Springer Japan KK

K. Horikoshi, K. Tsujii (Eds.)

Extremophiles in Deep-Sea Environments

With 122 Figures

Springer

Koki Horikoshi
Director General
The DEEPSTAR Group
Japan Marine Science and Technology Center (JAMSTEC)
2-15 Natsushima-cho, Yokosuka, Kanagawa 237-0061, Japan

Kaoru Tsujii
Team Leader
The DEEPSTAR Group
Japan Marine Science and Technology Center (JAMSTEC)
2-15 Natsushima-cho, Yokosuka, Kanagawa 237-0061, Japan

ISBN 978-4-431-68008-6

Library of Congress Cataloging-in-Publication Data

Extremophiles in deep-sea environments / K. Horikoshi, K. Tsujii (eds.).
p. cm.
Includes bibliographical references.
ISBN 978-4-431-68008-6 ISBN 978-4-431-67925-7 (eBook)
DOI 10.1007/978-4-431-67925-7
1. Extreme environments—Microbiology. 2. Benthos—Microbiology.
3. Submarine trenches. I. Horikoshi, Koki. II. Tsujii, Kaoru.
QR100.9 .E96 1999
579'.1777—dc21

99-36537

Printed on acid-free paper

Originally published by Springer Verlag Tokyo Berlin Heidelberg New York in 1999
Softcover reprint of the hardcover 1st edition 1999

Typesetting: Moriyama, Japan

SPIN: 10728618

Preface

On February 28, 1996, Japanese researchers working on the DEEP STAR program at the Japan Marine Science and Technology Center (JAMSTEC) broke their own world record by sending the unmanned submersible *Kaiko* (the Japanese word for "trench") to 10 898 meters beneath the Pacific Ocean. *Kaiko* sent back video images of life in the depths of the Mariana Trench, the deepest point in the world's oceans. A brief but excited fax from the mother ship described the scene captured on *Kaiko*'s camera: "The bed of the Mariana Trench was filled with a fine mud of reddish brownish particles. There were no rocks or cracks at all and it resembled a desert. However, very unusual organisms were observed here and there." In the following fax, they described types of sea urchin, jellyfish moving fairly fast, and the excrement of sea organisms. My colleagues also saw a fast-moving shrimp about three centimeters long. There was also a kind of sea cucumber, which was the same size as the jellyfish. *Kaiko* scooped out samples of mud to bring to the surface. Now we would be able to isolate living creatures from the deepest point of the Mariana Trench.

Not so many years ago, almost all biologists believed that life could survive only within so-called moderate environments; so when microbiologists looked around for interesting bacteria and other life-forms, they attempted to isolate microorganisms only from moderate environments.

About 30 years ago, I was visiting Florence, Italy. I was looking at the Renaissance buildings, which were very different from Japanese architecture. About 500 years ago, no Japanese could possibly have imagined this Renaissance culture. "There might be a whole new world of microorganisms in different unexplored cultures," I thought. "Could there be an entirely unknown domain of microorganisms at alkaline pH?" The acidic environment was being studied, probably because most food is acidic. However, hardly any work had been done in the alkaline region.

Science, just as much as the arts, relies upon a sense of romance and intuition. Upon my return to Japan, I prepared an alkaline medium containing 1% sodium carbonate, placed a small amount of soil in it, and incubated it overnight at 37°C. To my surprise, various microorganisms flourished in all 30 test tubes. Here was a new alkaline world which was utterly different from the neutral world discovered by Pasteur. I named these microorganisms which thrive in alkaline environments "alkaliphiles," and conducted systematic microbial physiological studies of them. At the same time I focused my interest on the enzymes produced by these alkaliphilic microorganisms as well as on their molecular genetics. The results showed that these microorganisms, which are completely different from any previously reported, were widely distributed on the globe and they produced new products. This was my first encounter with a kind of extremophile and led to my subsequent work with alkaliphiles.

The five-year Superbug Project of the Japanese ERATO program was launched in 1984 to search for extremophiles that grow in extreme environments and to try to use their unique properties to establish a "New Biotechnology." This research project resulted in many significant scientific discoveries. The Superbug Project was a reasonably successful ERATO project, in that some new, young scientists were trained and several successful patents have come out of the project. For example, one discovery was a very strong enzyme that can function in high alkalinity (pH 12) while degrading modified cellulose. This has already been incorporated in a Japanese detergent called "Attack." Another discovery has exceeded all conceptions of what constitutes a viable environment, an unusual, feisty life-form that can thrive in a 50% solution of very toxic toluene, in sharp contrast to ordinary organisms which perish in a solution of just 0.1%. We were able to isolate the DNA from the toluene-resistant bacterium and introduce it into toluene-sensitive mutants. It should ultimately be possible to isolate the special genes and transfer solvent-tolerance.

In more recent times, many organisms have been found in extreme environments. Moreover, some of them cannot survive in so-called moderate environments. For example, thermophilic bacteria grow in environments with extremely high temperatures, but will not grow at 20°C to 40°C. Some alkali-loving bacteria cannot grow in a nutrient broth at pH 7.0, but flourish at pH 10.5. Barophilic microorganisms isolated from the Mariana Trench absolutely require higher pressures (higher than 80 MPa) for growth. If a moderate environment for conventional organisms such as *Escherichia coli* were superimposed on that for extremophiles, the "moderate" environment would be an extreme environment. Thus, the idea of an extreme environment is relative, not absolute. Clearly we have been too anthropocentric in our thinking. We should therefore extend our consideration to other environments in order to isolate and cultivate new microorganisms.

A new 15-year research program called International DEEP STAR (Deep-sea Environment Exploration Program, Science and Technology for Advanced Research) was launched in October 1990 in Yokosuka through JAMSTEC. I have been placed in charge of this program and would like to expand the sources of microorganisms for study from the surface of the globe to the deep sea. JAMSTEC has two submarines (*Shinkai 2000* and *Shinkai 6500*), two unmanned submersibles (*Dolphin 3K* and *Kaiko*), and can collect samples from the deepest parts of the oceans. The remotely controlled *Kaiko* was used to collect samples from the deepest point of the Mariana Trench as described above. When I accepted the DEEP STAR post, my classmate, former President of JAMSTEC Isao Uchida told me: "JAMSTEC cannot study fish, because they belong to the Ministry of Agriculture and Fisheries, and it can't study the structure of the sea or ports, since this is under the Ministry of Construction." JAMSTEC decided to turn to microorganisms "because they belong to nobody." So I had good scientific and economic reasons for directing a major project to study microbial life at the bottom of the sea. One potential goal is to collect new information on the origins of life.

The program at Yokosuka turns out to be rather unusual in that 5 to 20 million dollars a year has been spent since 1991. At JAMSTEC there are 30 researchers working on deep-sea research, and from 1992 to 1998 another 45 researchers from ten countries came to work with us. Engineers and scientists from this project spent two years constructing an incredible facility that allows them to collect microorganisms under deep-sea hydrostatic pressure and to grow them continuously in culture, transferring them constantly under high pressure. These scientists also have developed the capability of working at very high temperatures, to better preserve and study deep-sea hydrothermal-vent bacteria.

The groups in DEEP STAR are focusing on the following areas:

1. Organic solvent-tolerant microorganisms. It took two years to isolate an organic solvent-tolerant bacterium in 1989. However, in 1991 our group developed a sophisticated method to isolate organic solvent-tolerant bacteria from deep-sea samples. Using this method, the isolation frequency of solvent-tolerant microorganisms was about 100 times higher than that of the isolation previously reported. Some of these microorganisms can completely digest crude oil in sea water. This finding has paved the way to removing oil spills. These organic solvent-tolerant microorganisms have considerable potential for application in bioreactors as solvent-tolerant microbial catalysts.
2. Hyperthermophiles. More than 50 strains of hyperthermophiles have been isolated from hydrothermal vents in the Okinawa area. One of these hyperthermophiles is *Pyrococcus horikoshii*, isolated in collaboration with The Center of Marine Biotechnology of the University of Maryland. The whole chromosomal DNA sequence of this strain was determined by the National Institute of Technology and Evaluation in 1998. Several enzymes in hyperthermophiles have been investigated for industrial applications.
3. Psychrophiles, bacteria that grow at very low temperatures. These bacteria are a good resource to isolate novel enzymes that are active at lower temperatures.
4. Barophiles, bacteria that grow only under hydrostatic pressure. DEEP STAR researchers have succeeded in isolating many obligately barophilic bacteria that function only under high pressure and have isolated the genes responsible for barophily. Phylogenetic studies of the isolates suggested that the isolated strains belong to a new sub-branch of the gamma group. It was discovered that the *lac* promoter was activated under high pressure in the absence of inducers. Al-

though the enzymes produced by these barophiles have not been studied, it is highly likely that it will be possible to clone novel genes from barophiles and to express them in conventional microorganisms for industrial applications. Furthermore, the physiology of *Saccharomyces cerevisiae* has been extensively studied under elevated hydrostatic pressures to establish barophysiology.

5. Whole genome analysis of alkaliphilic *Bacillus halodurans* C-125, a bacterium producing alkaline protease, alkaline cellulase, and alkaline xylanase. Alkaliphily and the environmental adaptation mechanism will be clarified from the whole DNA sequence determined by "shotgun sequencing."
6. Work is proceeding on preservation and culture collection, and research has begun on the genetics of these marine bacteria.

DEEP STAR researchers have successfully moved to the second phase (1998–2005) with the following results.

1. By the end of 1998 about 250 articles had been published in respected journals.
2. In January 1998, DEEP STAR organized the International Congress on Extremophiles at Yokohama. About 350 participants, including 150 scientists from abroad, discussed extremophiles.
3. The science journal *Extremophiles* was launched in 1997 by Springer-Verlag. I was appointed the first Chief Editor.

This book is written to commemorate our eight-year studies of deep-sea microorganisms. Some of the work has not been published in scientific journals for various reasons: some because it is not yet ready for submission, some for patent and other reasons. However, these personal reports should be useful in the study of deep-sea microorganisms. All the individuals involved have worked very hard to develop deep-sea microbiology, including barophysiology. I am extremely grateful for their cooperation in this project during the first phase.

I would also like to pay special thanks to former Presidents I. Uchida, and W. Ishizuka and President T. Hirano of JAMSTEC, who kindly gave me the opportunity to expand the new world of extremophiles.

K. Horikoshi

Contents

Preface V

Part 1 Biodiversity in Deep-Sea Environments

1 Isolation and Characterization of Microorganisms from Deep-Sea Mud
Hideto Takami 3

2 Molecular Analyses of the Sediment and Isolation of Extreme Barophiles from the Deepest Mariana Trench
Chiaki Kato 27

3 Taxonomy and Biotransformation Activities of Deep-Sea Actinomycetes
Alan T. Bull, Joy A. Colquhoun, and Stephen C. Heald 39

4 Microbial Diversity in the Sediments Collected from Cold-Seep Areas and from Different Depths of the Deep-Sea
Lina Li and Chiaki Kato 55

Part 2 Adaptation and Response of Extremophiles Toward Extreme Deep-Sea Conditions

5 Barophiles (Piezophiles)
Chiaki Kato ... 91

6 Thermophiles
Juan M. González ... 113

7 Deep-Sea Psychrophiles
Tetsuo Hamamoto ... 155

8 Organic Solvent-Tolerant Microorganisms
Akira Inoue ... 175

9 Microbial Processes Associated with a Methane Seep Tubeworm
Takeshi Naganuma ... 211

Part 3 New Approaches and Future Scope for Deep-Sea Microbiology

10 Barophysiology (Piezophysiology)
Fumiyoshi Abe and Chiaki Kato ... 227

11 Genome Analysis of Facultatively Alkaliphilic *Bacillus halodurans* C-125
Hideto Takami ... 249

12 Future Scope
Kaoru Tsujii ... 285

Appendix: International Congress on Extremophiles '98 at Yokohama ... 293

Appendix: List of Publications of the First DEEPSTAR Project: October 1, 1990 to September 30, 1998 ... 301

Subject Index ... 311

Contributors

Koki Horikoshi
The DEEPSTAR Group, Japan Marine Science and Technology Center,
2-15 Natsushima-cho, Yokosuka, Kanagawa 237-0061, Japan

Hideto Takami
The DEEPSTAR Group, Japan Marine Science and Technology Center,
2-15 Natsushima-cho, Yokosuka, Kanagawa 237-0061, Japan

Chiaki Kato
The DEEPSTAR Group, Japan Marine Science and Technology Center,
2-15 Natsushima-cho, Yokosuka, Kanagawa 237-0061, Japan

Alan T. Bull
Department of Biosciences, University of Kent,
Canterbury, Kent, CT2 7NJ, UK

Lina Li
The DEEPSTAR Group, Japan Marine Science and Technology Center,
2-15 Natsushima-cho, Yokosuka, Kanagawa 237-0061, Japan

Juan M. González
Center of Marine Biotechnology, University of Maryland Biotechnology Institute,
701 E. Pratt St., Baltimore, MD 21202, USA

Tetsuo Hamamoto
Biotechnology Transfer Services,
4-14-9 Kamiogi, Suginami-ku, Tokyo 167-0043, Japan

Akira Inoue
The DEEPSTAR Group, Japan Marine Science and Technology Center,
2-15 Natsushima-cho, Yokosuka, Kanagawa 237-0061, Japan

Takeshi Naganuma
Faculty of Applied Biological Science, Hiroshima University,
1-4-4 Kagamiyama, Higashi-Hiroshima, Hiroshima 739-8528, Japan

Fumiyoshi Abe
The DEEPSTAR Group, Japan Marine Science and Technology Center,
2-15 Natsushima-cho, Yokosuka, Kanagawa 237-0061, Japan

Kaoru Tsujii
The DEEPSTAR Group, Japan Marine Science and Technology Center,
2-15 Natsushima-cho, Yokosuka, Kanagawa 237-0061, Japan

1 Biodiversity in Deep-Sea Environments

1 Isolation and Characterization of Microorganisms from Deep-Sea Mud

Hideto Takami

The most concentrated and widespread occurrences of organisms are generally in "moderate" environments with approximately neutral pH, temperatures around 20°-37°C, pressures near 0.1 MPa, and adequate concentrations of nutrients and saline. In contrast, the deep-sea is an extreme environment with especially high hydrostatic pressure and low temperature. Microorganisms living there presumably have developed particular characteristics that allow them to thrive in such an environment. Bacteria have been isolated from deep-sea mud and from benthic organisms such as amphipods and sea cucumbers in the bathypelagic zone (Yayanos 1979; Yayanos et al. 1981). However, little information is available on bacterial diversity in sediments of the deep-sea floor because most marine biologists have focused on barophilic and psychrophilic inhabitants of the deep-sea environment (Yayanos 1995).

On March 2, 1996, the 3 m long unmanned submersible Kaiko touched the bottom of the Challenger Deep in the Mariana Trench and successfully scooped out a mud sample, the first obtained at a depth of 10 897 m. We isolated thousands of microbes from this deep-sea mud and found that the microbial flora there was composed of actinomycetes, yeasts, and a range of bacteria including various extremophilic bacteria (Takami et al. 1997). In studies aimed at further exploring the microbial diversity in various deep-sea environments, we attempted to isolate and characterize a number of bacteria from deep-sea mud collected by means of the manned submersibles Shinkai 2000 and 6500 (Takagawa et al. 1989). We recorded considerable bacterial diversity and the occurrence of extremophilic bacteria at several deep-sea sites located near the southern part of Japan.

Key words. Deep-sea, Mariana Trench, Phylogenetic tree, 16S ribosomal RNA, Extremophiles, α-Amylase

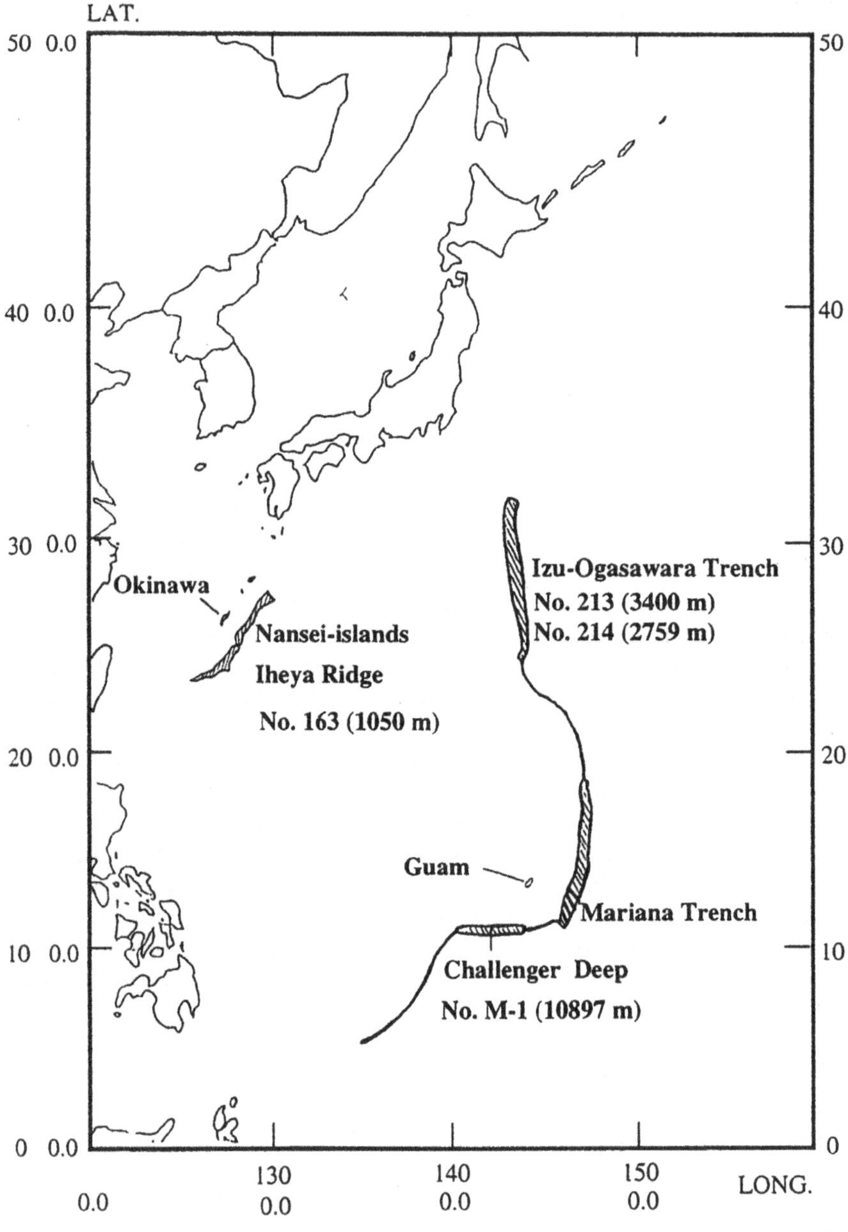

Fig. 1. Deep-sea sites for collection of mud samples. Sediment sample No. 163 was collected from the Nankai-islands Iheya Ridge by means of the manned submersible Shinkai 2000. Sediment samples No. 213 and No. 214 were obtained from the Izu-Ogasawara Trench (30°07.05'N, 139°58.42'E and 29°04.2'N, 140°43.3'E, respectively) by means of the manned submersible Shinkai 6500. Deep-sea mud (M-1) from the Challenger Deep region of the Mariana Trench was collected as described previously (Takami et al. 1997)

1.1 Collection of Deep-Sea Mud

Samples of deep-sea sediment were collected from the Nankai Islands Iheya Ridge (1050 m depth; 27°47.18'N, 126°54.15'E) by means of the manned submersible Shinkai 2000 and from the Izu-Ogasawara Trench (2759 m depth, 30°07.05'N, 139°58.42'E; and 3400 m depth; 29°04.2'N, 140°43.3'E) by means of the manned submersible Shinkai 6500, using cylinder mud samplers (Fig. 1).

The 3 m long submersible Kaiko touched the bottom of the Challenger Deep and successfully scooped out a mud sample, the first in the world obtained at such a great depth of 10 897 m as shown in Fig. 2B,C,D. A sterile 50-ml Falcon tube filled with a fine mud of grayish brownish particles was tightly inserted in a tube holder and brought to the sea surface without being contaminated by upper-ocean bacteria (Fig. 2A,D).

Fig. 2. Collection of a mud sample at a depth of 10 897 m from the Challenger Deep. The *white arrow* points at the sampling tube holder (**A**). This holder is usually set in the box shown at the right side of the *white arrow* (**A**). The *white arrowhead* and *open arrowhead* show a 50-ml Falcon tube and sample inlet, respectively (**A**, **C**, **D**). The deep sea mud was scooped out by a manipulator, as shown in alphabetical order (**B**, **C**, **D**). The sampling tube filled with the mud was tightly inserted in a sampling tube holder (**A**)

1.2 Isolation of Microorganisms

1.2.1 *Mariana Trench*

The mud sample was diluted two-fold with Marine broth 2216 (Difco Laboratories) (Zobell 1941); and 100–200 μl (5–10 mg as dry weight) of the suspension was spread on the Marine agar or half-strength nutrient agar plates used as a basal medium. In addition, modified Marine agar plates supplemented with 1% skim milk or 1% potato starch, which had varied conditions of pH (pH 3, 7, or 10) and concentrations of sodium chloride (0%, 2%, or 15%) were used for isolation. The alkaline and acidic media contained 1% sodium carbonate and 50 mM citric acid, respectively. The alkaline or acidic source 10-fold concentrated was autoclaved separately and then added to the plain or the modified Marine agar medium. The agar plates were incubated at 4°–75°C at atmospheric pressure (0.1 MPa) or at 100 MPa for 1–4 weeks. Cultivation techniques at high pressure were as described earlier (Abe and Horiko-

Table 1. Isolation of extremophilic bacteria from several deep-sea sites

Category	Isolation conditions	Origin no. (depth in meters)	Bacterial recovered (colonies g^{-1} dry sea mud)
Alkaliphile	pH 9.7 ± 0.3 25°C, 0.1 MPa	163 (1050)	$3.0 - 6.1 \times 10^2$
		214 (2759)	$0.2 - 2.3 \times 10^4$
		213 (3400)	0.9×10^2
		M1 (10 897) [a]	$0.4 - 1.2 \times 10^3$
Thermophile	55°C pH 7.3 ± 0.2, 0.1 MPa	163 (1050)	$0.8 - 2.3 \times 10^2$
		214 (2759)	$1.1 - 7.8 \times 10^2$
		213 (3400)	$1.0 - 6.0 \times 10^2$
		M1 (10 897) [a]	$0.6 - 3.5 \times 10^3$
Psychrophile	4°C ph 7.3 ± 0.2, 0.1MPa	163 (1050)	$0.8 - 5.3 \times 10^2$
		214 (2759)	$1.4 - 7.8 \times 10^2$
		213 (3400)	1.0×10^2
		M1 (10 897) [a]	2.0×10^2
Halophile	15%-NaCl, 25°C ph 7.3 ± 0.2, 0.1MPa	163 (1050)	4.6×10^2
		214 (2759)	3.6×10^3
		213 (3400)	0.9×10^2
		M1 (10 897) [a]	— [b]
Acidophile	ph 3.7 ± 0.2 25°C, 0.1MPa	163 (1050)	—
		214 (2759)	—
		213 (3400)	—
		M1 (10 897) [a]	—
Nonextremophile	25°C, 0.1MPa pH 7.3 ± 0.2	163 (1050)	$0.5 - 6.6 \times 10^3$
		214 (2759)	$0.2 - 1.1 \times 10^5$
		213 (3400)	$8.1 - 9.4 \times 10^2$
		M1 (10 897) [a]	$0.2 - 2.3 \times 10^5$

[a] Takami et al. (1997).
[b] No growth obtained.

shi 1995). Various nonextremophilic bacteria were isolated from the mud sample with a frequency of 2.2×10^4 to 2.3×10^5 colonies per gram of dry sea mud under the conditions described in Table 1.

This is the first report of the isolation of viable bacteria from the bottom of the Challenger Deep. It is a significant result in relation to understanding the nature of the microbial flora in the deepest sea mud. Previously, by means of microscopic observations, Zobell confirmed the existence of viable bacterial cells in mud collected at a depth of 10 400 m from the Philippine Trench, but none of these bacteria was cultured. We have obtained thousands of isolates from the Challenger Deep. Some were found to be producers of enzymes such as proteases and amylases.

Although barophilic, halophilic, and acidophilic bacteria have not yet been detected in mud samples from the Challenger Deep, various other types of extremophilic bacteria were isolated at 0.1 MPa (Table 1). Among the extremophiles found were many alkaliphiles and thermophiles, which would be expected to thrive in an extreme environment different from the in situ conditions of high hydrostatic pressure and low temperature in the Challenger Deep. The population of psychrophilic isolates showed a tendency to be smaller than those of alkaliphiles and thermophiles. In addition, we successfully isolated filamentous fungi and actinomycetes at 0.1 MPa with almost the same frequency (2.0×10^2 per gram of dry sea mud) as that of psychrophilic bacteria.

1.2.2 Other Deep-Sea Sites Located in Southern Japan

Halophiles, psychrophiles, thermophiles, and alkaliphiles were isolated at 0.1 MPa from the deep-sea mud collected at various sites (Fig. 1), although acidophilic bacteria were not detected in the mud samples in this study or the previous study (Takami et al. 1997, 1999). There was no substantial difference in the frequency of isolating thermophilic and psychrophilic bacteria among the three deep-sea mud samples examined in this study (Table 1). The population of alkaliphilic isolates from deep-sea site 214 (0.2×10^4 to 2.3×10^4) was larger than those from the other deep-sea sites, and the frequency of isolation of alkaliphiles did not vary depending on the depth of the sampling site. On the other hand, halophilic bacteria were isolated from mud samples collected at depths of 1050 m (no. 163), 2759 m (no. 214), and 3400 m (no. 213) under the conditions described in Table 1 whereas halophiles were not detected in our previous study (Takami et al. 1997). Some were found to be alkaliphilic halophiles growing at pH 9.3. The halophilic strain HTE831 grew on Marine agar plates containing 2% (0.34 M) to 23.4% (4 M) NaCl, and strain HTC018 grew on Marine agar plates containing 15% (2.6 M) NaCl but not those containing 2% NaCl. Other halophiles and alkaliphilic halophiles isolated grew well on Marine agar plates containing 2%–15% NaCl (data not shown).

Various nonextremophilic bacteria also were isolated at 0.1 MPa from the mud samples at a frequency of 0.5×10^3 to 6.6×10^3 (no. 163) to 0.2×10^5 to 1.1×10^5 (no. 214) colonies per gram of dry sea mud under the conditions described in Table 1. The frequency of isolating nonextremophilic bacteria did not change depending on the depth of habitat. Some of them were found to be producers of enzymes such as proteases and amylases. In this study, it was found that the frequency of isolating extremophilic and nonextremophilic bacteria at 0.1 MPa was not directly related to the depth of sampling sites. We presume that the frequency of isolating bacteria from the deep-sea sediment at 0.1 MPa may be influenced by other factors, such as the size and kind of sediment particles, nutrient conditions in the sediment, and the speed of the bottom current.

1.3 Growth Patterns of Deep-Sea Isolates Under High Pressure

1.3.1 Mariana Isolates

Seventeen isolates were selected at random as candidates of ordinary or extremophilic bacteria. Each isolate was basically inoculated on the Marine agar plate and overlaid with low melting agarose (1% v/v) of the same medium. Half-strength nutrient agar was used for the last three strains in Table 2. The agar plate was slipped into a plastic bag and sealed with the heat sealer. The bacteria were cultivated under proper temperature and pH conditions at 30 or 60 MPa for 2 weeks. All isolates were cultivated at 4°C under 100 MPa for 2 weeks. The 17 isolates described in Table 2 were cultivated under high hydrostatic pressure. Thermophilic bacteria did not grow at pressures higher than 30 MPa during 2 weeks of cultivation, whereas strain HTA462 grew slightly at 30 MPa. Neutrophilic strain HTA064 showed better growth at 60 MPa than at 30 MPa and also grew slightly even at 100 MPa—same as the strain HTA208. Alkaliphiles were more sensitive at 60 MPa than neutrophiles, although three of them grew slightly at 100 MPa, same as neutrophiles. Psychrophilic bacteria showed a tendency similar to that of mesophilic bacteria, both neutrophiles and alkaliphiles, in their growth pattern under high hydrostatic pressure. In this experiment, it was found that some of the Mariana isolates grew at 100 MPa although slowly, which is one of in situ conditions in the Mariana Trench, although no barophilic or barotolerant bacteria have been isolated after 2 weeks of cultivation under 100 MPa, mentioned above.

We have obtained many isolates widely distributed in the bacterial taxon, although various microbes other than those capable of growth under the laboratory conditions employed exist in the mud of the deepest sea floor. The microbes present in the aquatic environment are free-living or attached. The attached microbes are usually found with the flocculate aggregates of particulate matter constituting "marine snow" (Silver et al. 1978; Kirchman 1993; Olsen et al. 1994). Theoretical settling rates of the particulate matter, as well as phytoplankton, range from 1.0 to 0.1 m per day or 5000 m in 1-50 years (Jannasch and Taylor 1984). Accordingly, almost all isolates obtained in this study would have been brought down with the particulate matter to the bottom of the Challenger Deep over the course of a long time.

Table 2. Growth pattern of Mariana isolates under high hydrostatic pressure

Strain no.	Temp. (°C)	pH	Hydrostatic pressure (MPa)		
			30	60	100 [a]
Mesophile					
Neutrophile					
HTA064	25	7.6	++	+++	+
HTA208	25	7.6	++	++	+
Alkaliphile					
HTA212	25	9.5	++	–	–
HTA275	25	9.5	++	+	+
HTA459	25	9.5	++	+	+
HTA474	25	9.5	+	+	+
Psychrophile					
HTA554	8	7.6	++	+	+
HTA557	8	7.6	++	++	+
HTA563	8	7.6	++	++	+
HTA580	8	7.6	++	+	+
HTA608	8	7.6	++	++	+
Thermophile					
HTA426	55	7.6	–	–	–
HTA454	55	7.6	–	–	–
HTA462	55	7.6	+	–	–
HTA1416	55	7.6	–	–	–
HTA1417	55	7.6	–	–	–
HTA1418	55	7.6	–	–	–

–, no growth; +, slight growth; ++, growth; +++, good growth.
[a] All isolates were cultivated at 4°C under 100 MPa.

1.3.2 Isolates from Other Deep-Sea Sites

The 25 isolates described in Table 3 were cultivated at 4°C (the in situ temperature of the deep-sea floor) under high hydrostatic pressure to examine the growth patterns of the isolates obtained from various depths (Takami et al. 1999). As shown in Table 3, the alkaliphilic bacteria did not grow at pressures higher than 30 MPa during a 4-week period of cultivation, although two strains (HTB113 and HTB119) grew slightly at 30 MPa. Two moderately halophilic isolates, HTB068 and HTB069, grew well at 30 MPa; but another one, HTE830, showed no growth at the same hydrostatic pressure. Two of three alkaliphilic halobacteria did not grow at pressures higher than 30 MPa, the same as was observed with alkaliphiles, whereas strain HTB122 grew at 30 MPa (Table 3). Eight of fourteen neutrophilic strains grew at 30 MPa, and of these only three strains grew slightly at 100 MPa. Alkaliphiles were more sensitive to inhibition by pressure at 30 MPa than the neutrophiles, although two of them (HTB113 and HTB119) grew slightly, the same as observed in our previous study on isolates from deep-sea mud of the Mariana Trench (Takami et al. 1997). It was found that some Mariana isolates grew slowly at 100 MPa, which is similar to the in situ pressure conditions in the Mariana Trench; and they presumably had become adapted to such hydrostatic pressure over a long time. In the present study, most of the isolates from the deep-sea mud collected at depths of 1050, 2759, and 3400 m were sensitive to hydrostatic pressures higher than 30 MPa. Thus, the bacteria living in the deep-sea sediment may be well adapted to the in situ hydrostatic pressure of their deep-sea environment.

Table 3. Growth patterns of deep-sea isolates under high hydrostatic pressure

Strain no. (origin no.)	pH	Growth under 0.1–100 MPa			
		0.1	30	60	100
Alkaliphile					
HTB110 (214)	9.5	++	–	–	–
HTB113 (214)	9.5	+++	+	–	–
HTB119 (214)	9.5	+++	+	–	–
HTB127 (214)	9.5	++	–	–	–
HTB132 (214)	9.5	+++	–	–	–
Halophile					
HTE830 (163)	7.4	++	–	–	–
HTB068 (214)	7.4	+++	+++	–	–
HTB069 (214)	7.4	+++	+++	–	–
Alkaliphilic halophile					
HTB 122 (214)	9.3	+++	++	–	–
HTB 123 (214)	9.3	++	–	–	–
HTB 124 (214)	9.3	++	–	–	–
Neutrophile					
HTE777 (163)	7.6	++	–	–	–
HTE807 (163)	7.6	+	–	–	–
HTE811 (163)	7.6	+	–	–	–
HTE846 (163)	7.6	+++	++	–	–
HTE856 (163)	7.6	+	+	+	+
HTB006 (214)	7.6	+++	++	+	–
HTB021 (214)	7.6	+++	++	+	+
HTB035 (214)	7.6	+++	–	–	–
HTB080 (214)	7.6	+++	+++	–	–
HTB095 (214)	7.6	+++	+++	+	+
HTC004 (213)	7.6	+	–	–	–
HTC010 (213)	7.6	++++	+++	–	–
HTC011 (213)	7.6	+++	++	–	–
HTC028 (213)	7.6	++	–	–	–

–, no growth; +, slight growth; ++, growth; +++, good growth; ++++, very good growth.
All isolates were incubated at 4°C under high hydrostatic pressure.

1.4 16S rDNA Sequencing and Analysis

1.4.1 Mariana Isolates

Twenty-eight strains were selected among the Mariana isolates for 16S rDNA analysis. Each strain was cultivated on an agar plate under proper growth conditions, as shown in Table 4. One loop of cell pellet was suspended in 50 µl of GTE solution (25

Table 4. Mariana isolates used for 16S rDNA analysis

Strain no.	Growth conditions			Remarks
	Medium	Temp. (°C)	pH	
Mesophile				
Neutrophile				
HTA064	Marine agar	25	7.6	Protease producer
HTA208	Marine agar	25	7.6	
HTA437	Marine agar	25	7.6	Protease producer
HTA456	Marine agar	25	7.6	Amylase producer
HTA473	Marine agar	25	7.6	Amylase producer
HTA484	Nutrient agar	25	7.6	
HTA506	Nutrient agar	25	7.6	
HTA527	Nutrient agar	25	7.6	
Alkaliphile				
HTA212	Marine agar	25	7.6 – 10.5	Amylase producer
HTA218	Marine agar	25	7.6 – 9.5	Amylase producer
HTA275	Marine agar	25	7.6 – 9.5	Amylase producer
HTA333	Marine agar	25	7.6 – 10.5	
HTA459	Marine agar	25	7.6 – 10.5	
HTA474	Marine agar	25	7.6 – 10.5	
Psychrophile				
HTA554	Nutrient agar	4 – 25	7.6	
HTA557	Nutrient agar	4 – 25	7.6	
HTA563	Nutrient agar	4 – 25	7.6	
HTA580	Nutrient agar	4 – 25	7.6	
HTA608	Sabouraud's agar [a]	4 – 25	7.6	
Thermophile				
HTA426	Marine agar	55 – 75	7.6	
HTA454	Marine agar	45 – 65	7.6	
HTA462	Marine agar	55 – 75	7.6	
HTA1415	1/2 Nutrient agar [b]	45 – 70	7.6	
HTA1416	1/2 Nutrient agar	45 – 70	7.6	
HTA1417	1/2 Nutrient agar	45 – 70	7.6	
HTA1418	1/2 Nutrient agar	55 – 70	7.6	
HTA1420	1/2 Nutrient agar	45 – 70	7.6	
HTA1422	1/2 Nutrient agar	45 – 70	7.6	

[a] Sabouraud's agar containing glucose (4%) and NaCl (2%).

[b] Half-strength of nutrient broth.

mM Tris-HCl, 50 mM glucose, and 10 mM EDTA) containing lysozyme (1 mg/ml). The cell suspension was incubated at 37°C for 20 min, and 5 μl of 10% sodium dodeyl sulfate (SDS) solution was added to the cell suspension to extract chromosomal DNA. The DNA was purified by phenol/chloroform treatment and then used as a template DNA for polymerase chain reaction (PCR) amplification of 16S rDNA. The following prokaryote-specific primers were used for gene amplification and sequencing : EU10F (5'-AGAGTTTGATCCTGGCTCAG-3'), EU1500R (5'-GGTTACCTTGTTACGACTT-3'), EU500F (5'-GTGCCAGCAGCCGCGG-3'), EU500R (5'-GTATTACCGCGGCTGCTG-3'), EU1100F (5'-AAGTCCCGCAACGAGCGCA-3'), and EU1100R (5'-TTGCGCTCGTTGCGGGACT-3'). The amplified fragments were purified with Suprec (Takara Shuzo Co., Japan) and sequenced by using *LI-COR* DNA sequencer model 400L (LI-COR, Nebraska, USA). 16S rDNA sequences were aligned by using the Clustal multiple-alignment program (Clustal W) (Higgins and Sharp 1988). Sites involving gaps were excluded from all analyses. A phylogenetic tree was inferred by the neighbor-joining (NJ) method (Saitou and Nei 1987). The sequences, which were determined in this study, were aligned with previously deposited sequences in the GeneBank and Ribosomal Database of the University of Illinois at Urbana-Champaign.

The total 16S rDNA sequences of Mariana isolates shown in Table 4 were determined. These sequences were aligned and compared with the 16S rDNA sequences of 27 reference strains categorized as actinomycetes, low guanine and cytosine (G+C) gram-positive bacteria, proteobacteria α-subdivision, and γ-subdivision (Olsen et al. 1994). Evolutionary distances between Mariana isolates and reference strains were computed, and phylogenetic tree was constructed using the NJ algorithm and the pairwise evolutionary distances (Fig. 3).

Five of six alkaliphiles were filed into actinomycetes. The 16S rDNA sequences of alkaliphilic strains HTA474 and HTA459 were 99% similar to that of *Diezia maris* and 98% similar to that of *Aureobacterium testaceum*, respectively. On the other hand, the 16S rDNA sequence of HTA333 categorized as proteobacteria gamma subdivision was almost identical to that of *Brevundimonas diniminuta.* Thermophiles filed into low G+C gram-positive bacteria were distant from reference strains used in this study, whereas the 16S rDNA sequences of HTA426 were similar to that of *Bacillus kaustophilus* or *Bacillus thermoleovorans*. Four of five psychrophiles were categorized as proteobacteria γ-subdivision. These four psychrophiles, which constitute two subclasses categorized may be members of genus *Acinetobacter*. Another psychrophile, HTA563, categorized as a low G+C gram-positive bacterium was 99% identical to *Staphyloccocus epidermidis* on the basis of 16S rDNA sequence. Neutrophiles were widely distributed in the bacterial taxon shown in Fig. 3. Two strains, HTA437 and HTA506, which constitute a subclass, may be members of the genus *Bacillus*. The 16S rDNA sequences of HTA484 and HTA208 were 98% identical to that of *Bacillus cereus* and 99% identical to that of *Pseudomonas stutzeri*, respectively.

For further characterization of Mariana isolates, the morphological, physiological, and biochemical properties of isolates listed in Table 4 were investigated accord-

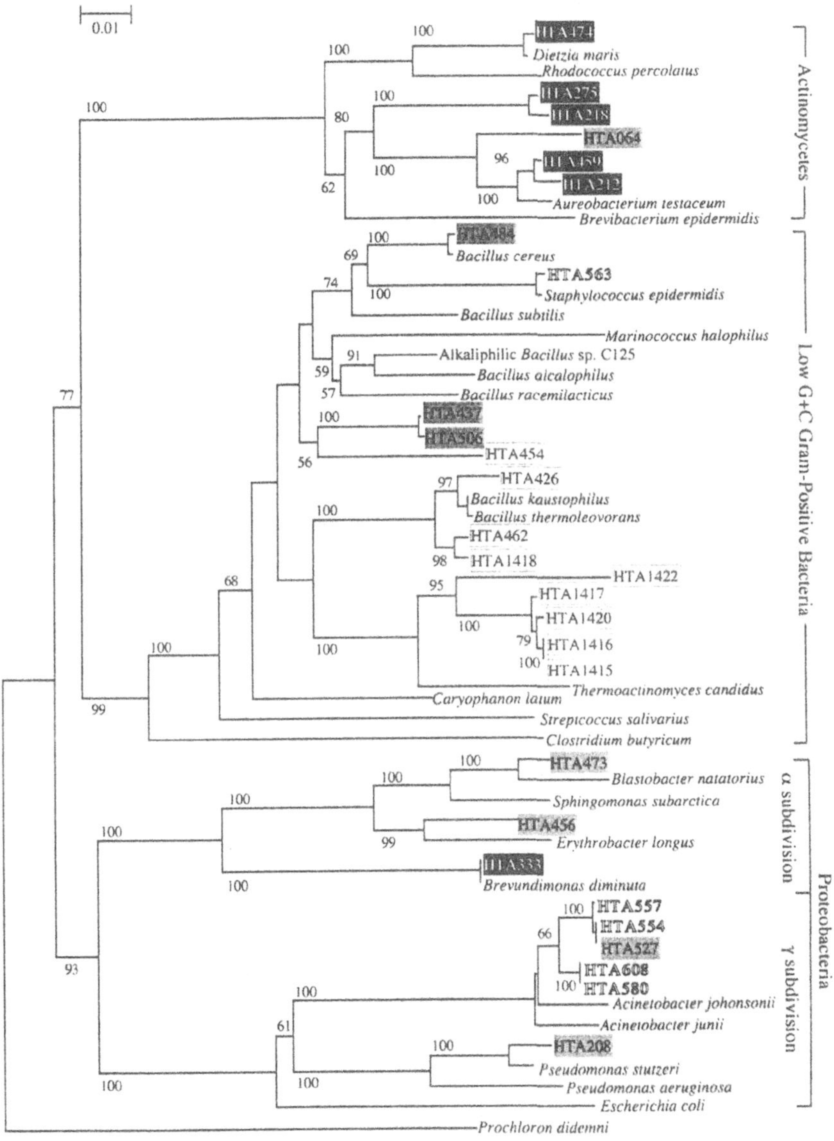

Fig. 3. Unrooted phylogenetic tree showing the relation of Mariana isolates to reference organisms. The numbers indicate the percentages of bootstrap samples, derived from 10 000 samples, that supported the internal branches (Silver et al. 1978). Bootstrap probability values less than 50% were omitted from this figure. Bar 0.01 K_{nuc} unit. ABC, alkaliphile; ABC, neutrophile; ABC, thermophile; ABC, psychrophile

Table 5. Identification of Mariana isolates

Strain no.	Isolation conditions	16S rDNA search (BLASTN·similarity%)	Proposed name by identification
Eubacteria			
HTA208	Marine broth, 1% starch, pH 7.6, 25°C	*Pseudomonas putida* (97.3)	*Pseudomonas* sp. (**new species?**)
HTA212	Marine broth, 1% starch, pH 7.6, 25°C	*Aureobacterium testaceum* (97) **P**	*Aureobacterium* sp. (**new species?**) (amylase producer)
THA333	Marine broth, 1% casein, pH 7.6, 25°C	*Brevundimonas diminuta* (99)	*Brevundimonas diminuta*
HTA423	Marine broth, 1% casein, pH 7.6, 55°C	*Bacillus stearothermophilus* (98.7) **P**	*Bacillus* sp. group 5 (**new species?**)
HTA426	Marine broth, 1% casein, pH 7.6, 55°C	*Bacillus kaustophilus* (99.4)	*Bacillus* sp. group 5
HTA427	Marine broth, 1% starch, pH 7.6, 55°C	*Bacillus subtilis* (99.9)	*Bacillus subtilis* (protease producer)
HTA437	Marine broth, 1% casein, pH 7.6, 25°C	*Bacillus chonii* (96.1)	*Bacillus* sp. group 1
HTA456	Marine broth, 1% starch, pH 7.6, 25°C	*Erythrobacter longus* (94.9)	*Erythrobacter longus* ? (α-subgroup)
HTA462	Marine broth, 1% casein, pH 7.6, 55°C	*Bacillus thermocatenulatus* (98.5)	*Bacillus* sp. group 5
HTA473	Marine broth, 1% starch, pH 7.6, 25°C	*Blastomonas natatoria* (94.8)	(α-subgroup) (**new species?**)
HTA474	Marine broth, 1% casein, pH 9.5, 25°C	*Dietzia maris* (99)	*Dietzia maris*
HTA484	Nutrient broth, 2% NaCl, pH 7.6, 25°C	*Bacillus cereus* (98.7)	*Bacillus cereus* group
HTA554	Nutrient broth, 0.1% glucose pH7, 8°C	*Acietobacter johnsonii* (99.6%)	*Acietobacter johnsonii*
HTA557	Nutrient broth, 0.1% glucose pH7, 8°C	*Acietobacter johnsonii* (99.6%)	*Acietobacter johnsonii*
HTA608	Sabouraud's 4% glucose pH 7.6, 8°C	*Acietobacter johnsonii* (99.4%)	*Acietobacter johnsonii*
HTA1415	1/2 Nutrient broth, pH 7.6, 55°C	*Bacillus thermoleovorans* (92.3)	*Bacillus* sp. group 5 (**new species?**)
HTA1422	1/2 Nutrient broth, pH 7.6, 55°C	*Bacillus stearothermophilus* (92.3)	*Bacillus* sp. group 5 (**new species?**)
MS300	Marine broth, 1% starch, pH 7.6, 8°C	*Pseudomonas stuzeri* (98)	*Pseudomonas stuzeri* (G4 amylase producer)
Eukaryotae			
HTA550	Nutrient, 0.1% glucose,1% casein, pH 7, 25°C	—	*Penicillium lagena*
HTA563	Nutrient broth, 0.1% glucose, pH 7, 8°C	*Rhodotorlula graminis* (98)	*Rhodotorlula mucilaginosa*

P, partial sequence.

ing to the methods described in *Bergey's Manual of Systematic Bacteriology* (Sneath et al. 1986). As shown in Table 5, eight strains (HTA333, HTA427, HTA474, HTA554, HTA557, HTA608, MS300, HTA484) were identified as known species, and another six strains may be categorized as new species. We attempt DNA-DNA hybridization between closely related strains to determine whether they are new species. On the other hand, two eukaryotic isolates were identified as *Penicillium lagena* and *Rhodotorula mucilaginosa* by conventional taxonomic methods and phylogenetic analysis based on 18S rDNA sequences (Table 5).

1.4.2 Isolates from Other Deep-Sea Sites

Twenty-four strains were selected on the basis of cell and colony morphology from among the deep-sea isolates categorized as alkaliphiles, halophiles, alkaliphilic halophiles, and neutrophiles for 16S rDNA analysis (Takami et al. 1999). Each strain was cultivated on an agar plate under appropriate growth conditions as shown in Table 6.

One loop of cell pellet was suspended in 50 μl of 25 mM Tris-HCl, 50 mM glucose, 10 mM EDTA, and 1 mg ml^{-1} lysozyme. The cell suspension was incubated at 37°C for 20 min, and 5 μl of 10% SDS solution was added to the cell suspension to extract chromosomal DNA. The DNA was purified by phenol/chloroform treatment and then used as template DNA for PCR amplification of 16S rDNA. Total 16S rDNA sequences were determined for the deep-sea isolates shown in Table 6. These sequences were aligned and compared with the 16S rDNA sequences of 40 reference strains and 28 Mariana isolates (Takami et al. 1997) categorized as actinomycetes, low G+C Gram-positive bacteria, and proteobacteria of the α–, β– and γ-subdivisions (Olsen et al. 1994). Evolutionary distances among the deep-sea isolates in this study, the Mariana isolates, and the reference strains were computed, and a phylogenetic tree was constructed using the NJ algorithm (Fig. 4).

The 16S rDNA sequences of alkaliphilic strain HTB113 and neutrophilic strain HTB096, both categorized as actinomycetes, were 99.6% similar to each other and showed 99.1% and 98.8% similarity, respectively, to that of the Mariana isolate HTA212. The 16S rDNA sequence of neutrophilic strain HTE856 showed 99% and 98.4% similarity to those of *Bacillus cereus* and Mariana isolate HTA484, respectively. Halophilic isolate HTE831, which grew on medium containing 0.34–4.00 M NaCl, was positioned close to *Bacillus subtilis* in the phylogenetic tree (Fig. 4).

The halophilic isolate HTC018 and the neutrophilic isolate HTB091 were categorized as members of the b subdivision of proteobacteria, which were not found among the Mariana isolates (Takami et al. 1997) based on 16S rDNA sequence analysis. The 16S rDNA sequences of HTC018 and HTB091 were 99.4% similar to each other and showed 99.3% and 99.4% similarity, respectively, to that of *Janthiobacterium lividum*, which is categorized as a member of the proteobacteria β-subdivision (O'Sullivan et al. 1990). Two halophilic strains (HTB122 and HTB123), alkaliphilic halophilic strain HTB069, and alkaliphilic strain HTB138, which were categorized as members of the proteobacteria γ-subdivision, constituted two subclasses; and they were quite distantly related to the Mariana isolates (Takami et al. 1999). On the other hand, these isolates were comparatively related to another subclass constituted by two alkaliphilic strains (HTB111 and HTB143) and *Halomonas meridiana* (Dobson et al. 1993). The 16S rDNA sequence of strain HTB148 was 99% similar to that of *Marinobacter aquaeolei*, which is phylogenetically distant from any other isolates. Three alkaliphilic isolates (HTB110, HTB139, HTB147) were found to be closely related to *Pseudomonas pseudoalcaligenes* (Moore et al. 1990). Finally, three neutrophilic isolates (HTB010, HTB019, HTB021), which constituted a subclass based on the results of 16S rDNA sequence analysis, were distantly related to the Mariana isolates but were related to *Curacaobacter baltica.*

We have thus obtained many isolates that differ in taxonomic affiliation, including some that belong to the proteobacteria β-subdivision and some of which belong

Table 6. Deep-sea isolates selected for 16S rDNA analysis

Strain no. (origin no.)	Growth conditions		Remarks
	Medium [a]	pH	
Alkaliphile			
HTB110 (214)	3M	9.6	
HTB111 (214)	3M	9.6	
HTB113 (214)	3M	9.6	
HTB138 (214)	MC	9.6	
HTB139 (214)	MC	9.6	
HTB143 (214)	MS	9.6	
HTB147 (214)	MS	9.6	
Halophile			
HTE831 (163)	15M	7.4	
HTB148 (214)	15M	7.4	
HTB069 (214)	15M	7.4	
HTC018 (213)	15M	7.4	
Alkaliphilic halophile			
HTB122 (214)	15M	9.3	
HTB123 (214)	15M	9.3	
Neutrophile			
HTE856 (163)	MS	7.6	Amylase productor
HTB082 (214)	3M	7.6	
HTB010 (214)	3M	7.6	
HTB015 (214)	3M	7.6	
HTB019 (214)	3M	7.6	
HTB021 (214)	MC	7.6	Protease productor
HTB028 (214)	MC	7.6	
HTB091 (214)	MS	7.6	Amylase productor
HTB095 (214)	MS	7.6	Amylase productor
HTB096 (214)	MS	7.6	
HTC042 (213)	MS	7.6	

All isolates were incubated at 25°C under atmospheric pressure.
[a] 3M, Marine broth containing 3% NaCl; 15M, Marine broth containing 15% NaCl; MC, Marine broth containing 1% skim milk; MS, Marine broth containing 1% starch.

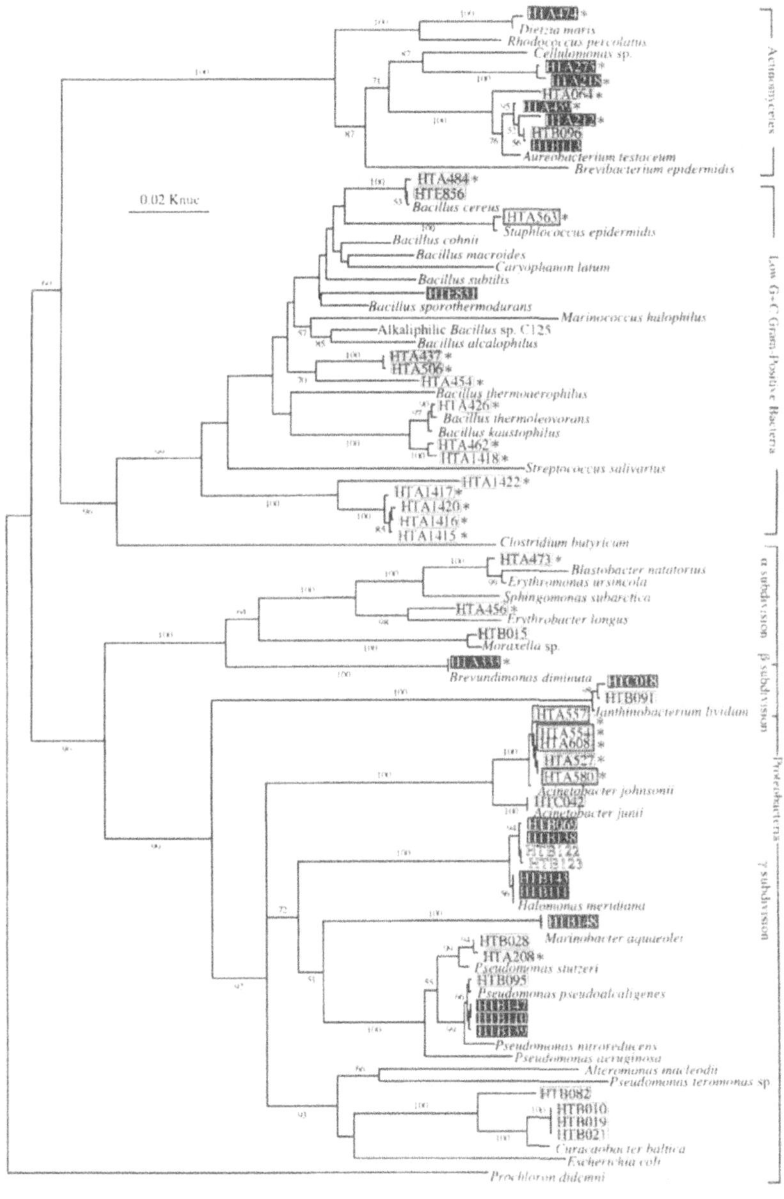

Fig. 4. Unrooted phylogenetic tree showing the relationship of bacteria isolated from the deep-sea to reference organisms. The numbers indicate the percentages of bootstrap samples, derived from 10 000 samples that supported the internal branches. Bootstrap probability values less than 50% were omitted from this figure. ABC, alkaliphile; ABC, halophile; ABC, neutrophile; ABC, thermophile; ABC, psychrophile; ABC, alkaliphilic halophile. * Isolates from the Mariana Trench (Takami et al. 1997)

to new subclasses not found among the Mariana isolates (Fig. 4). These findings demonstrate that various types of extremophilic bacteria are widely distributed in the deep-sea mud at depths of 1050-10 897 m near the southern part of Japan (Fig. 1).

1.5 Exploring Unique Enzyme Producers from the Deep-Sea for Industrial Application

As mentioned above, we have isolated thousands of microbes from sediments collected at several deep-sea sites from 1000 to 11 000 m depth to identify the microbial diversity in the deep-sea environment and showed that some of isolates can be categorized as new species. These results indicate that there is good possibility that many unknown microbes with unique and useful properties for human life thrive in the deep-sea environment. Therefore, we attempted to explore unique new enzyme producers using deep-sea isolates. We first focused on amylase producers because oligosaccharides, especially G3–G5, are widely used as a sweetener, nutrient, substrate for the blood α-amylase assay, and so on.

1.5.1 Screening Amylase Producers

The collected sea mud was spread onto Marine broth plates containing 1% potato starch and incubated at 4°, 25°, or 55°C. The amylase producers were detected as those showing formation of a halo around the colony. We isolated 39 amylase producers; and among them, strain MS300 formed the largest halo. Strain MS300 was found to be an aerobic, gram-negative, motile, rod-shaped bacterium. Test for oxidase and denitrification were positive. We identified MS300 as a member of species *Pseudomonas* on the basis of these data (Kobayashi et al. 1998).

1.5.2 Purification of Amylase Produced by *Pseudomonas* sp. MS300

MS300 was grown aerobically in the medium containing 10 g of potato starch (Difco Laboratories USA), 10 g of yeast extract (Difco), 10 g of polypeptone (Wako, Japan), and 10 g of NaCl per liter (PYP medium) or Marine broth 2216 (Difco) for 2 days at 25°C. The sodium dodecyl sulfate-polyacrylamide gel electrophoresis (SDS-PAGE) of culture supernatant showed only one prominent protein band, with its size estimated at 55 kDa (data not shown). A zymogram prepared using the culture supernatant showed four bands of amylase activity of which two were dominant amylases (A, B) and the other two were minor amylases (C, D). The pI of the major bands (amylase A and B) were determined to be 5.0 and 4.7, respectively. Purification of the two major amylases (A, B) was performed as follows. After removing

Table 7. Purification of the amylase produced by MS300

Step	Volume (ml)	Protein (mg)	Activity (U)	Specific activity (U/mg)
Culture supernatant	1800	286	1.6×10^5	550
$(NH_4)_2SO_4$ precipitation	152	176	9.1×10^4	439
DEAE-Toyopearl I	132	110	9.1×10^4	824
DEAE-Toyopearl II	8	26	2.8×10^4	1074
HPLC (Protein-Pak 300 × 2)				
F-1 (amylase A)	1.2	0.30	463	1544
F-2 (amylase B)	1.6	0.31	530	1775

cells by centrifugation, ammonium sulfate was added to the culture supernatant at 40% of saturation, and the mixture was held at 4°C for 16 h. After centrifugation (10 000 g, 30 min, 4°C) of the mixture, the precipitation was dialyzed against 10 mM Tris-HCl buffer (pH 7.5) at 4°C and applied to a diethylamino ethyl(DEAE)-Toyopearl column (i.d. 2.5×6 cm, 30 ml) equilibrated with 10 mM Tris-HCl (pH 7.5). The amylases were eluted with a linear gradient of 0–0.1 M NaCl in the same buffer. The fractions containing amylase activity were collected. These fractions were applied to a DEAE-Toyopearl column (i.d. 1.7×7.4 cm, 13 ml) equilibrated with 10 mM Tris-HCl (pH 8.5). The enzymes were eluted with a stepwise gradient of NaCl concentration of 40, 80, and 120 mM. The fractions containing amylase activity were collected. The amylase-containing fraction was applied to 2×Protein-Pak 300 column (Millipore Co.) and eluted with 25 mM 3-[n-morpholino]propanesulphonic acid (MOPS)-NaOH buffer (pH 7.0). The purity of each fraction was checked by means of a zymogram. The two major protein species (amylase A and B) could be separated in pure forms. We purified the two major amylases to the extent that a single band was obtained by SDS-PAGE in each instance, and it summarized amylase activities at each purification step (Table 7).

The N-terminal amino acid sequences of the two major amylases were found to be ADLAGKSPAGVRYHGGDEII. Similarly, the N-terminal amino acid sequences of the two minor amylases were identical. The nucleotide sequence of the G4 amylase (*amy* P) from *P. stutzeri* MO-19 has been reported (Fujita et al. 1989). The first 20 amino acids of the N-terminal sequence of amylase A and B showed 80% similarity with that of G4 amylase from *P. stutzeri* MO-19. However, other properties of this G4 amylase have not been reported.

1.5.3 Enzyme Profiles

Substrate Specificity and Product by Enzyme Reaction

We purified and characterized the two major amylases produced by strain MS300, which made the largest halo on the starch-containing medium among all of the Mariana isolates. MS300 was identified as a member of *Pseudomonas* species. This strain was found to produce two major amylases (A and B) and two minor amylases. Both amylases A and B generated maltotetraose from starch and did not react with α-, β- or γ-cyclodextrin (CD). The maltotetraose produced by these amylases from potato starch was the α-anomer (data not shown). Therefore, amylases A and B were identified as α-maltotetraohydrase (G4 amylase). G4 amylases from *Pseudomonas stutzeri* strains have been reported (Schmidt and John 1979). Robyt and Ackerman (1971) reported that *P. stutzeri* produced seven G4 amylases that differed in molecular mass and isoelectric point. Sakano et al. (1983) reported two G4 amylases that had the same molecular mass but different isoelectric points. Those G4 amylase isozymes had almost the same properties, and their N-terminal amino acid sequences were identical. Strain MS300 was found to produce four G4 amylases.

Effects of Temperature and pH on the Activity and Stability of Amylases

The optimum temperature for amylase A and B activity was 50°C in each instance, and these enzymes were inactivated by a 10-min incubation at 60°C (pH 7.0) (Fig. 5). Both enzymes maintained their full activities after incubation for 30 min at temperatures below 40°C at pH 7.0 (Fig. 5).

In the presence of 5 mM of $CaCl_2$, the activity at 60°C was about 1.3-fold (amy-

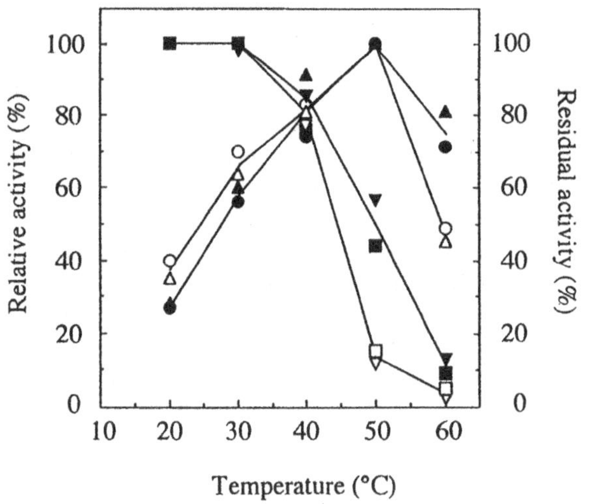

Fig. 5. Effects of temperature on the activity of amylases A and B. The relative activity of amylase A (*circles*) and amylase B (*triangles*) at each temperature in the presence or absence of 5 mM $CaCl_2$ was calculated with the activity observed at 50°C taken as 100%. The residual activity of amylase A (*inverted triangles*) and amylase B (*squares*) in the presence or absence of 5 mM $CaCl_2$ was measured after incubation for 30 min at each temperature. All solid symbols indicate the relative or residual activity of enzymes in the presence of 5 mM $CaCl_2$

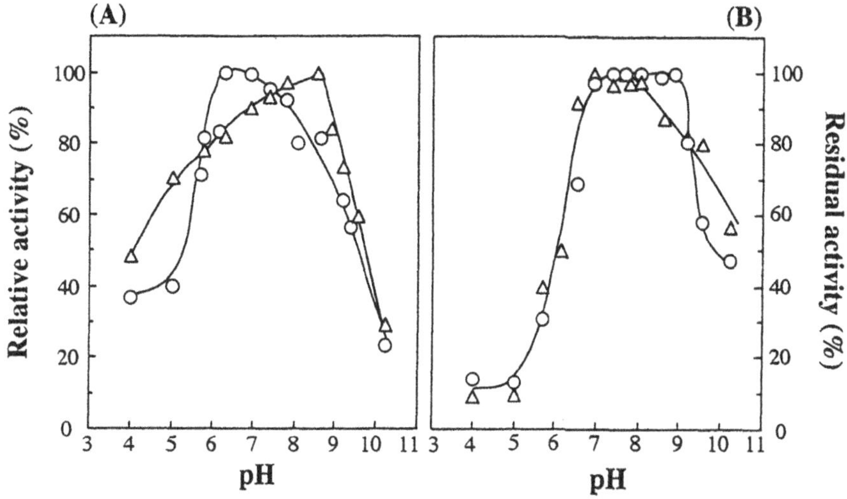

Fig. 6. Effects of pH on the activity of amylases A and B. **A, B** Relative activity and residual activity of amylase A (*circles*) and amylase B (*triangles*) under various pH conditions. The relative activity was calculated with the maximum activity taken as 100%. The residual activity was measured after incubation under various pH conditions for 30 min. Reaction mixtures were buffered with 50 mM of acetate-NaOH buffer (pH 4.0- 5.7), phosphate-NaOH buffer (pH 5.7- 6.3), 3-[n-morpholino]propanesulphonic acid (MOPS)-NaOH buffer (pH 6.5-7.8), Tris-HCl buffer (pH 7.8- 8.6), *N*-cyclohexyl-2-aminoethanesulfonic acid (CHES)-NaOH (pH 8.7- 9.5), or glycine-NaOH-NaCl buffer

Table 8. Effect of metal ions on amylase activity

Metal ions [a]	Relative activity (%)	
	Amylase A	Amylase B
Na^+	105	105
K^+	110	105
Ni^{2+}	41	38
Zn^{2+}	1	3
Co^{2+}	50	45
Fe^{2+}	90	85
Ca^{2+}	103	98
Cu^{2+}	2	0
Mn^{2+}	89	81
Hg^{2+}	1	3
Mg^{2+}	98	91
Pb^{2+}	19	24
Al^{3+}	83	82
Fe^{3+}	23	13
None	100	100

[a] All metal ions were added as the chloride salt at a concentration of 1 mM.

lase A) and 1.7-fold (amylase B) higher than that in the absence of calcium ions. The optimum pH for activity at 40°C was pH 6.3 for amylase A and pH 7.8 for amylase B (Fig. 6A). Amylase A was stable over the pH range 7.0–8.0, and amylase B was stable over the pH range 7.0–9.0 upon incubation at 40°C for 30 min (Fig. 6B).

The optimum pH for activity of amylase A and B differed. Furthermore, the pI, optimum temperature, and pH for amylase A and B activity were different from those of the G4 amylases from *P. stutzeri* (Figs. 5, 6AB). Although N-terminal amino acid sequences of amylase A and B were identical, their pI and properties were different. C-terminal amino acids of these enzymes may be changed. Thus, our findings indicate that amylase A and B are novel types of G4 amylases.

Effect of Metal Ions on Activity

The purified enzymes were treated in a starch solution (0.1%) with 1 mM metal chloride. As shown in Table 8, both amylases A and B were completely inhibited by 1 mM of Zn^{2+}, Cu^{2+}, and Hg^{2+} . Upon addition of 1 mM Mn^{2+}, Fe^{2+}, and Al^{3+}, slight inhibition (81%-90%) was observed (Table 8). None of the metal ion tested could selectively inhibit one of the enzymes.

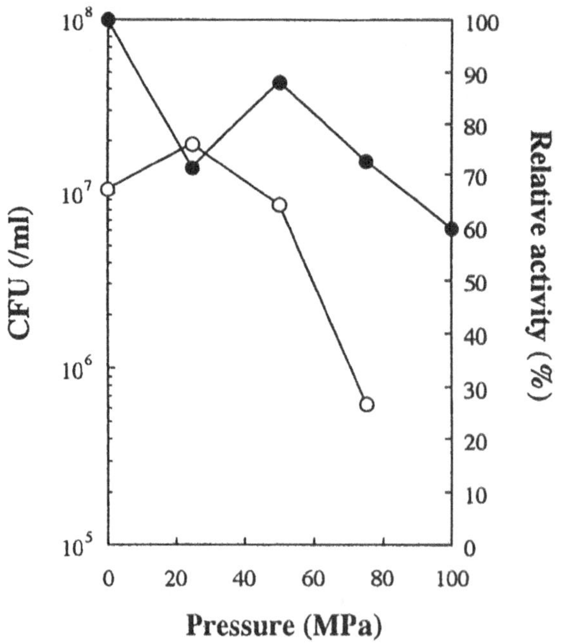

Fig. 7. Effects of pressure on the growth and amylase production by strain MS300. The cells were incubated at 4°C under various pressure conditions for 2 weeks. The number of colony forming units (CFU, *open circles*) was determined by counting the colonies that grew on Marine broth plates after incubation. The cell density before incubation was 4.0×10^4 CFU/ml. Amylase activity (*filled circles*) was measured by the 2-dinitrosalitiric reaction (Sumner 1925). Relative activity was calculated with maximum activity (0.1 MPa) taken as 100%

Effect of Hydrostatic Pressure on the Growth and Amylase Production of MS300

Strain MS300 was incubated at 4°C under various hydrostatic pressures for 2 weeks, and the number of colony-forming units (CFU) was found to increase at pressures less than 75 MPa, starting with the initial CFU of 4×10^5 /ml. Amylase activity was detected in the culture supernatant at pressures below 100 MPa (Fig. 7).

Optimum growth occurred at 25 MPa, and the number of viable cells was about twice that observed at 0.1 MPa. The CFU at 100 MPa could not be accurately determined because of cell aggregation. The initial amylase activity was not detected because the cells were washed before inoculation of the medium. The amount of amylase activity in the supernatant decreased under high hydrostatic pressure. The amylase activity at 75 MPa was 73% that at 0.1 MPa, although the CFUs at 75 MPa was 6% that at 0.1 MPa. Amylase activity was also detected at 100 MPa despite cell aggregation. The specific activity of purified enzymes was not influenced by hydrostatic pressure less than 100 MPa.

The microbial diversity in the Mariana trench has been reported as described above, but activity of the microbes there is still unknown. Mariana isolate MS300 could grow and produce amylase at pressures below 75 MPa in Marine broth (Fig. 7). The amylase activities of culture supernatant decreased with increasing pressure (Fig. 7). However, amylase activities per viable cell were increased at 60 and 80 MPa. The growth curve under high hydrostatic pressure was different from that under atmospheric pressure. Although MS300 cells aggregated at pressures below 100 MPa, amylase activity was detected in the culture supernatant, and turbidity of the culture began to increase when the culture was decompressed to 0.1 MPa (data not shown). MS300 was active at a very low level at pressures below 100 MPa.

Marine broth provides growth conditions far different from in situ conditions of the Mariana Trench. Our findings suggest that MS300 may be active at the sea bottom in the Mariana Trench. Although we have not fully established ideal methods for isolating and cultivating deep-sea microorganisms, the increased CFUs and turbidity have been monitored as an indicator of bacterial activity (Kato and Bartlett 1997). When the microorganisms grow at a very low rate, it is difficult to apply such an indicator to estimate bacterial growth. In the case of enzyme producers, not only the CFU but also the amount of enzyme activity present in the culture supernatant can be applied as an indicator of bacterial activity. Thus, enzyme producers are considered ideal candidates for studying a deep sea ecosystem.

Table 9. Effect of hydrostatic pressure on amylase production

Pressure (MPa)	Activity (U/ml)	Growth (OD_{660} [a])
0.1	3.6	2.0
30	4.5	2.1
60	4.8	1.9

[a] OD_{660} was measured after 72 h of incubation

Influence of High Hydrostatic Pressure on Amylase Productivity

MS300 cells grown to the stationary phase in YP medium were harvested and washed with same medium. Thereafter, the cells were suspended in fresh YP medium, and the cell suspension was adjusted to an optical density of 2.0 at 660 nm (OD_{660}). The culture was incubated at 25°C under pressure conditions of 0, 30, and 60 MPa for 3 days. The amount of amylase produced by MS300 was 3.7 U/ml in 50 ml of the culture under 0.1 MPa. As shown in Table 9, amylase productivity increased as the pressure was elevated. At 30 and 60 MPa, MS300 produced about 1.3 and 1.4 times, respectively, as much amylase as that observed under 0.1 MPa. The OD_{660} of these cultures did not change during the course of incubation. We investigated whether high hydrostatic pressure enhanced enzyme production of strain MS300 at the same cell density and found that this strain produced much more amylase under high pressure conditions (Table 9).

Strain MS300 could grow and produce amylase under 75 MPa (Fig. 7). It produced much larger amounts of amylase under high hydrostatic pressure than at atmospheric pressure (Table 9). The amylase activity in culture at 75 MPa was 73% that at atmospheric pressure despite lower cell density (Fig. 7). The mechanism responsible for the increase in amylase production at high pressure is still unknown. There are some reports that some deep-sea bacteria have pressure-regulated operons (Kato et al. 1995, 1996, 1997). High hydrostatic pressure influences phospholipid membrane fluidity (DeLong and Yayanos 1985) and causes a change in phospholipid composition of membranes (Kaneshiro and Clark 1995). These changes may promote enzyme secretion.

The deep-sea environment is thought to support a very low level of biological activity, and the variety of microorganisms are thought to be limited. However, we have isolated the novel type G4 amylase producer from this extreme environment. We are convinced that the deep sea is a valuable source of new microbial resources. Also, we can expect that the deep sea environment will provide many novel microbial biocatalysts.

References

Abe F, Horikoshi K (1995) Hydrostatic pressure promotes the acidification of vacuoles in *Saccharomyces cerevisiae*. FEMS Microbial Lett 130:307-312

DeLong EF, Yayanos AA (1985) Adaptation of the membrane lipids of a deep-sea bacterium to changes in hydrostatic pressure. Science 228:1101-1103

Dobson SJ, McMeekin TA, Franzmann, PD (1993) Phylogenetic relationships between some members of the genera Deleya, Halomonas, and Halovibrio. Int J Syst Bacteriol 43:665-673

Fujita M, Torigoe K, Tsusaki K, Kubota M, Sakai S, Tsujisaka Y (1989) Cloning and nucleotide sequence of the gene (amyP) for maltotetraose-forming amylase from *Pseudomonas stutzeri* MO-19. J Bacteriol 171:1333-1339

Higgins DG, Sharp PM (1988) Clustal: a package for performing multiple alignment on a microcomputer. Gene 73:237-244

Jannasch HW, Taylor CD (1984) Deep-sea microbiology. Annu. Rev. Microbiol. 38:487- 514

Kaneshiro SM, Clark DS (1995) Pressure effects on the composition and thermal behavior of lipids from the deep-sea thermophile *Methanococcus jannaschii*. J Bacteriol 177:3668-3672

Kato C, Bartlett DH (1997) The molecular biology of barophilic bacteria. Extremophiles 1:111-116

Kato C, Ikegami A, Smorawinska M, Usami R, Horikoshi K (1997) Structure of genes in a pressure-regulated operon and adjacent regions from a barotolerant bacterium strain DSS12. J Mar Biotechnol 5:210-218

Kato C, Smorawinska M, Sato T, Horikoshi K (1995) Cloning and expression in *Escherichia coli* of a pressure-regulated promoter region from a barophilic bacterium, strain DB6705. J Mar Biotechnol 2:125-129

Kato C, Smorawinska M, Sato T, Horikoshi K (1996) Analysis of a pressure-regulated operon from the barophilic bacterium strain DB6705. Biosci Biotechnol Biochem 60:166-168

Kirchman, DL (1993) Particulate detritus and bacteria in marine environments. In : Ford TE (ed) Aquatic microbiology. Blackwell Scientific , Cambridge, pp 321-341.

Kobayashi H, Takami H, Inoue A, Horikoshi K (1998) Deep-sea, new resource for screening of biocatalysts. In: Kieslich K et al.(eds) New frontiers in screening for microbial biocatalysts. Elsevier Science, Amsterdam, pp 259-264

Moore ERB, Mau M, Arnscheidt A, Bottger EC, Hutson RA, Collins MD, van de Peer Y, de Wachter R, O'Sullivan J, McCullough J, Johnson JH, Bonner DP, Clark JC, Dean L, Trejo WH (1990) Janthinocins A, B and C, novel peptide lactone antibiotics produced by *Janthiobacterium lividum*. I. Taxonomy, fermentation, isolation, physicochemical and biological characterization. J Antibiot 43:913-919

Olsen GJ, Wose CR, Overbeek R (1994) The winds of (evolutionary) change: breathing new life into microbiology. J Bacteriol 176:1-6

O'Sullivan J, McCullough J, Johnson JH, Bonner DP, Clark JC, Dean L, Trejo WH (1990) Janthinocins A, B and C, novel peptide lactone antibiotics produced by *Janthiobacterium lividum*. I. Taxonomy, fermentation, isolation, physicochemical and biological characterization. J Antibiot 43:913-919

Robyt J, Ackerman R (1971) Isolation, purification and characterization of a maltotetraose-producing amylase from *Pseudomonas stutzeri*. Arch Biochem Biophys 145:105-109

Saitou N, Nei M (1987) A neighbor-joining method: a new method for reconstructing phylogenetic trees. Mol Biol Evol 44:406-425

Sakano Y, Kashiyama E, Kobayashi T (1983) Purification of α-maltotetraose-forming exo-amylase of *Pseudomonas stutzeri*: two-forms of the amylase and their enzymatic properties. Agric Biol Chem 47:1761-1768

Schmidt J, John M (1979) Starch metabolism in *Pseudomonas stutzeri*. I. Studies on maltotetraose-forming amylase. Biochim Biophys Acta 566:88-89

Silver MW, Shanks AL, Trent JD (1978) Marine snow: microplankton habitat and source of small-scale patchness pelagic populations. Science 201:371-374

Sneath PHA, Mair NS, Sharpe ME, Holt JG (1986) Bergey's manual of systematic bacteriology, vol 2. Williams & Wilkins, Baltimore

Sumner JB (1925) A more specific reagent for the determination of sugar in urine. J Biol Chem 47:393-395

Takagawa S, Takahashi K, Sano T, Mori Y, Nakanishi T, Kyo M (1989) 6500 m Deep manned research submersible *Shinkai* 6500 system. Oceans 89:741-746

Takami H, Inoue A, Fuji F, Horikoshi K (1997) Microbial flora in the deepest sea mud of the Mariana Trench. FEMS Microbiol Lett 152:279-285

Takami H, Kobata K, Nagahama T, Kobayashi H, Inoue A, Horikoshi K (1999) Biodiversity in deep-sea sites near the south part of Japan. Extremophiles 3:97-102

Yayanos AA (1979) Isolation of a deep-sea barophilic bacterium and some of its growth characteristics. Science 205:808-810

Yayanos AA (1981) Obligately barophilic bacterium from the Mariana Trench. Proc Natl Acad Sci USA 78:5212-5215

Yayanos AA (1995) Microbiology to 10,500 meters in the deep sea. Annu Rev Microbiol 49:777-805

Yayanos AA, Dietz AS, Boxtel RV (1981) Obligately barophilic bacterium from the Mariana Trench. Proc Natl Acad Sci USA 78:5212-5215

ZoBell CE (1941) Studies on marine bacteria. I. The cultural requirements of heterotrophic aerobes. J Mar Res 4:42-75

2 Molecular Analyses of the Sediment and Isolation of Extreme Barophiles from the Deepest Mariana Trench

Chiaki Kato

The Mariana Trench, Challenger Deep (11°22'N. 142°25'E), is the deepest ocean bottom in the world. Sediment samples have been obtained from this extreme environment by the unmanned submersible Kaiko (Kyo et al. 1995) in three dives(nos. 21, 22, and 23) at a depth of 11 000 m (Kato et al. 1997). It seems likely that these were the first sediment samples recovered from the world's deepest point with no microbiological contamination. In 1960 Jacques Piccard and D. Walsh dived to the world's deepest point aboard the manned submersible Trieste. They reported seeing several deep-sea animals and fishes there (Piccard and Dietz 1961). Unfortunately, that dive was not for scientific purposes, and no films were made or samples collected, so their story could not be confirmed scientifically.

The Mariana Trench is characterized by cold, dark, remoteness, with the highest pressure environment in the ocean bottom. Relatively little is known about the natural bacterial communities in the sediment there because of the need to use uncommon instruments to keep deep-sea organisms at high pressure and to develop new methods to maintain and cultivate these organisms in the laboratory. Two isolates, *Pseudomonas bathycetes* and an obligately barophilic bacterium strain MT41, have been isolated from the Mariana Trench (Morita 1976; Yayanos et al. 1981). Apparently, *P. bathycetes* is not an organism adapted to the deep-sea environment, as its generation time is 33 days under such conditions (2°C, 100 MPa) (Schwarz and Colwell 1975). This bacterium grows at atmospheric pressure and at temperatures as high as 25°C (Pope et al. 1975). Bacterium strain MT41 was isolated from a deep-sea amphipod (*Hirondellea gigas*) obtained from the Mariana Trench using an insulated trap (Yayanos et al. 1981). This bacterium could grow only under pressure

---

Key words. Deep-sea, Mariana Trench, Phylogenetic tree, 16S Ribosomal RNA, Extremophiles, Barophilic bacteria conditions greater than 50 MPa; it is closely related to the genus *Colwellia* (Yayanos et al. 1981; DeLong et al. 1997).

The Mariana Trench is a challenging point in the field of ecology and evolution. To understand the natural bacterial communities in sediment from the Mariana Trench, we have undertaken a molecular analysis of the sediment and isolation of unique microorganisms in it, which we describe in the following sections.

2.1 Molecular Phylogenetic Analyses of the Sediment

2.1.1 PCR Amplification of DNA from the Mariana Trench

The DNA fragments encoding the rRNA small-subunit were amplified by the polymerase chain reaction (PCR) using the primers Eubac27F and Eubac1492R (for bacterial 16S rRNA), Arch21F and Arch958R (for archaeal 16S rRNA), and EukF and EukR (for eukaryal 18S rRNA) from the purified DNA extracted from the Mariana sediment. The results are shown in Fig. 1 (Kato et al. 1997). DNA fragments corresponding to the rRNA small-subunit sequences of bacteria and archaea were clearly amplified, but no DNA corresponding to the eukaryal rRNA small-subunit was amplified. These amplified fragments were then cloned and sequenced.

2.1.2 Phylogenetic Relationships of Amplified Bacterial 16S rDNA Sequences

The amplified bacterial 16S rDNA sequences were determined and compared with sequences in the DNA databases. The results suggested that at least three kinds of

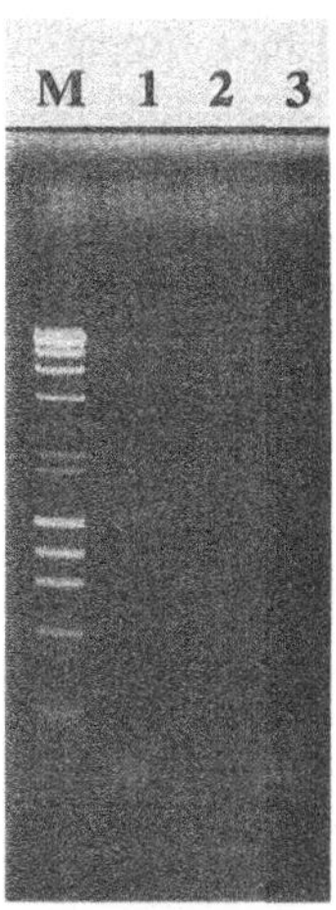

M: λ *Hind*III · ϕX174 *Hae*III marker

1: Eubacteria 16S rDNA

2: Archaea 16S rDNA

3: Eukaryote 18S rDNA

Fig. 1. Gel electrophoresis of polymerase chain reaction (PCR) products amplified with 16S and 18S rDNA primers. Lane M, molecular size markers (*Hind*III-digested λ DNA and *Hae*III digested ϕX174 DNA); lanes 1–3, amplified fragments of eubacterial 16S ribosomal DNAs, archaeal 16S ribosomal DNAs, and eukaryotes 18S ribosomal DNAs from the DNA purified from the Mariana sediment

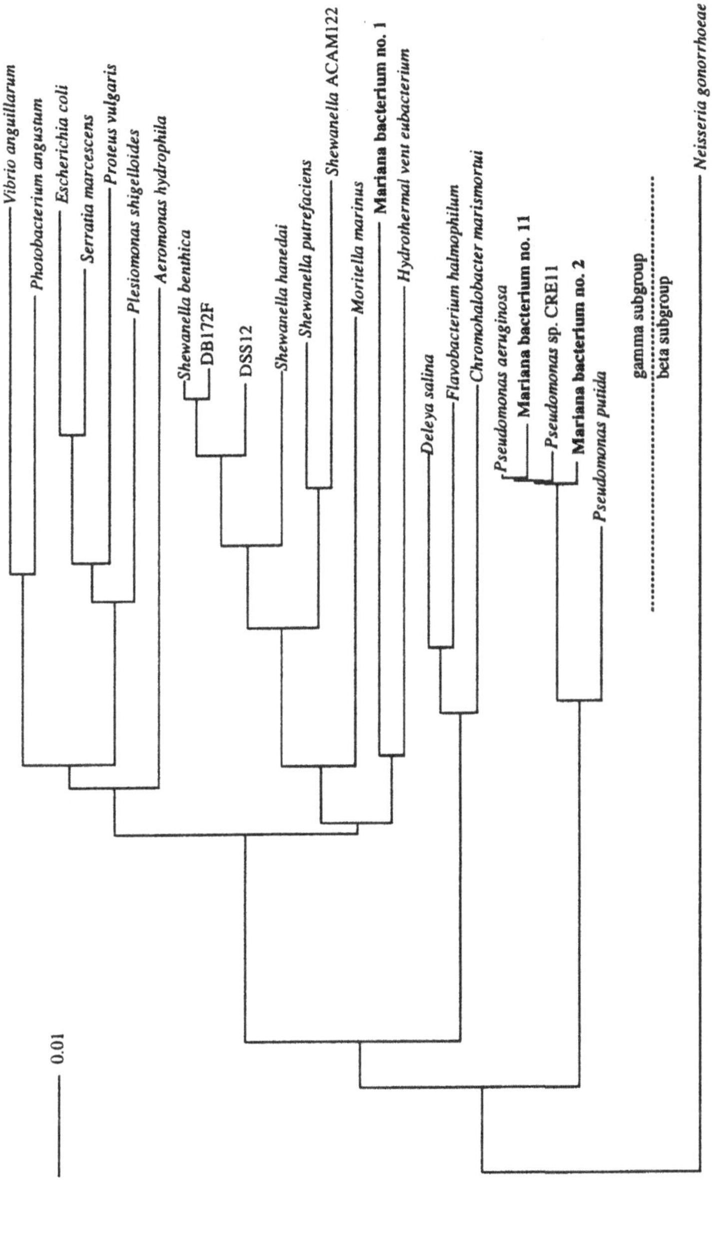

Fig. 2. Phylogenetic tree showing the relationship of amplified eubacterial 16S rDNA within the Proteobacterium gamma subgroup using the neighbor-joining method. The strains DSS12 and DB172F are barophilic bacteria isolated in our laboratory from deep-sea sediment samples collected at a depth of 5100–6500 m (Kato et al. 1995, 1996). *Neisseria gonorrhoeae* (Proteobacteria β-subgroup) 16S rDNA sequence was used as an out-group to make the tree

sequences from the bacterial domain were present in DNA isolated from the sediment and belonged to the γ-subdivision of Proteobacteria. The phylogenetic relationships between the amplified sequences and this subgroup of bacteria are shown in Fig. 2 (Kato et al. 1997). Two of the amplified sequences (Mariana bacterium no. 2 and no. 11) were closely related to the *Pseudomonas* branch, and another one (Mariana bacterium no. 1) belonged to the branch containing *Shewanella* and barophilic bacteria (Kato et al. 1995, 1996). The first bacterium isolated from the Mariana Trench was *Pseudomonas bathycetes* (Morita 1976), so it seems reasonable that members of the genus *Pseudomonas* were identified as microbial residents of the sediment.

2.1.3 Phylogenetic Relationships of Amplified Archaeal 16S rDNA Sequences

The amplified archaeal 16S rDNA sequences were also determined and compared with the DNA databases. The phylogenetic relationships between these amplified sequences and archaea are shown in Fig. 3 (Kato et al. 1997). The results indicate that at least three archaeal sequences were present in DNA from the sediment and belonged to the planktonic archaea group I reported by DeLong (1992). These planktonic archaea were unculturable strains cloned from ocean waters, and group I belongs to one of the major branches of archaea; however, group I shares no close evolutionary relationship with previously cultured archaea (DeLong 1992; Stein et al. 1996). These unculturable archaea are widely distributed in the ocean. Recently, the group I archaea were also identified from the midgut of the deep-sea holothurians as a paraphyletic component (McInerney et al. 1995). It is still unknown whether the group I archaea have some physiological functions as a symbiont in such deep-sea animals. Our results indicate that the planktonic archaea exist at the sea bottom of the Mariana Trench, and these archaea probably arrived from the middle ocean with planktonic particles that settled there like a marine snow (Kato et al. 1997).

2.2 Isolation and Characterization of the Extremely Barophilic Bacteria

2.2.1 Analysis of Barophilic Isolates

Six obligately barophilic strains that were able to grow well at 100 MPa but not able to grow at atmospheric pressure were isolated from the Mariana Trench sediment. The pressure-regulated operons particularly identified in *Shewanella* barophilic strains (Li et al. 1998) were amplified by PCR from DNAs extracted from three strains (DB21MT-2, DB21MT-4, DB21MT-7) but were not amplified from DNAs from the other three strains, (DB21MT-5, DB21MT-6, DB21MT-9). Sequence analysis sug-

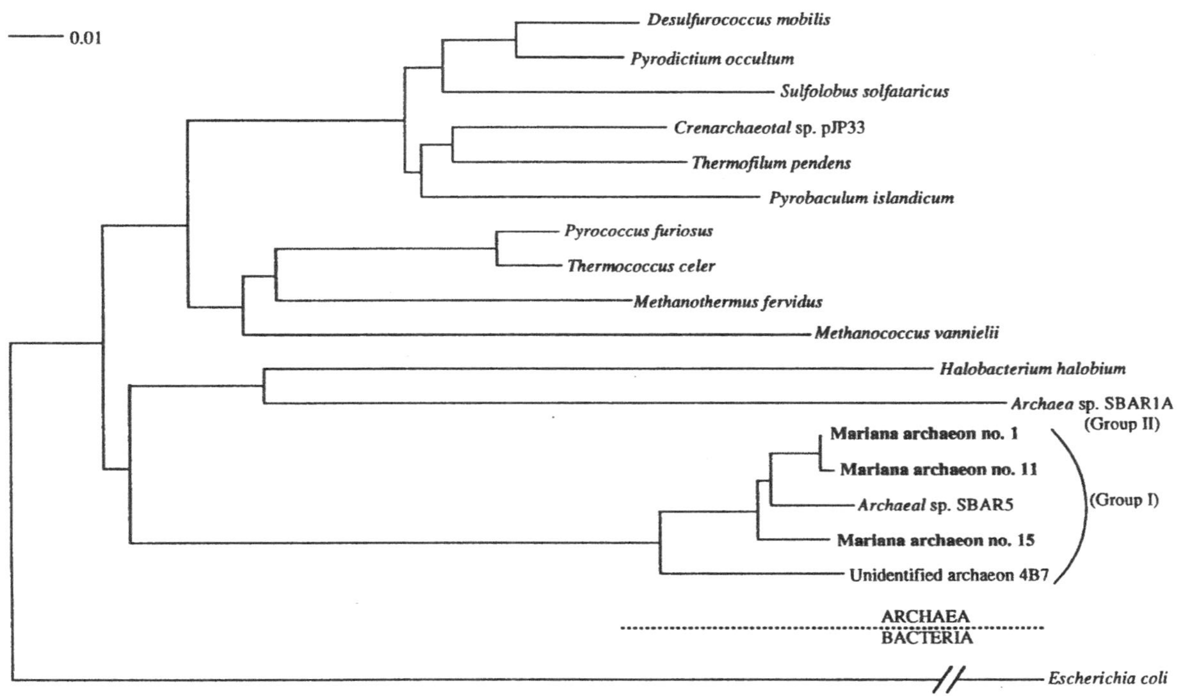

Fig. 3. Phylogenetic tree showing the relationship of amplified archaeal 16S rDNA within the Archaeal branch using the neighbor-joining method. *Escherichia coli* (eubacteria) 16S rDNA sequence was used as an out-group of the tree. *Archaea* sp. designated as clone SBAR5 and clone SBAR1A were reported by DeLong (1992), and unidentified archaeon 4B7 was reported by Stein et al. (1996). *Crenarchaeotal* sp. was an unculturable archaeon isolated from a Yellowstone National Park hot spring environment and designated as clone pJP41 (Barns et al. 1994)

gested that two different 16S rDNA sequences were present, one in the three isolates with the pressure-regulated operons and the other in the three isolates without these operons. Therefore, two barophilic strains, DB21MT-2 and DB21MT-5, were chosen for the following studies.

2.2.2 Growth Properties Under Hydrostatic Pressures

The cells of strain DB21MT-2 were 2.0 × 0.8–1.0 μm in size, and those of strain DB21MT-5 were a little bigger, 2.5×1.0 to 3.0×1.0 μm. Both of these strains were

A 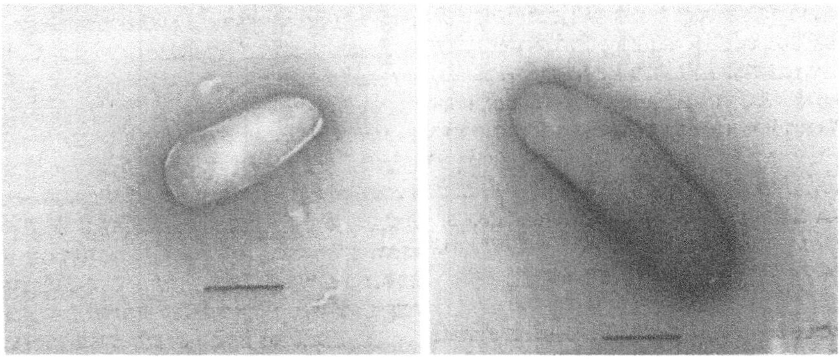 B

Fig. 4. Electron micrographs (negative staining) of extremely barophilic bacteria (strains DB21MT-2 [A] and DB21MT-5 [B]). Bar 1 μm

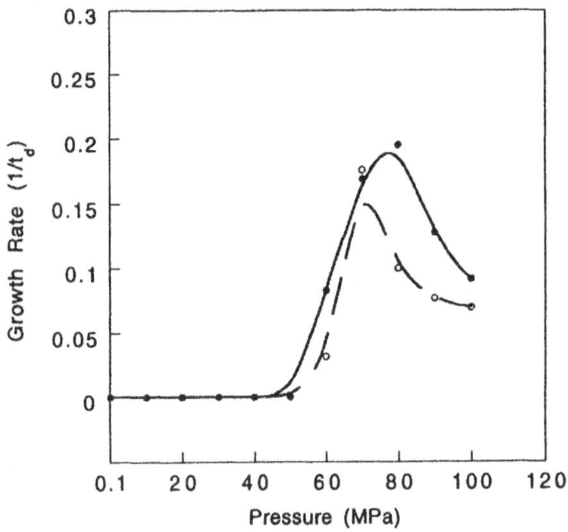

Fig. 5. Growth properties of the extreme barophiles at elevated hydrostatic pressure. *Dotted line*, DB21MT-2; *solid line*, DB21MT-5. t_d indicates the doubling time (hour)

found to have a single polar flagellum (Fig. 4). As shown in Fig. 5, the optimal pressure conditions for growth of strains DB21MT-2 and DB21MT-5 were 70 MPa and 80 MPa, respectively. Neither strain was able to grow at pressures below 50 MPa, but both were able to grow well at higher pressures, even at 100 MPa. Therefore, strains DB21MT-2 and DB21MT-5 are defined as "extremely" barophilic bacteria (Kato et al. 1998).

2.2.3 Phylogenetic Analysis of Extremely Barophilic Isolates

A phylogenetic tree was constructed based on the 16S rDNA sequences of the extremely barophilic isolates and the members of bacteria in the γ-subdivision of Proteobacteria (Kato et al. 1998). As illustrated in Fig. 6, strain DB21MT-2 is closely related to *Shewanella benthica* and other barophilic strains belonging to the "*Shewanella* barophiles branch." This result corresponds well with the findings shown above, as this strain was found to have the pressure-regulated operon identified only in members of this branch (Li et al. 1998). The other isolated strain (DB21MT-5) is

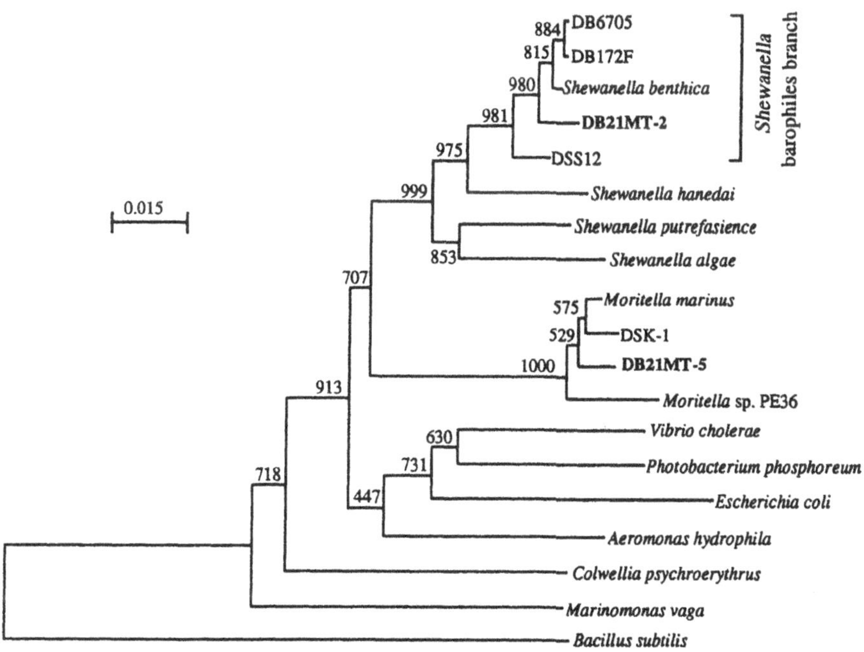

Fig. 6. Phylogenetic tree showing the relationships between the isolated bacteria within the γ-Proteobacteria. Bar represents the average number of nucleotide substitutions per site. Bootstrap values are shown for frequencies per 1000 times

Table 1. Levels of homology for chromosomal DNAs of the extremely barophilic isolates and closely related strains

(a) Strain DB21MT-2

Strain	% Homology with DNA from						
	S. han	*S. put*	*S. ben*	172F	6705	MT-2	DSS12
S. han	100	7	7	5	13	10	16
S. put	15	100	6	6	11	9	9
S. ben	12	7	100	79	88	80	47
172F	16	9	90	100	96	77	39
6705	14	10	91	79	100	84	46
MT-2	20	21	81	79	78	100	48
DSS12	11	13	47	43	45	40	100

S. han, *Shewanella hanedai*; *S. put*, *S. putrefaciens*; *S. ben*; *S. benthica*; 172F, strain DB172F; 6705, strain DB6705; MT-2, strain DB21MT-2; DSS12, strain DSS12.

(b) Strain DB21MT-5

Strain	% Homology with DNA from				
	S. put	*S. ben*	*M. mar*	DSK1	MT-5
S. put	100	8	9	10	7
S. ben	7	100	11	14	12
M. mar	9	8	100	33	28
DSK1	16	4	17	100	22
MT-5	10	8	32	41	100

M.mar, *Moritella marinus*; DSK1, strain DSK1; MT-5, strain DB21MT-5.

closely related to *Moritella marina* and the barotolerant strain DSK1 reported by Kato et al. (1995).

2.2.4 DNA–DNA Hybridization Analysis

The results of DNA–DNA hybridization analysis comparing the isolated strains, DB21MT-2 and DB21MT-5, and the related bacterial species suggested from the 16S rDNA phylogenetic tree (Fig. 6) are shown in Table 1 (Kato et al. 1998). A high level of DNA–DNA relatedness between strain DB21MT-2, the previously reported barophilic isolates DB6705 (Kato et al. 1995) and DB172F (Kato et al. 1996), and *Shewanella benthica* was evident (77%-84%), indicating that these strains might be members of the same species (Wayne et al. 1987). The hybridization values obtained for strain DB21MT-5 and the *Moritella* and *Shewanella* reference strains were significantly lower than that accepted as the phylogenetic definition of a species

Table 2. Fatty acid compositions of the isolated strains and type strains [a]

Strains	Fatty acids							
	14:0	14:1	15:0	16:0	16:1	18:1	20:5 (EPA)	22:6 (DHA)
DB21MT-2	4		5	22	31	19	18	
S. benthica	21		6	18	45	1	10	
DB21MT-5	25	12		9	45	1		8
M. marina	16	1	1	21	45	2		12

Results are percents.
EPA, eicosapentaenoic acid; DHA, docosahexaenoic acid.
[a] Type strains were grown at atomospheric pressure and 4°C (*S. benthica*) or at 15°C (*M. marina*).

described by Wayne et al. (1987). In addition to the phylogenetic analysis, this finding suggested that strain DB21MT-5 is new species of the genus *Moritella*. Hence we have reported this strain as *Moritella yayanosii* sp. nov. (Nogi and Kato 1999).

2.2.5 Fatty Acids Analysis

The whole-cell fatty acid composition of the isolated barophilic strains DB21MT-2 and DB21MT-5 and that of selected reference strains, *S. benthica* and *M. marinus*, respectively, are shown in Table 2 (Kato et al. 1998). The dominant fatty acid in the isolated barophiles was hexadecenoic acid (16:1), and the long-chain polyunsaturated fatty acids (PUFAs) present in significant amounts were eicosapentaenoic acid (EPA) (20:5) in strain DB21MT-2 and docosahexaenoic acid (DHA) (22:6) in strain DB21MT-5. The fatty acid profiles of these isolated strains DB21MT-2 and DB21MT-5 were basically similar to those of the reference species *S. benthica* and *M. marinus*, respectively, particularly the PUFA patterns. Around 70 % of the membrane lipids in both extreme barophiles were unsaturated fatty acids; and compared with the reference strains, higher amounts of octadecenoic acid (18:1) in strain DB21MT-2 and tetradecenoic acid (14:1) in strain DB21MT-5 were evident.

2.2.6 Features of the Mariana Extremely Barophilic Bacteria

Extremely barophilic bacteria—defined as bacteria unable to grow at pressures below 50 MPa but able to grow well at 100 MPa—were isolated from the sediment of Mariana Trench Challenger Deep at a depth of 10 898 m (Kato et al. 1998). One of the isolates, strain DB21MT-2, which shows optimal growth at a pressure of 70 MPa, belongs to the species *Shewanella benthica*, based on 16S rDNA sequence comparisons and DNA–DNA hybridization analysis (Fig. 6; Table 1a). This strain particularly produces EPA as a component of its membrane lipids (Table 2). All of the barophilic *Shewanella* strains isolated from the deep-sea produce EPA. Another isolate, strain DB21MT-5, which shows optimal growth at a pressure of 80 MPa, is closely related to the genus *Moritella* based on 16S rDNA sequence comparisons, but DNA–DNA hybridization results suggest that this strain does not belong to the same species as known *Moritella* strains (Fig. 6; Table 1b). This strain produces DHA as a component of its membrane lipids (Table 2). *M. marinus* also produces DHA (Table 2); thus, strain DB21MT-5 may share some properties with this species, except for its pressure requirements for growth. The *Moritella* strains reported are not obligately barophilic; they are barotolerant (strain DSK1) and facultatively barophilic (strain PE36) deep-sea bacteria.

Production of EPA is one of the particular characteristics of deep-sea adapted *Photobacterium* species, such as *Photobacterium profundum* (Nogi et al. 1998), and the incidence of such PUFAs in barophilic genera suggests that PUFAs may play an important role in maintenance of membrane fluidity at low temperature and high pressure (DeLong and Yayanos 1986).

The first obligately barophilic strain isolated from the Mariana Trench was strain MT41 (Yayanos et al. 1981), which was unable to grow at pressures below 50 MPa; its optimal pressure for growth was around 100 MPa (Yayanos 1986). Thus this strain is also an extremely barophilic bacterium. It is not related to the genus *Shewanella* or *Moritella* but is closely related to the genus *Colwellia* (DeLong et al. 1997). Deming et al. (1988) reported that optimal pressure for growth of the obligately barophilic bacterium strain BNL-1 was 93 MPa at 10°C and 74 MPa at 2°C. Strain BNL-1 was designated *Colwellia hadaliensis*, and this strain is similar to extreme barophiles. *Shewanella* sp. strains PT99, PT48 (DeLong and Yayanos 1986), and DB172F (Kato et al. 1996) are also obligate barophiles, showing optimal growth at pressures of 69, 62, and 70 MPa, respectively. However, these *Shewanellas* are also able to grow at pressures below 50 MPa, so they are not extremely barophilic strains. A facultatively barophilic strain PE36 has been reported (Yayanos 1986) among the members of the genus *Moritella*,but no obligately barophilic strains are known.

Therefore this is the first evidence of the existence of an obligately barophilic *Moritella* sp. from the deep-sea environment, and it is an extremely barophilic bacterium. The strain DB21MT-5 was studied for more details taxonomically and was named as *Moritella yayanosii* sp. nov. (Nogi and Kato 1999)(see Chapter 5).

References

Barns SM, Fundyga RE, Jeffries MW, Pace NR (1994) Remarkable archaeal diversity detected in a Yellowstone National Park hot spring environment. Proc Natl Acad Sci USA 91:1609-1613

DeLong EF (1992) Archaea in coastal marine environments. Proc Natl Acad Sci USA 89:5685-5689

DeLong EF, Yayanos AA (1986) Biochemical function and ecological significance of novel bacterial lipids in deep-sea procaryotes. Appl Environ Microbiol 51:730-737

DeLong EF, Franks DG, Yayanos AA (1997) Evolutionary relationships of cultivated psychrophilic and barophilic deep-sea bacteria. Appl Environ Microbiol 63:2105-2108

Deming JW, Somers LK, Straube WL, Swartz D G, MacDonell MT (1988) Isolation of an obligately barophilic bacterium and description of a new genus, *Colwellia* gen. nov. Syst Appl Microbiol 10:152-160

Kato C, Li L, Nogi Y, Nakamura Y, Tamaoka J, Horikoshi K (1998) Extremely barophilic bacteria isolated from the Mariana Trench, Challenger Deep at a depth of 11,000 m. Appl Environ Microbiol 64:1510-1513

Kato C, Li L, Tamaoka J, Horikoshi K (1997) Molecular analyses of the sediment of the 11 000 m deep Mariana Trench. Extremophiles 1:117-123

Kato C, Masui N, Horikoshi K (1996) Properties of obligately barophilic bacteria isolated from a sample of deep-sea sediment from the Izu-Bonin Trench. J Mar Biotechnol 4:96-99

Kato C, Sato T, Horikoshi K (1995) Isolation and properties of barophilic and barotolerant bacteria from deep-sea mud samples. Biodiv Conserv 4:1-9

Kyo M, Miyazaki E, Tsukioka S, Ochi H, Amitani Y, Tsuchiya T, Aoki T, Takagawa S (1995) The sea trial of "KAIKO," the full ocean depth research ROV. Oceans'95 3:1991-1996

Li L, Kato C, Nogi Y, Horikoshi K (1998) Distribution of the pressure-regulated operons in deep-sea bacteria. FEMS Microbiol Lett 159:159-166

McInerney JO, Wilkinson M, Patching JW, Embley TM, Powell R (1995) Recovery and phylogenetic analysis of novel archaeal rRNA sequences from a deep-sea deposit feeder. Appl Environ Microbiol 61:1646-1648

Morita RY (1976) Survival of bacteria in cold and moderate hydrostatic pressure environments with special reference to psychrophilic and barophilic bacteria. In: Gray TRG, Postgate JR (eds) The survival of vegetative microbes. Cambridge Univ. Press, Cambridge pp.279-298

Nogi Y, Kato C (1999) Taxonomic studies of extremely barophilic bacteria isolated from the Mariana Trench and description of *Moritella yayanosii* sp. nov., a new barophilic bacterial isolate. Extremophiles 3:71-77

Nogi Y, Masui N, Kato C (1998) *Photobacterium profundum* sp. nov., a new, moderately barophilic bacterial species isolated from a deep-sea sediment. Extremophiles 2:1-7

Piccard J, Dietz RS (1961) Seven miles down. Putnum, New York

Pope DH, Smith WP, Swartz RW, Landau JV (1975) Role of bacterial ribosomes in barotolerance. J Bacteriol 121:664-669

Schwarz JR, Colwell RR (1975) Macromolecular biosynthesis in *Pseudomonas bathycetes* at deep-sea pressure and temperature, session 104, abstr. K.94. In: Abstracts, 75th Annual Meeting of the American Society for Microbiology. American Society for Microbiology, Washington DC, p162

Stein JL, Marsh TL, Wu KY, Shizuya H, DeLong EF (1996) Characterization of uncultivated prokaryotes: isolation and analysis of a 40-kilobase-pair genome fragment from a planktonic marine archaeon. J Bacteriol 178:591-599

Wayne LG, Brenner DJ, Colwell RR, Grimont PAD, Kandler O, Krichevsky MI, Moore LH, Moore WEC, Murray RGE, Stackbrandt E, Starr MP, Truper HG (1987) Report of the Ad Hoc Committee on reconciliation of approaches of bacterial systematics. Int J Syst Bacteriol 37:463-464

Yayanos AA (1986) Evolutional and ecological implication of the properties of deep-sea barophilic bacteria. Proc Natl Acad Sci USA 83:9542-9546

Yayanos AA, Dietz AS, Boxtel RV (1981) Obligately barophilic bacterium from the Mariana Trench. Proc Natl Acad Sci USA 78:5212-5215

3 Taxonomy and Biotransformation Activities of Deep-Sea Actinomycetes

Alan T. Bull, Joy A. Colquhoun, and Stephen C. Heald

Microbial diversity as a source of innovation for biotechnology is a major interest of our laboratory (Bull et al. 1992), and it is in this context that the deep-sea environment has proved such an attractive one to explore. Despite the fact that 60% of the Earth's surface is covered by seas of depths exceeding 2000 m, the inventory of microorganisms in the deep-sea is very incomplete. Recent surveys of the open oceans, hydrothermal vents, and the deepest ocean floor (Jollivett 1996; DeLong 1997; Takami et al. 1997) have revealed unsuspected diversity and novelty in their microbiota. Our research has focused on bacteria of soft deep-sea sediments.

In this section we report the results of taxonomic and biotransformation studies aimed at detecting novelty among deep-sea bacteria. Actinomycetes were chosen for this study because (1) they have been reported previously in marine ecosystems, (2) the taxonomy of terrestrial actinomycetes is well established and provides an excellent background against which to assess novelty in the deep-sea, and (3) they have singular value from a biotechnological perspective. Although reports of actinomycetes in marine sediments go back several decades, for the most part the reference to actinomycetes was incidental, or the sediments were from shallow in-shore locations (Goodfellow and Haynes 1984; Jensen et al. 1991; Takizawa et al. 1993). Moreover, the few deep-sea sediments examined contained relatively low numbers of actinomycetes (Weyland 1969; Goodfellow and Haynes 1984; Helmke and Weyland 1984), and it has not been established whether these were truly indigenous organisms or simply residual populations from terrestrial wash-in. Consequently we have analyzed sediments obtained from the complete range of bathyal, abyssal, and hadal depths and have used taxon-designed selective procedures for isolating

Key words. Deep-sea, Mariana Trench, Phylogenetic tree, 16S Ribosomal RNA, Actinomycetes, Selective isolation, Pyrolysis-mass spectrometry, Nitrile transformations, High pressure streptomycetes, micromonosporae, and mycolate (mycolic acid-containing) actinomycetes. We have assessed the terrestrial wash-in of actinomycetes by detecting the presence of thermoactinomycetes in sediment samples (Attwell and Colwell 1984); and profiling isolates in terms of their temperature, salinity, and pressure requirements and tolerances has provided evidence for an indigenous actinomycete community in deep-seas.

This phase of our deep-sea actinomycete program has focussed on culturable, aerobic, barotolerant populations. To circumscribe large numbers of isolates rapidly we have deployed pyrolysis mass spectrometry as a means of detecting phenotypic novelty and evaluated its reliability by making polyphasic taxonomic analyses of the clusters generated. The potential of deep-sea actinomycetes as a source of novel biotransformations was explored with respect to their nitrile-metabolizing activities. Such activities are exploitable for chemical syntheses and for biotreatment of toxic environmental chemicals, and they are well documented in terrestrial bacteria (Wyatt and Linton 1988).

3.1 Culturable Actinomycetes

Culturable bacteria, isolated on 10% tryptic soy agar fomulated with artificial seawater (Takizawa et al. 1993), were recovered from all of the deep-sea sites examined (Table 1). There was no obvious correlation between population density and depth. The highest counts were obtained from both shallow sites (Sea of Japan, Sagami and Suruga Bays) and certain abyssal and hadal sites in the Ryukyu and Japan Trenches; but even at the extreme depth of the Challenger Deep nearly 10^3 bacteria per milliliter of wet sediment were recovered. Numbers of the target actinomycetes (Table 1) in most cases were approximately 10^3/ml wet sediment (i.e., less than 1% of the total culturable bacteria), but in one case (Japan Trench, 6455m) actinomycetes constituted nearly 9% of this total. These numbers are comparable to those recorded in relatively shallow marine sediments [Chesapeake Bay (Takizawa et al. 1993); coastal sediments in the United Kingdom (Goodfellow and Haynes 1984)] and the few deep-sea sediments that have been investigated (Weyland 1969; Goodfellow and Haynes 1984). In the sediments we have examined, the dominant actinomycetes appear to belong to mycolate taxa.

The highest counts of *Thermoactinomyces* strains also were found in shallow sea sediments (Colquhoun et al. 1998b), with maximum numbers recovered at 289 m from the Sea of Japan (1.7 x 10^5 CFU/ml wet sediment). Such data are consistent with the ingress of terrigenous microorganisms into marine ecosystems, and the population sizes are similar to those reported for estuarine waters and sediments (Attwell and Colwell 1984). At deep-sea sites the *Thermoactinomyces* numbers were several orders of magnitude lower, ranging from 10% to less than 1% of the total culturable actinomycetes, or were completely absent [Okinawa Trough, Ryukyu and Mariana Trenches (Colquhoun et al. 1998b)]. Thus, the deep-sea sites included in this study varied in the extent to which they were sinks for terrestrial material.

Table 1. Total culturable bacteria and abundance of actinomycetes in deep-sea sediments [a]

Sampling location and depth (m)	Total cfu/ml wet sediment	Actinomycetes/ml wet sediment		
		Mycolate	Other	Total
Japan Sea				
289	2.5×10^6	2.0×10^3	2.3×10^3	4.3×10^3
Sagami Bay				
1168	1.9×10^6	2.0×10^2	3.2×10^2	5.2×10^2
Suruga Bay				
1151	1.3×10^6	6.7×10^2	4.0×10^3	4.7×10^3
1487	1.2×10^7	0	1.0×10^3	1.0×10^3
1948	ND	2.4×10^3	3.3×10^3	5.7×10^3
Okinawa Trough				
1393	2.0×10^3	3.0×10^2	1.0×10^3	1.3×10^3
Archipelago, SE Okinawa				
2859	1.0×10^3	0	0	0
Ryukyu Trench				
5110	2.7×10^7	0	0	0
5425	3.5×10^6	0	0	0
6600	5.0×10^2	0	0	0
Izu-Bonin Trench				
2679	3.0×10^6	5.0×10^3	3.3×10^2	5.3×10^3
4056	ND	1.0×10^3	0	1.0×10^3
4644	2.0×10^5	0	0	0
4739	1.2×10^6	0	0	0
6390	2.0×10^5	1.0×10^3	0	1.0×10^3
6499	1.1×10^6	2.0×10^3	3.3×10^2	2.3×10^3
Japan Trench				
4418	2.1×10^5	1.3×10^3	3.3×10^2	1.6×10^3
6048	2.0×10^5	3.3×10^2	0	3.3×10^2
6142	1.1×10^7	0	0	0
6300	1.3×10^6	0	0	0
6372	8.3×10^4	0	0	0
6455	1.5×10^4	2.0×10^3	0	2.0×10^3
6475	7.4×10^6	0	0	0
Mariana Trench				
10898	8.0×10^2	0	0	0

ND, not determined.

[a] Enumeration of culturable bacteria was made on 10% trypic soy agar formulated with artificial seawater (Takizawa et al. 1993); incubations were made at atmospheric pressure and 30°C. Isolation of target actinomycetes was made using the following selective media:(1) *Rhodococcus*: Münz medium containing 1%(v/v) paraffin (Nesterenko et al. 1978), M3 medium (Rowbotham and Cross 1977); (2) *Micromonospora*: starch-casein-nitrate containing novobiocin, and cellulose-asparagine media (Goodfellow and Haynes 1984); (3) *Streptomyces*: starch-casein-nitrate medium (Küster and Williams 1964). All selective media were supplemented with cycloheximide (100 mg ml^{-1}) to suppress fungal growth.

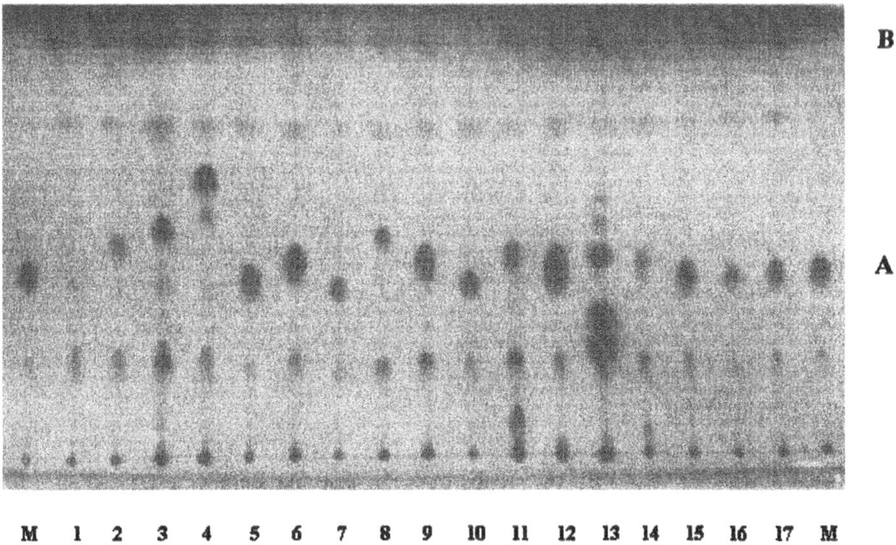

Fig. 1. Representative mycolic acid thin-layer chromatography (TLC) profiles of type strains and deep-sea mycolate actinomycetes.* **A** Mycolic acid; **B** Nonhydroxylated fatty acid methyl esters. *1*, negative control (nonmycolate actinomycete); *2*, *Nocardia asteroides*; *3*, *Gordonia bronchialis*; *4*, *Tsukamurella paurometabola*; *5*, *Dietzia maris*; *6*, *Rhodococcus rhodochrous*; *7*, *R. marinonascens*; *8*, IBT2679/051; *9*, IBT2679/068; *10*, OKT1393/078; *11*, SGB1168/117; *12*, SRB1948/A07; *13*, SRB1948/E04; *14*, SJS0289/JS3; *15*, IBT6499/047; *16*, JTS6455/242b; *17*, JTS4418/260b.*Cell wall mycolic acids were extracted and analyzed by TLC (Minnikin 1988). Detection was made with molybdophosphoric acid (5% w/v) followed by development of the TLC plates at 180°C for 10 min

Preliminary grouping of isolates was performed on the basis of colony characteristics, microscopy, and in the case of putative rhodococci mycolic acid profiles. *Rhodococcus* type strains and deep-sea isolates produced a single mycolic acid spot on thin-layer chromatography (TLC) plates (R_f 0.39-0.47) and were distinguishable from strains of *Gordonia* (R_f 0.52) and *Tsukamurella* (R_f 0.63) and from mycobacteria (e.g., Suruga Bay isolate SRB1948/E04) that produced multiple mycolic acids (Fig. 1) (Colquhoun et al. 1998a).

3.2 Taxonomy of Deep-Sea Mycolata

Curie-point pyrolysis mass spectrometry (PyMS) was used as the preferred means for sorting the putative rhodococci into clusters. The advantages of PyMS in this context are its speed, sensitivity, and capacity to handle large numbers of strains.

Before deciding on the use of PyMS, it was validated against other methods for sorting the deep-sea rhodococci by comparison with an extensive numerical taxonomic (NT) analysis. PyMS and NT analyses, made independently in different laboratories, produced complete congruence of clusters (Colquhoun et al. unpublished data). Our subsequent strategy for circumscribing putative rhodococcal isolates was to use the following polyphasic taxonomic approach: (1) PyMS to define the range of phenotypic diversity, (2) 16S rDNA sequencing to establish phylogenetic relationships, and (3) DNA:DNA pairing to resolve questions of species identity. The following selection of data illustrate the scope and efficacy of the approach.

The dendrogram shown in Fig. 2 was produced from the PyMS analysis of 39 deep-sea isolates recovered from a range of sites and depths, and 16 type strains of mycolate actinomycetes (Colquhoun et al. 1998a). Three major (A, C, D) and two minor (B, E) clusters were defined by this initial analysis. Intra-cluster variation is small, suggesting a close relationship between strains within such clusters. The percentage similarity values are derived from Mahanalobis distances for each pair of strains; and although these values define only a portion of interstrain discrimination (Magee 1994), when the dataset contains reference strains it is possible to establish the relative closeness of unknown to known organisms. It is important to point out, however, that even small intra-cluster discriminations of the sort seen in Fig. 2 may be highly relevant for biotechnology screening purposes (Bull et al. 1992).

The major clusters shown in Fig. 2 were progressively reanalyzed by removing outlying strains from the dataset. Thus cluster C, which contained isolates from various depths in the Izu Bonin and Japan Trenches, was differentiated into four subclusters. Subcluster C2 isolates were grouped with *Dietzia maris,* whereas the Japan Trench isolates (JTS) were clearly differentiated on the basis of depth (C4, 4418 m; C1, 6048 m; C3, 6455 m) (Colquhoun et al. 1998b).

Fifteen deep-sea isolates representative of the PyMS clusters, together with a number of PyMS outliers, were subjected to BioLog (BioLog, Haywood, CA, USA) analysis. Also included in the analysis were (1) Izu Bonin Trench isolates (IBT2679/024, IBT2679/068) that had been stored at -70°C in glycerol for 2 years; and (2) a number of recently isolated terrigenous strains of *Rhodococcus*. The results of this analysis are shown in Fig. 3 (Colquhoun et al. 1998a). Whereas clusters generated by BioLog analysis were much more heterogeneous compared to those by PyMS, clusters A, C, and D were recognizable. The poorer resolving power of BioLog clearly reflects the smaller number and narrower range of phenotypic characters on which it is based. The storage of isolates at -70°C appeared not to affect their phenotypic properties, as evidenced by the close grouping of glycerol-stored and agar-passaged replicates of the same strain (cf. IBT2679/024g and IBT2679/024, and IBT2679/068g and IBT2679/068, respectively)(Fig. 3). BioLog analysis of recently isolated terrigenous rhodococci was made to check that the PyMS clustering of deep-sea strains away from the type strains did not simply reflect type strain degeneration during prolonged laboratory cultivation. The terrigenous strains were mainly collected into a large heterogeneous group at the 80% similarity level that contained members of PyMS clusters A, B, and C; they clustered closely with terrigenous type

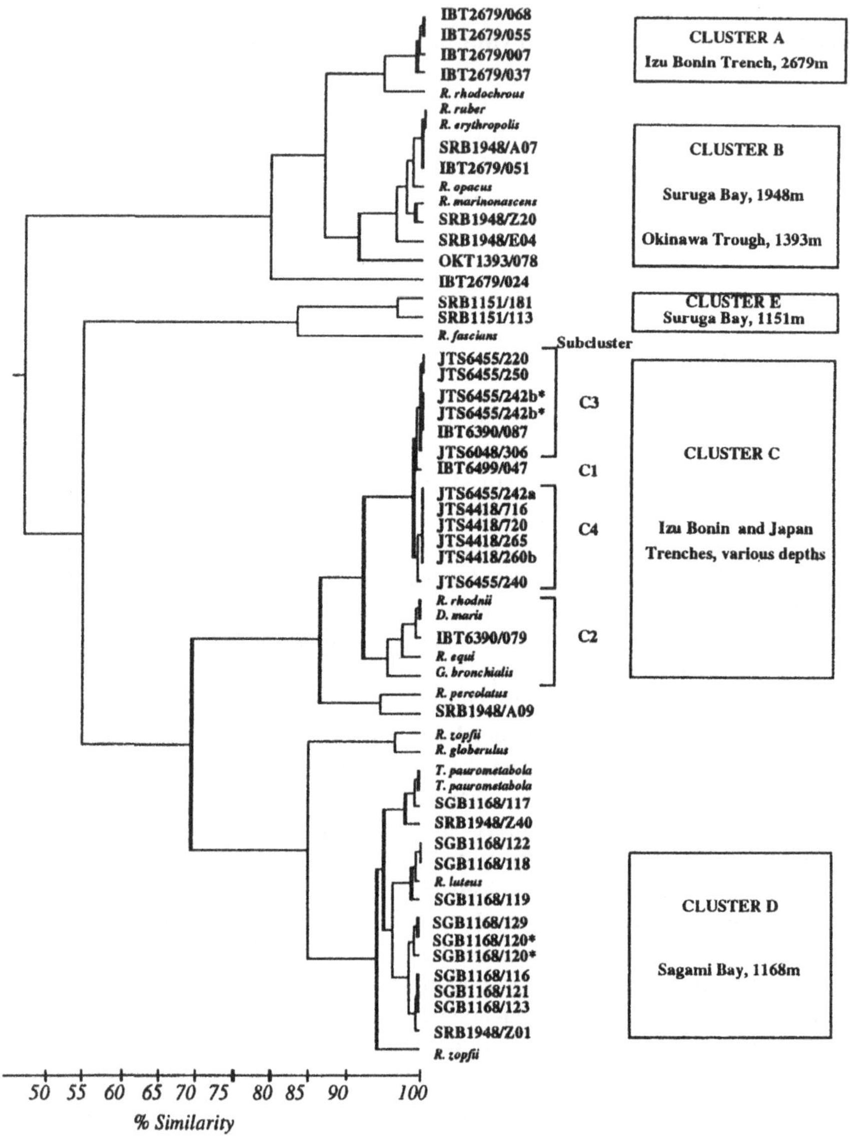

Fig. 2. Dendrogram derived from PyMS data for deep-sea and type strain actinomycetes isolated from diverse sampling sites. Deep-sea isolates, type strains of *Rhodococcus*, and other mycolate actinomycete genera were grown on GYE agar (Gordon and Mihm 1962) for 5 days at 30°C. Samples for pyrolysis were prepared and analyzed with a Horizon RAPyD 400 pyrolysis mass spectrometer (Horizon Instruments, Heathfield, UK) using methods developed for actinomycete classification (Sanglier et al. 1992). Triplicate samples were analyzed for each isolate and type strain. Ranked masses were subjected to principal components and canonical variate analyses using the GENSTAT statistical package (Numerical Algorithms Group, Oxford, UK). Similarities were calculated as Mahalanobis distances for each pair of strains, and clusters were defined using the unweighted pair group method with arithmetic averages (UPGMA; Sneath and Sokal 1973)

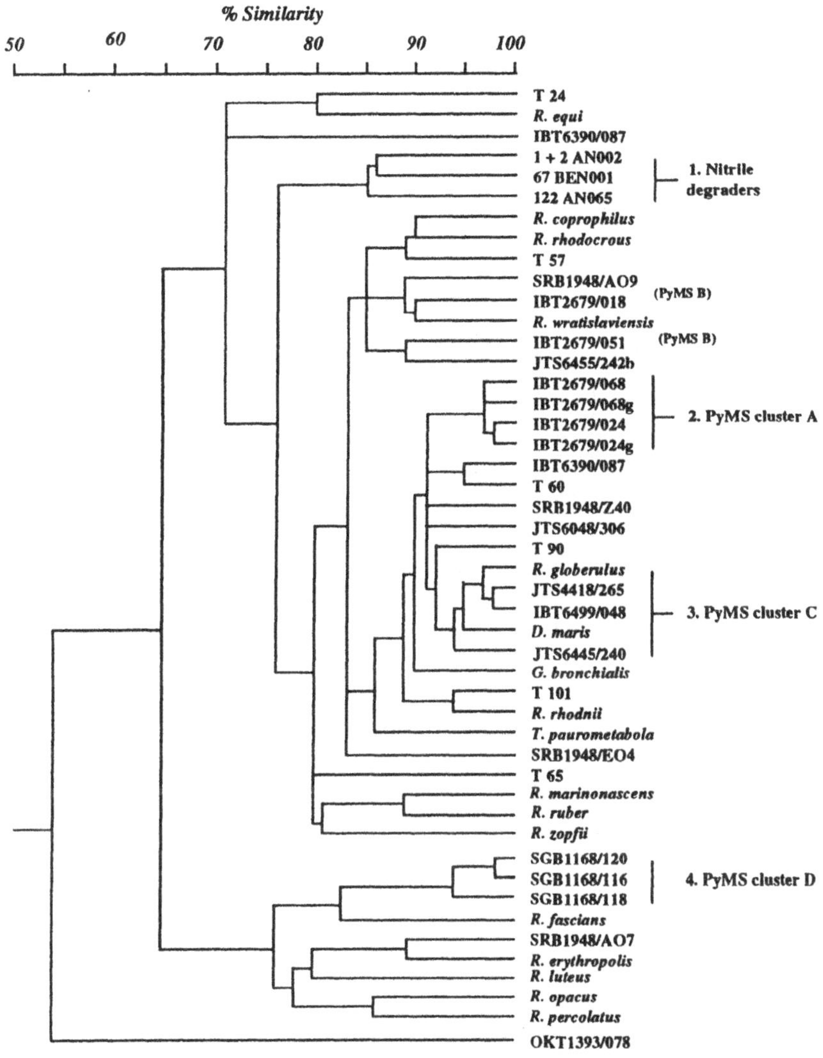

Fig. 3. Dendrogram derived from UPGMA cluster analysis of BioLog data for representative deep-sea and terrestrial actinomycetes and type strains: T-series rhodococci recently isolated from terrestrial sites. Substrate utilization were tested against the selected strains' ability to oxidize a panel of 95 carbon sources (BioLog, Haywood CA, USA). Tests were run in duplicate, and the data were subjected to cluster analysis using the UPGMA algorithm (MicroLog3 software)

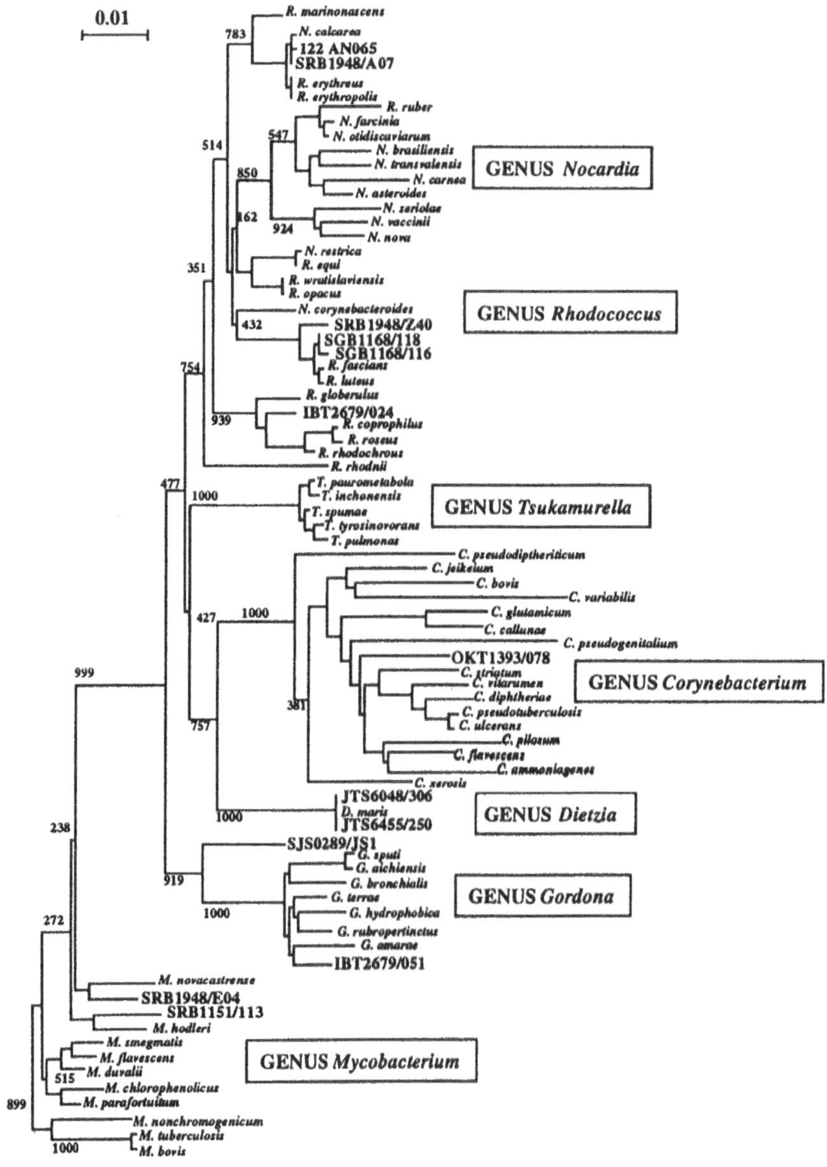

Fig. 4. Phylogram based on a comparison of 16S rDNA sequences of mycolate actinomycete type strains and selected deep-sea isolates. Bootstrap values are given for the likelihood of evolution from each major node. Bar indicates inferred substitution per 100 nucleotides. The 16S rDNA sequences of the isolated strains have been deposited in the DDBJ, EMBL, and GenBank nucleotide sequence databases. Strains names and accession numbers are as follows: IBT2679/024; AB010902, IBT2679/051; AB010903, JTS6048/306; AB010904, JST6455/250; AB010905, OKT1393/078; AB010906, SGB1168/116; AB010907, SGB1168/118; AB010908, SJS0289/JS1; AB010909, SRB1151/113; AB010910, SRB1948/A07; AB010911, SRB1948/E04; AB010912, SRB1948/Z40; AB010913

strains or remained separated from their nearest neighbors, suggesting that the discrimination of the deep-sea isolates was not artifactual. BioLog analysis recovered the nitrile-transforming rhodococci as a minor cluster whose identity was supported by 16S rDNA sequencing (see below).

Sequencing of the 16S rDNA gene was done on representatives of the major PyMS clusters together with a number of deep-sea isolates of indeterminate PyMS grouping (SRB1151/113, SRB1948/A07, SRB1948/E04, OKT1393/078, IBT2679/051, SJS0289/JS1) according to the procedure reported previously (Kato et al. 1997). The phylogenetic relationships between the deep-sea isolates and 65 type strains of seven genera of mycolata are shown in Fig. 4 (Colquhoun et al. 1998a). *Corynebacterium*, *Gordonia*, *Mycobacterium*, and *Tsukamurella* formed phylogenetically distinct clades, whereas it is clear that the status of the *Rhodococcus-Nocardia* taxonomic group requires further resolution. Our analysis confirms the view that *Nocardia* constitutes a clade within the *Rhodococcus* radiation (Rainey et al. 1995).

The 16S rDNA analysis provides strong evidence for the occurrence of novel mycolate taxa in the deep-seas. The Okinawa Trough isolate, strain OKT1393/078, appears to be a new species of *Corynebacterium*; and isolates from the Izu Bonin Trench (IBT2679/051, cluster B) and Suruga Bay (SRB1151/113, cluster E, and SRB1948/E04) are probably new *Gordonia* and *Mycobacterium* species, respectively. The status of isolate SJS0289/JS1 from the Sea of Japan is especially interesting, and its phylogenetic depth is suggestive of a new genus separate from the *Gordonia* clade. The 16S rDNA data confirmed that among PyMS cluster C (including JTS6048/306 and JTS6455/250) are strains of the authenticated deep-sea actinomycete *Dietzia maris*. Finally, the 16S rDNA data point to PyMS cluster A and D isolates being terrigenous actinomycetes that have entered the marine environment by wash-in. DNA:DNA pairing experiments according to the procedure of Ezaki et al. (1989) confirmed cluster D isolates as *Rhodococcus luteus*. The PyMS cluster A isolates (IBT2679/024 and IBT2679/068), which 16S rDNA data place with *R. rhodochrous* and *R. coprophilus*, were identical to each other on the basis of DNA pairing analyses, and the low similarity (20%) proved that they were not *R. rhodochrous*. They showed 40% similarity to *R. coprophilus*, a value well below the 70% similarity accepted as defining a genomic species (Wayne et al. 1987). The 40% value is equivocal and could indicate a distinct genomic species or a DNA group within the *R. coprophilus* taxon; however, it does caution too rigid reliance on 16S rDNA evidence for circumscribing species. This point has been stressed by Fox et al. (1992), who commented that, although 16S sequences are valuable for distinguishing genera and well-resolved species, recently diverged species, of the sort we suspect among deep-sea rhodococci, may not be recognizable.

3.3 Nitrile-Transforming Activities of Deep-Sea Rhodococci

We have isolated a number of bacteria from marine sediments by direct and batch enrichment culture on acetonitrile or benzonitrile. All of these strains contained mycolic acids with mobilities on TLC plates similar to those of the reference strains of *Rhodococcus*, suggesting that they were likely to be members of this genus. Strains 122-AN065 and 67-BEN001, recovered from depths greater than 5000 m (Ryukyu Trench 5425 m and Japan Trench 6475 m, respectively), grew consistently well on the nitriles on which they were isolated. 16S rDNA sequencing of strain 122-AN065 showed it to have 99.8% sequence homology with *Nocardia calcarea* and 99.7% homology with *Nocardioides simplex* (Rainey et al. 1995), which was reclassified *R. erythropolis*. Strain 67-BEN001 had a 16S rDNA restriction length fragment polymorphism (RLFP) pattern identical to that of 122-AN065 and is likely to have a similar sequence identity.

Strain 122-AN065 grew on both aliphatic (acetonitrile) and aromatic (benzonitrile) nitriles, although growth was slower on the latter compound. The specific growth rate on 20 mM acetonitrile at 30°C was 0.21 h^{-1} compared to 0.02 h^{-1} on acetonitrile at 4°C and 0.08 h^{-1} on 10 mM benzonitrile at 30°C. Growth was fastest at acetonitrile concentrations of 50 and 100 mM, it was severely inhibited at concentrations in excess of 500 mM and ceased completely at 2 M (Heald et al. paper in preparation).

Strain 67-BEN001 also was able to grow on aromatic and aliphatic nitriles as sole carbon sources. Growth occurred at concentrations between 1 and 30 mM benzonitrile but not at 50 mM, the maximum solubility of the substrate in minimum medium. The maximum specific growth rate on benzonitrile minimum medium supplemented with yeast extract (5 mg l^{-1}) was 0.12 h^{-1}.

Acetonitrile-grown, resting cells of strain 67-BEN001 metabolized acetonitrile and benzonitrile with production of the corresponding acid and ammonia. During these transformations amides were not detected in the reaction mixtures, although the strain was able to metabolize acetamide and benzamide, indicating the presence of amidase activity. Biotransformation data suggested that strain 122-AN065 had a nitrile hydratase-amidase system in which amidase activity was higher than nitrile hydratase activity when the strain was grown on a nitrile substrate; however, when growth was on succinate, the amidase activity was lower than that of the hydratase; thus, amide accumulated transiently. Strain 67-BEN001, when grown on benzonitrile, acetonitrile, glycerol, or benzonitrile plus yeast extract (5 mg l^{-1}), released ammonia from both acetonitrile and benzonitrile. Acetonitrile was transformed at a faster rate than benzonitrile. The transformation of acetamide by these bacteria again was indicative of the presence of amidase activity. Resting cell suspensions also were able to metabolize benzamide. When strain 67-BEN001 was grown on benzonitrile without yeast extract, amide did not accumulate and the rate of benzonitrile transformation was approximately 10-fold slower than that in the presence of yeast extract.

Table 2. Hydrolysis rates of nitrile and amide substrates by deep sea *Rhodococcus* strains 67-BEN001 and 122-AN065

Substrate	Strain 67-BEN001					Strain 122-AN065	
		Nitrogen release after growth on					
	Growth	Benzonitrile	Glycerol	Benzonitrile/yeast	Acetonitrile	Growth	Nitrogen release after growth on acetonitrile
Acetonitrile (C2)	Y	184	124	112	403	Y	100
Acrylonitrile	—	—	—	—	—	N	64
4-Aminobenzonitrile	—	—	—	—	—	N	18
Benzonitrile	Y	100	100	100	100	Y	50
Butyronitrile (C4)	Y	940	364	630	—	Y	840
Chloroacetonitrile	N	438	270	—	—	N	76
Crotononoitrile	N	64	17	—	—	Y	36
3-Cyanopyridine	N	126	—	—	—	N	0
Fumaronitrile	—	—	—	—	—	N	0
Isovaleronitrile	—	143	71	212	—	Y	—
Malononitrile	—	—	—	—	—	N	0
Methoxyacetonitrile	N	—	189	—	—	N	38
D,L-Mandelonitrile	—	—	—	—	—	N	0
2-Nitrobenzonitrile	—	—	—	—	—	N	20
Propiononitrile (C3)	Y	942	257	435	—	Y	70
Succinonitrile (C6)	Y	—	—	—	—	Y	379
Valeronitrile (C5)	Y	1636	362	—	—	Y	815
Acetamide	Y	420	124	630	—	Y	> 4000
Acrylamide	—	—	103	—	—	—	—
Benzamide	Y	—	68	—	—	—	—
Lactamide	Y	—	322	—	—	—	1124
Salicylamide	—	—	0	—	—	—	37

Rates of hydrolysis were normalized to the rate of hydrolysis of benzonitrile (nmol min^{-1} mg protein) by strain 67-BEN001 (grown on benzonitrile, 162; glycerol, 292; benzonitrile/yeast, 200; acetonitrile, 302) and acetonitrile by strain 122-AN065 (grown on acetonitrile, 380).
—, not determined; Y, growth on compound; N, no growth on compound.
Values in parentheses indicate carbon number of straight-chain aliphatic compounds.

Cell free extracts of strains 122-AN065 and 67-BEN001 grown on acetonitrile and benzonitrile respectively were electrophoresed on native polyacrylamide gel electrophoresis (PAGE) gels and stained with anthranilonitrile. A single distinct fluorescent band, visible under ultraviolet (UV) light, was produced by both strains as a consequence of conversion of the nitrile to fluorescent amide or acid products. This result was indicative of a single nitrile-metabolizing enzyme system present in each strain, although it did not rule out the possibility of the coexistence of a nitrile hydratase-amidase and nitrilase that co-migrated or that the activity of one of these enzymes was lost during extraction.

The nitrile metabolizing enzyme of strain 67-BEN001 has been distinguished from the amidase activity by ion exchange chromatography (Heald et al. paper in preparation). The nitrile active fraction when electrophoresed by sodium dodecyl sulfate(SDS)-PAGE was shown to comprise two 25 kDa bands; this is the same size as the α- and β- subunits of nitrile hydratase (Yamada and Kobayashi 1996). High-performance liquid chromatography (HPLC) analysis of benzonitrile transformation mixtures produced by the nitrile active fraction revealed benzamide as the predominant product, confirming the presence of a nitrile hydratase-amidase system in this bacterium.

The nitrile transforming activities of both deep-sea strains had broad substrate specificities (Table 2) with a preference for aliphatic nitriles (Colquhoun et al. 1998a). Valeronitrile (C5) and butyronitrile (C4) were the preferred substrates and were metabolized much faster than either acetonitrile or benzonitrile. Strain 67-BEN001 also had a high rate of hydration of propiononitrile (C3). Dinitriles (succininitrile), chloroacetonitrile, and a number of amides were also metabolized. The two deep-sea rhodococci characterized in this study differ from the only other reported marine nitrile transforming *Rhodococcus* (*R. erythropolis* strain BL1) (Langdahl et al. 1996) in that the latter was isolated from coastal sediments and was unable to metabolize benzonitrile.

3.4 Effects of Simulated Deep-Sea Conditions on Growth of Deep-Sea Rhodococci

The effect of NaCl concentration, temperature, and pressure (over ranges similar to those of the deep-sea environment) on growth were investigated. The marine sediments from which these strains were isolated had total halide concentrations ranging from 2.5% to 3.2%. Strains grew on 20 mM acetonitrile in liquid minimum medium at salinities from 0% to 4% (w/v).

The deep-sea abyssal depths are at a constant temperature of 2° – 4°C (Yayanos 1995). All nitrile transforming bacteria were isolated at 30°C. Strain 122-AN065 grew on acetonitrile at 4°C, although the growth rate was only 10% of that at 30°C. Acetonitrile transformations rates were approximately 10-fold lower at 4°C. Strain 67-BEN001 also grew on benzonitrile at 4°C, but transformation rates were more than 30-fold lower than at 30°C.

The effect of elevated hydrostatic pressure on growth in glucose-yeast extract

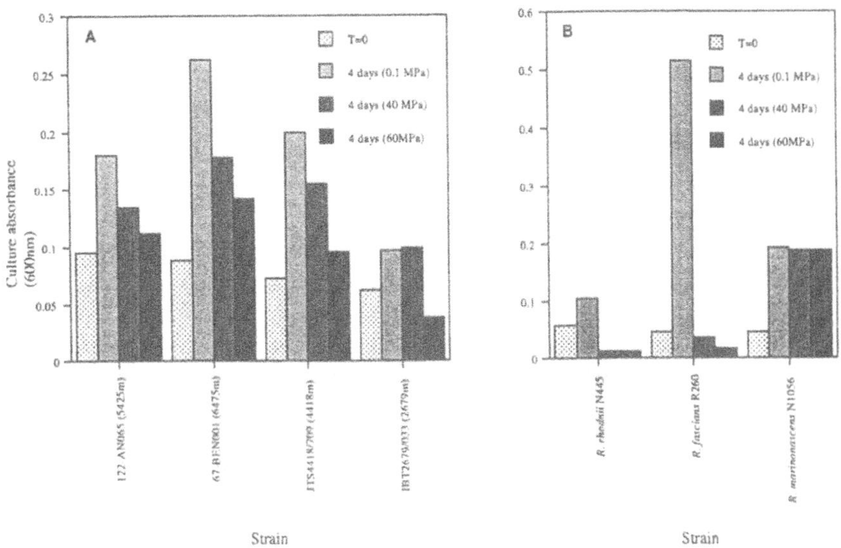

Fig. 5. Growth of deep-sea rhodococci (**A**) and selected type strains of *Rhodococcus* (**B**) at pressures of atmospheric pressure (1 atm), and high pressures (40 and 60 MPa). Control cultures were incubated at atmospheric pressure. All experiments were made for 4 days at 30°C. T = 0, culture absorbance of the inoculum. Growth at hyperbaric pressures was in 1.5-ml sterile autosampler vials containing 0.5 ml of the oxygen carrier Fluorinet FC-72 (Sigma) without a gaseous headspace. Prior to use, Fluorinet was degassed under vacuum, oxygenated for 2 h, and filter-sterilized. Vials were incubated in hydraulically pressurized vessels constructed from Inconel 625 high strength nickel-chromium alloy (Stott and Herbert 1986)

medium of deep-sea strains from a range of depths is shown in Fig. 5 (Colquhoun et al. 1998a). Strains 122-AN065 and 67-BEN001 isolated from 5425 and 6475 m, respectively, grew at a pressure of 40 MPa. At 60 MPa growth still occurred, but the isolate from the shallower depth was more sensitive to this higher pressure. The terrestrial strains *R. rhodnii* and *R. fascians* did not grow at 60 or 40 MPa, and after incubation at elevated pressures cell lysis occurred. Interestingly, a culture collection type strain, *R. marinonascens* (Helmke and Weyland 1984), originally isolated from marine sediments down to 2000 m in depth but that had been maintained at atmospheric pressure for many years, also was resistant to high pressure and grew at 40 and 60 MPa. The specific responses of the deep-sea rhodococci to conditions found in the marine environment strongly suggests that they are marine in origin. This conclusion is supported by the low numbers or complete absence of *Thermoactinomyces* in the deep ocean trenches.

The diversity of actinomycetes in deep-sea sediments and their ability to synthesize biotechnologically important enzymes and bioactive metabolites such as receptor antagonists (Silva et al. unpublished data) argue that this poorly researched environment is a propitious one in which to search for novel, exploitable microorganisms, activities, and products.

3.5 Conclusion of Biotransformation Activities of Deep-Sea Actinomycetes

Deep-sea soft sediments from trench systems and depths in the north-western Pacific Ocean ranging from less than 300 to 10 897 m have been analyzed for three target genera of actinomycetes: *Micromonospora*, *Rhodococcus*, and *Streptomyces*. Only culturable strains, recovered at atmospheric pressure on selective isolation media, have been examined to date. Maximum recoveries of culturable bacteria were more than 10^7/ml wet g sediment, but actinomycetes comprised a small proportion of this population (usually less than 1%). The target actinomycetes were isolated at all depths except from the Mariana Trench sediments. Actinomycete colonies were defined initially on the basis of colony morphologies, and preliminary identification was then made by chemotaxonomic tests. Pyrolysis mass spectrometry (PyMS) of deep-sea mycolic acid-containing actinomycetes gave excellent correspondence with numerical (phenetic) taxonomic analyses and subsequently was adopted as a rapid procedure for assessing taxonomic diversity. PyMS analysis enabled several clusters of deep-sea rhodococci to be distinguished that are quite distinct from all type strains. 16S rRNA gene sequence analysis has revealed that several of these marine rhodococci have sequences similar to certain terrestrial species of *Rhodococcus* and to *Dietzia*. There is evidence for the intrusion of terrestrial runoff into these deep trench systems, and the inconsistency of the phenotypic and molecular taxonomies may reflect recent specification events in actinomycetes under the high-pressure conditions of the deep-sea. The results of DNA:DNA pairing experiments point to the novelty of *Rhodococcus* strains recovered from hadal depths in the Izu Bonin Trench.

Biotransformation studies of deep-sea bacteria have focused on nitrile compounds. Nitrile-metabolizing bacteria, closely related to rhodococci, have been isolated that grow well at low temperature, high salt concentrations, and high pressures, suggesting that they are of marine origin or have adapted to the deep-sea environment.

References

Attwell RW, Colwell RR (1984) Thermoactinomycetes as indicators for estuarine and marine waters. In: Ortiz-Ortiz L, Bojalil LF, Yakoleff V (eds) Biological, biochemical and biomedical aspects of actinomycetes. Academic, Orlando, pp 441-452

Bull AT, Goodfellow M, Slater JH (1992) Biodiversity as a source of innovation in biotechnology. Annu Rev Microbiol 46:219-251

Colquhoun JA, Heald SC, Li L, Tamaoka J, Kato C, Horikoshi K, Bull AT (1998a) Taxonomy and biotransformation activities of some deep-sea actinomycetes. Extremophiles 2:269-277

Colquhoun JA, Mexson J, Goodfellow M, Ward AC, Horikoshi K, Bull AT (1998b) Novel rhodococci and other mycolata from the deep-sea. Antonie Van Leeuwenhoek 74:27-40

DeLong EF (1997) Marine microbial diversity: the tip of the iceberg. Trends Biotechnol 15:203-

207

Ezaki T, Hashimoto Y, Yabuuchi E (1989) Fluorometric deoxyribonucleic acid-deoxyribonucleic acid hybridization in microdilution wells as an alternative to membrane filter hybridization in which radioisotopes are used to determine genetic relatedness among bacterial strains. Int J Syst Bacteriol 39:224-229

Fox G, Wisotskey JD, Jurtshuk P (1992) How close is close: 16S rRNA sequence identity may not be sufficient to guarantee species identity. Int J Syst Bacteriol 42:166-170

Goodfellow M, Haynes JA (1984) Actinomycetes in marine sediments. In: Ortiz-Ortiz L, Bojalil LF, Yakoleff V (eds) Biological, biochemical and biomedical aspects of actinomycetes. Academic, Orlando, pp 453-472

Gordon RE, Mihm JM (1962) Identification of *Nocardia caviae* (Erikson) nov.comb. Ann NY Acad Sci 98:628-636

Helmke E, Weyland H (1984) *Rhodococcus marinonascens* sp.nov., an actinomycete from the sea. Int J Syst Bacteriol 34:127-138

Jensen PR, Dwight R, Fenical W (1991) Distribution of actinomycetes in near-shore tropical marine sediments. Appl Environ Microbiol 57:1102-1108

Jollivett D (1996) Specific and genetic diversity at deep-sea hydrothermal vents: an overview. Biodiv Conserv 5:1619-1653

Kato C, Li L. Tamaoka J, Horikoshi K (1997) Molecular analyses of the sediment of the 11 000 m deep Mariana Trench. Extremophiles 1:117-123

Küster E, Williams ST (1964) Selection of media for isolation of streptomycetes. Nature 202:928-929

Langdahl BR, Bisp P, Ingvorsen K (1996) Nitrole hydrolysis by *Rhodococcus erythropolis* BL1, an acetonitrile-tolerant strain isolated from a marine sediment. Microbiology 142:145-154

Magee JT (1994) Analaytical fingerprinting methods. In: Goodfellow M, O'Donnell AG (eds) Chemical methods in prokaryotic systematics. Wiley, Chichester, pp 523-553

Minnikin DE (1988) Isolation and purification of mycobacterial cell wall lipids. In: Hancock IC, Poxton IR (eds) Bacterial cell surface techniques. Wiley, Winchester, pp 125-135

Nesterenko OA, Kasumova SA, Kvasnikov EI (1978) Microorganisms of the *Nocardia* genus and the '*rhodochrous*' group in soils of the Ukranian SSR. Mikrobiologiia 47:866-870

Rainey FA, Burghardt J, Kroppenstedt RM, Klatte S, Stackebrandt E (1995) Phylogenetic analysis of the genera *Rhodococcus* and *Nocardia* and evidence for the evolutionary origin of the genus *Nocardia* from within the radiation of *Rhodococcus* species. Microbiology 141:523-528

Rowbotham TJ, Cross T (1977) Ecology of *Rhodococcus coprophilus* and associated actinomycetes in freshwater and agricultural habitats. J Gen Microbiol 100:231-240

Sanglier JJ, Whitehead D, Sadler GS, Ferguson EV, Goodfellow M (1992) Pyrolysis mass spectrometry as a method for the classification and selection of actinomycetes. Gene 115:235-242

Sneath PHA, Sokal RR (1973) Numerical taxonomy: the principles and practice of numerical classification. Freeman, San Francisco.

Stott JFD, Herbert BN (1986) The effects of temperature and pressure on sulphate-reducing bacteria and the action of biocides in oilfield water injection systems. J Appl Bacteriol 60:57-66

Takami H, Inoue A, Fuji F, Horikoshi K (1997) Microbial flora in the deepest sea mud of the Mariana Trench. FEMS Microbiol Lett 152:279-285.

Takizawa M, Colwell RR, Hill RT (1993) Isolation and diversity of actinomycetes in the

Chesapeake Bay. Appl Environ Microbiol 59:997-1002

Wayne LG, Brenner DJ, Colwell RR, Grimont PAD, Kandler O, Krichevsky MI, Moore LH, Moore WEC, Murray RGE, Stackbrandt E, Starr MP, Truper HG (1987) Report of the ad hoc committee on reconciliation of approaches of bacterial systematics. Int J Syst Bacteriol 37:463-464

Weyland H (1969) Actinomycetes in North Sea and Atlantic Ocean sediments. Nature (Lond) 223:858

Wyatt JM, Linton EA (1988) The industrial potential of microbial nitrile biochemistry. In: Evered D, Harnett S (eds) Cyanide compounds in biology. Wiley, Chichester, pp 32-48

Yamada H, Kobayashi M (1996) Review: nitrile hydratase and its application to industrial production of acrylamide. Biosci Biotechnol Biochem 60:1391-1400

Yayanos AA (1995) Microbiology to 10,500 meters in the deep sea. Annu Rev Microbiol 49:777-805

4 Microbial Diversity in the Sediments Collected from Cold-Seep Areas and from Different Depths of the Deep-Sea

LINA LI and CHIAKI KATO

Defining the diversity and distribution of natural microbial communities from sea sediments has been a long-standing challenge in the field of microbial ecology and evolution. It is generally estimated that about 95% of the organic matter produced photosynthetically in the surface of waters recycled in the upper 100-300 m (Jannasch and Taylor 1984). Only about 1% of photosynthetically produced organic carbon reaches the deep-sea floor. Therefore, the major nutritional characteristics of the deep-sea as a habitat for organisms are relatively low input of organic carbon and its consumption. The availability of nutrients, being geographically highly variable, and the elevated pressures characterized the deep sea sediments as potentially unique bacterial environments.

Deep-sea "cold-seep" areas have been found in subduction zones off the coast of Oregon, off the coast of Peru, and in the Japan Trench and the Nankai Trough (Kulm et al. 1986; Ohta and Laubier 1987; Sibuet et al. 1988). *Calyptogena* is a typical chemoautotrophic bivalve living in such environments. Many studies have revealed that chemosynthetic symbiont bacteria inhabit the gill surface of *Calyptogena* as symbionts and support its growth (Paull et al. 1984; Kennicutt et al. 1985, 1989). Studies of the microbial communities around the *Calyptogena* colonies could provide us with information about the microbial diversity and its relationship to *Calyptogena*, the specialized functional composition of bacteria groups represented, their utilization of energy sources, and their interactions.

To investigate the microbial community in deep-sea environments, we have isolated and characterized several deep-sea microorganisms (Kato et al. 1998a). So far only limited numbers of deep-sea microorganisms have been isolated in the labora-

Key words. Deep-sea, Phylogenetic tree, 16S ribosomal RNA, Microbial diversity, Ocean sediment, Cold-seep, *Calyptogena*, *Pseudomonas*, symbiont tory; most of the microorganisms could not be cultivated by standard techniques. Fortunately, descriptions of these uncultured microorganisms can now be obtained by means of molecular phylogenetic techniques based on analyses of small-subunit rRNA genes. We examined three groups of sediment samples to illustrate the microbial diversity in the deep-sea environments.

1. Two sediments from inside and beside the area of *Calyptogena* communities in the Japan Trench demonstrated that δ- and ε-Proteobacteria-related sequences are particularly abundant in this area. Archaea-related sequences are also present in these sediments (Li et al. 1999c).
2. Five sediments from core samples inside and outside the area of *Calyptogena* communities in the Nankai Trough indicated that microbial diversity is greater in the sediment inside the *Calyptogena* communities than in the sediments outside them, especially the δ- and ε-Proteobacteria-related sequences (Li et

Table 1. Deep-sea sediment samples

Sample/dive no.	Depth (m)	Sampling location	Dive date
Group 1			
JT1/6K[a]374-1	6367	Japan Trench (40°06' N, 144°11' E) (inside *Calyptogena* community)	6/7/97
JT2/6K374-3	6354	Japan Trench (40°06' N, 144°11' E) (beside *Calyptogena* community)	6/7/97
Group 2			
NK1/6K322-1	3843	Nankai Trough (33°39' N, 137°56' E) (inside *Calyptogena* community 0 cmd)	7/27/96
NK2/6K322-1	3843	Nankai Trough (33°39' N, 137°56' E) (inside *Calyptogena* community 7 cmd)	7/27/96
NK3/6K322-1	3843	Nankai Trough (33°39' N, 137°56' E) (inside *Calyptogena* community 12 cmd)	7/27/96
NK4/6K322-2	3843	Nankai Trough (33°39' N, 137°56' E) (outside *Calyptogena* community 5 cmd)	7/27/96
NK5/6K322-2	3843	Nankai Trough (33°39' N, 137°56' E) (outside *Calyptogena* community 12 cmd)	7/27/96
Group 3			
BD1/2K[b]870-01	1159	Suruga Bay (35°02' N, 138°37' E) (near a site containing manmade garbage)	6/11/96
BD2/2K871-01	1521	Suruga Bay (34°55' N, 138°39' E)	6/12/96
BD3/6K330-C1	2339	Ryukyu Trench (28°10' N, 130°23' E)	8/20/96
BD4/6K332-C2Up	3671	Ryukyu Trench (28°29' N, 131°11' E)	8/22/96
BD5/6K340-R	4031	Izu-Bonin Trench (30°55' N, 141°49'E) (near whale bone remains)	10/6/96
BD6/6K331-01	6482	Ryukyu Trench (27°53' N, 130°32' E)	8/21/96
BD7/6K273-1	6379	Japan Trench (40°06' N, 144°11' E) (inside *Calyptogena* community)	7/20/95

cmd, depth in centimeters from the sediment surface.
[a] 6K: *Shinkai* 6500.
[b] 2K: *Shinkai* 2000.

al. 1999a). A model of sulfur circulation in the *Calyptogena* communities was proposed from the results of items 1 and 2.

3. Seven sediment samples from different depths of the deep-sea at the range of 1159–6482 m were examined. The results demonstrated that some common bacterial sequences are present in all the sediments, microbial diversity is greater in the nutrient-rich area than in the normal area, and pressure seems not to influence the microbial diversity (Li et al. 1999b).

4.1 Experimental Procedure

4.1.1 Collection of Deep-Sea Sediment Samples

The deep-sea sediment samples were collected using JAMSTEC's submersible Shinkai 2000 or Shinkai 6500 from various points listed in Table 1. The collected samples were kept in 10% glycerol/artificial seawater (3.0 g NaCl, 0.7 g KCl, 5.3 g $MgSO_4 7H_2O$, 10.8 g $MgCl_2 6H_2O$,and 1.0 g $CaSO_4.2H_2O$ per liter) in liquid nitrogen (-180°C) immediately after being taken from the deep-sea. Sediments in group 1 were collected from inside and beside the *Calyptogena* community situated in the Japan Trench. Sediments in group 2 were collected from inside and outside the *Calyptogena* community situated in the Nankai Trough; and the subsamples were taken every 5 cm from the core sample on board ship. The sediment sample sites in group 3 in this study are from the normal cold sea area; only three are characterized by being near a manmade garbage site (BD1), a whale-bone remains site (BD5), and inside the area of the *Calyptogena* community (BD7).

The procedure for microbial diversity analysis by the molecular phylogenetic techniques employed in this study is shown in Fig. 1.

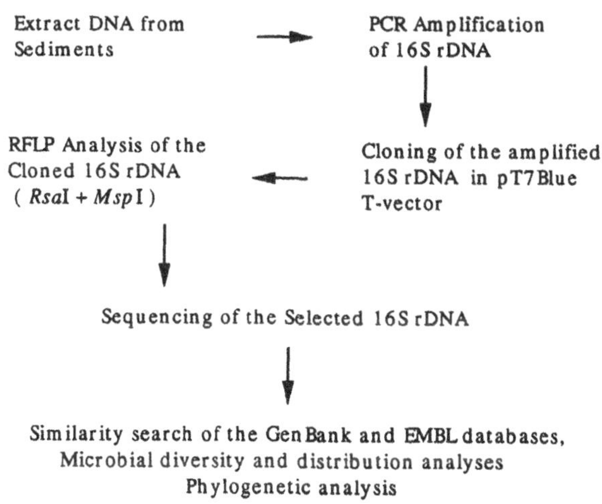

Fig. 1. Procedure for analysis of microbial diversity in deep-sea sediments

4.1.2 DNA Extraction

Total DNA was extracted from the sediment samples by a modification of the methods of Herrick et al. (1993). Approximately 20 g of the sediment (wet weight) was suspended in 50 ml of sterilized artificial seawater and centrifuged for 5 min at 1000 g and 4°C. The supernatant was pooled in an Erlenmeyer flask and maintained on ice until further centrifugation. That step was repeated three times. The supernatants were mixed and centrifuged for 30 min at 10 000 g and 4°C to collect the microbial cells. The pellets were resuspended in a lysozyme solution [150 mM NaCl, 100 mM Na_2EDTA (pH 8.0), lysozyme 15 mg/ml] and incubated at 37°C for 1 h with occasional agitation. A sodium dodecyl sulfate (SDS)-solution [100 mM NaCl, 500 mM Tris-HCl (pH 8.0), 10% SDS] was added; the reaction mixture was mixed gently by inversion and incubated at room temperature for 5 min. Three cycles of freezing in liquid nitrogen and thawing in a 65°C waterbath were conducted to release nucleic acids. A 7.5 M ammonium acetate solution was added to give a final concentration of 2.5 M; and then acid-washed polyvinylpolypyrrolidone (PVPP) (Sigma) was added to achieve 2% of the total mixture. The mixture was then centrifuged at 6000 g for 10 min. This process resulted in the precipitation of humic materials. Nucleic acids in the supernatant fraction were precipitated by the addition of 2.5 volumes of ice-cold ethanol and glycogen (Sigma) 20 µg/ml after incubation at -80°C for 2 h or more. The precipitate was collected by centrifugation at 4°C and washed with 70% ethanol. The dried pellet was resuspended in 10 mM Tris-HCl+1 mM EDTA (pH7.6)(TE). The crude DNA extract was subjected to further purification using Gene Clean II (Bio101 Co.).

4.1.3 Polymerase Chain Reaction Amplification

The polymerase chain reaction (PCR) amplification was performed with a DNA Thermal Cycler model 9600 (Perkin-Elmer/Cetus Co.) using 50 µl of the PCR reaction mixture under the conditions recommended by the enzyme manufacturers (Takara Co.). A total of 30 cycles of amplification was performed with temperate DNA denaturation at 95°C for 1.5 min, primer annealing at 55°C for 1.5 min, and primer extension at 72°C for 1.5 min (DeLong 1992). The oligonucleotide primer sequences designated by DeLong (1992) used in this study were as follows:

For bacterial 16S rDNA:

Eubac27F	AGA GTT TGA TCC TGG CTC AG
Eubac1492R	GGT TAC CTT GTT ACG ACT T

For archaea 16S rDNA:

Arch21F	TTC CGG TTG ATC CYG CCG GA
Arch958R	YCC GGC GTT GAM TCC AAT T

where Y is a C or T, and M is an A or C.

4.1.4 Purification and Cloning of PCR Products

Amplified DNA was purified from the reaction components using a DNA recovery filter SUPREC-02 (Takara Co.), and quantities were measured spectrophoto-metrically. The PCR products were cloned into pT7Blue(R)T-vector, according to the manufacturer's directions (Novagen). Clones containing the DNA inserts were identified by agarose gel electrophoresis in 1xTAE [40 mM Tris-acetate, 1 mM EDTA (pH 8.0)].

4.1.5 Restriction Fragment Length Polymorphism (RFLP) Analysis and Sequencing of the Cloned PCR Products

To screen the clones into similar clone types for subsequent sequence analysis, clones containing full-length 16S rDNA were subjected to restriction fragment length polymorphism (RFLP) analysis using tetrameric restriction enzymes (i.e., recognizes a 4-bp restriction site). Two tetrameric restriction enzymes were used, *Rsa*I (GT'AC) and *Msp*I (C'CGG). The restriction enzyme reactions were electrophoresed through a 3% agarose X (Nippon Gene, Toyama, Japan) gel in 1x TAE. The selected cloned 16S rDNA fragments were then PCR amplified and purified; they were subsequently sequenced by the dideoxynucleotide chain-termination method using a model 373S DNA Sequencer (Perkin Elmer/Applied Biosystems) according to the manufacturer's instructions. The primers used for sequencing are forward and reverse primers of pUC vectors (commercial; Toyobo, Osaka, Japan). The oligonucleotide synthesized primers are as follows:

Forward (-20)	5'-GTA AAA CGA CGG CCA GT-3'
Reverse	5'-AAC AGC TAT GAC CAT G-3'

For bacterial 16S rDNA:

357F	5'-TAC GGG AGG CAG CAG-3'
519R	5'-GTA TTA CCG CGG CTG CTG-3'
803F	5'-GAT TAG ATA CCC TGG TAG-3'
909R	5'-CCG TCA ATT CAT TTG AGT-3'
1114F	5'-GCA ACG AGC GCA ACC C-3'

Primer numbers correspond to the *Escherichia coli* 16S rDNA sequence (accession number, JO1695).

For archaeal 16S rRNA:

275F	5'-GAC TAT TAC AGG TAC GGG CTC TG-3'
677R	5'-CCA CAG GTG GTC ATC GAG AGA TC-3'
533F	5'-GTG CCA GCM GCC GCG GTA A-3'

The numbers correspond to the common archaeal 16S sequence (Robb et al. 1995).

4.1.6 Analysis of DNA Sequences and Phylogenetic Relationships

The sequences of 16S rDNAs determined were checked for similarities to DNA sequences in the GenBank and EMBL database, using the FASTA program of the GENETYX-MAC/CD program (version 37.0; Software, Tokyo, Japan). Sequences were aligned, and phylogenetic trees were constructed from a matrix of pairwise genetic distances by the maximum-parsimony algorithm and the neighbor-joining method using the Clustal W program (Saitou and Nei 1987).

4.2 Microbial Diversity in the Area of *Calyptogena* Communities in the Japan Trench

4.2.1 Clones of Bacterial and Archaeal 16S rDNA Sequences

The sediments analyzed in this study were taken from the deepest *Calyptogena* communities in the deep-sea (Fujioka and Murayama 1992). Samples were obtained in two places inside/beside the area of the *Calyptogena* communities. DNA was extracted directly from the sediments, and 16S rRNA genes were amplified by PCR using the primers Eubac27F and Eubac1492R for bacterial genes and Arch21F and Arch958R for archaeal genes. Both bacterial and archaeal 16S rDNA were amplified from the two sediment samples JT1 and JT2. The amplified DNAs were cloned in *Escherichia coli* with pT7Blue(R)T-vector. A total of 91 clones of bacterial 16S rDNA and 45 clones of archaeal 16S rDNA from the sediment samples were analyzed. On the basis of RFLP analysis, the 91 bacterial clones were classified into 33 sequences, and the 45 archaeal clones were classified into 8 sequences.

4.2.2 Phylogenetic Analysis of Bacterial 16S rDNA Sequences

Large-scale phylogenetic trees were constructed based on all of the sequenced clones and representative members of the major lines of bacteria. The trees were constructed on the basis of 1.5-kb aligned position corresponding to *Escherichia coli* positions 27-1492 of the 16S rDNA. The sequences from the two sediments fell into similar phylogenetic affiliations, and some sequences were common in the two sediments (Table 2). Hence all of the sequences were constructed as one tree and are described together here. The phylogenetic tree constructed demonstrates that the 33 sequences fell into three groups of phylogenetic affiliation: Proteobacteria, *Cytophaga,* and gram-positive bacteria (Fig. 2). There were 20 sequences of the 16S rDNA belonging to the Proteobacteria group. Of these 20 sequences, 5 fell in the δ-subdi-

Table 2. Distribution of bacterial 16S rDNA sequences cloned from the sediments of *Calyptogena* communities

Phylogenetic affiliation and sequenced clone no. (accession number)	No. of clones from sediment samples		
	JT1 [a]	JT2 [b]	Total
δ-Proteobacteria			
JTB16 (AB015240)	19	1	20
JTB20 (AB015241)	4	1	5
JTB36 (AB015242)		2	2
JTB38 (AB015243)		3	3
JTB133 (AB015244)	2		2
Total			32
α-Proteobacteria			
JTB131 (AB015245)	1	2	3
JTB260 (AB015246)		2	2
JTB359 (AB015247)		1	1
Total			6
γ-Proteobacteria			
JTB23 (AB015248)		5	5
JTB12 (AB015249)	6	2	8
JTB35 (AB015250)		1	1
JTB247 (AB015251)	6	1	7
JTB148 (AB015252)		1	1
JTB254 (AB015253)		1	1
JTB255 (AB015254)		1	1
JTB256 (AB015255)		2	2
Total			26
ε-Proteobacteria			
JTB129 (AB015256)	1		1
JTB146 (AB015257)		1	1
JTB315 (AB015258)	1		1
JTB360 (AB015259)		6	6
Total			9
Cytophaga			
JTB132 (AB015260)	2	1	3
JTB143 (AB015261)	1	1	2
JTB244 (AB015262)		1	1
JTB248 (AB015263)		1	1
JTB250 (AB015264)		1	1
JTB251 (AB015265)		1	1
JTB220 (AB015266)	1		1
Total			10
Gram-positive bacteria			
JTB31 (AB015267)		3	3
JTB32 (AB015268)		1	1
JTB138 (AB015269)		1	1
JTB215 (AB015270)	1	1	2
JTB243 (AB015271)		1	1
JTB326 (AB015272)		2	2
Total			10

[a] Inside the area of the *Calyptogena* community.
[b] Beside the area of the *Calyptogena* community.

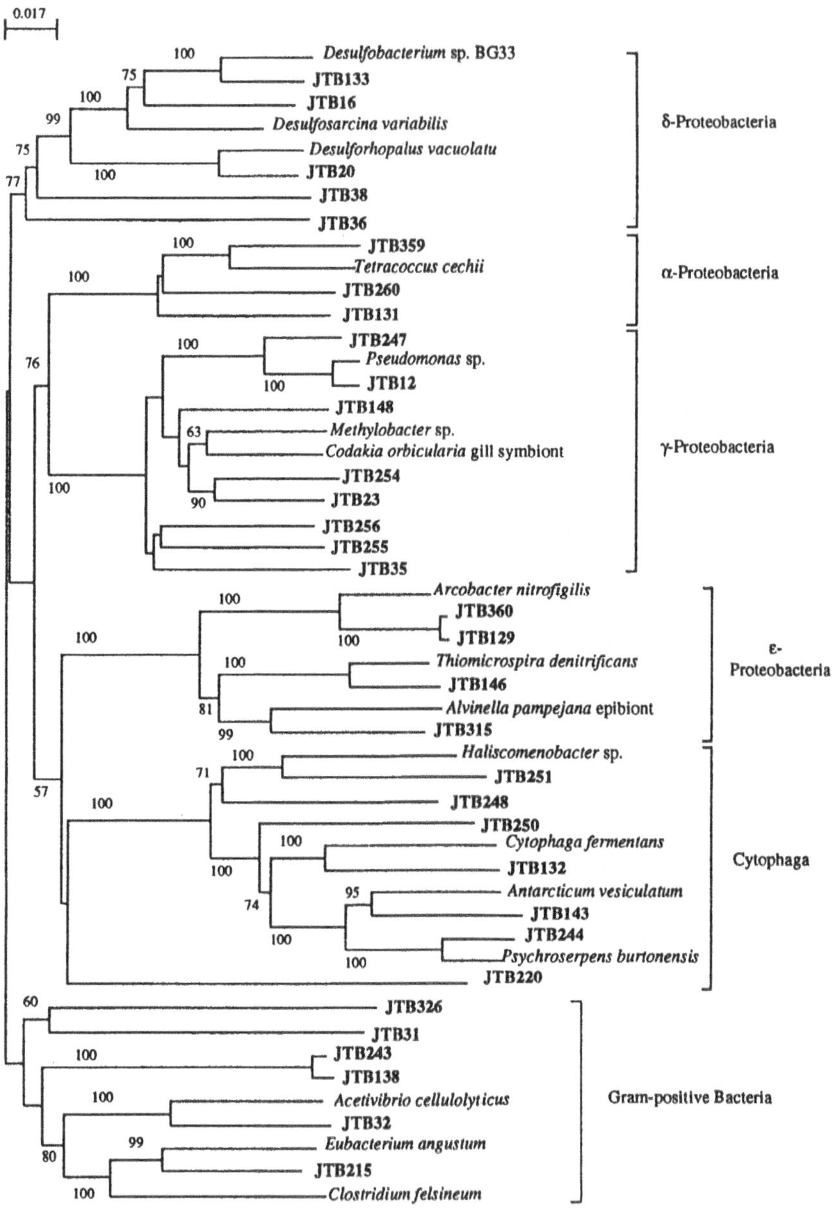

Fig. 2. Phylogenetic tree showing the relationship of 16S rDNA sequences amplified from sediments JT1 and JT2 to those of representatives of bacteria. The tree was constructed from a matrix of pairwise genetic distances by the maximum-parsimony algorithm and the neighbor-joining method. Representatives of the gram-positive bacteria were used as the outgroup. Bar: 0.017 substitutions per base position. Bootstrap values are show for frequencies at or above a threshold of 50%

vision of Proteobacteria, 3 in the α-subdivision, 8 in the γ-subdivision, and 4 in the ε-subdivision. There were six sequences belonging to *Cytophaga* and seven sequences belonging to gram-positive bacteria.

Among the 16S rDNA sequences belonging to the α-Proteobacteria, JTB260 cloned from sample JT2 showed 97.83% identity to that of *Bartonella vinsonii*; originally isolated from dogs (Kordick et al. 1996); JTB359 from sample JT2 showed 93.59% similarity to that of *Tetracoccus cechii*, which was commonly seen in activated sludge mixed liquor, arranged in tetrads. This bacterium is not able to grow phototrophically, nor can it reduce nitrate beyond nitrite nor grow anaerobically; but it is oxidase- and catalase-positive (Blackall 1997). JTB131 only has a 88.7% similarity with an unidentified bacterium in the α-subdivision, indicating that it might be a novel bacterium sequence. Eight sequences that fell in the γ-subdivision of Proteobacteria were analyzed. Sequences JTB12 (8 clones) and JTB247 (7 clones) cloned from both samples showed an identity of more than 99% homology with *Pseudomonas* sp. The genus *Pseudomonas* commonly existed in the sediment samples (as discussed later). Sequences JTB23 (5 clones), JTB148, JTB254, and JTB255 cloned from sample JT2 showed 90.0%-91.5% identity to the 16S rDNA of the *Codakia orbicularis* gill symbiont, which is a sulfur-oxidizing chemoautotrophic (thioautotrophic) bacterium (Distel et al. 1994). Sequence JTB35 showed 89.8% identity to the 16S rDNA of a methane oxidizing bacterium *Methylobacter* sp. JTB256 showed 90.5% identity to the 16S rDNA of *Coxiella burnetii;* this bacterium can cause poisoning in food and also was used as an antigen in diagnostic serological tests (Rasolofonirina 1989). Six *Cytophaga*-related sequences from both samples were cloned. These sequences showed an affiliation with the 16S rDNA of *Cytophaga fermentans, Cytophaga diffluens, Antarcticum vesiculatum, Psychroserpens burtonensis, Microcilla aggregans* and *Haliscomenobacter hydrossis* with identities of about 85.6-97.0%. Seven sequences fell into the group of gram-positive bacteria. JTB215 from both samples is related to *Eubacterium angustum*, showing 89.0% identity. JTB32 from sample JT2 is related to *Acetivibrio cellulolyticus* (90.8% identity), and JTB220 is related to *Anaeroplasma abactoclasticum* (85.1% identity). JTB31, JTB138, JTB243, and JTB326 clustered in the gram-positive group but showed very low identities to *Sanguibacter suarezii* (82.5%), *Desulfotomaculum orientis* (81.1%, 81.3%), and *Moorella glycerini* (81.8%), respectively, in this group. The α- and γ-Proteobacteria, *Cytophaga*, and gram-positive bacteria seem to exist commonly in the deep-sea environments; the *Codakia orbicularis* gill symbiont related sequences might be specially abundant in the *Calyptogena* communities.

Bacteria belonging to the δ- and ε-subdivisions of Proteobacteria were considered to be particularly abundant in the area of the *Calyptogena* communities in this study. A total of 32 clones related to δ-Proteobacteria 16S rDNA sequences were cloned from the two sediments and represented 34% of the total clones. The cloned δ-Proteobacteria-related sequences clustered mainly with those of the sulfate- and sulfur-reducing bacteria, which can utilize sulfate as an electron acceptor for energy-generating processes. Sequence JTB20 (five clones) cloned from both samples showed 93.6% identity to the 16S rDNA of *Desulforholalus vacuolatus*, which is a sulfate reducing bacterium with gas vacuoles isolated at 10°C from reduced mud obtained from a temperate estuary with thiosulfate and lactate as substrates. This strain is moderately psychrophilic with optimum growth at 18°-19°C. Propionate, lactate, and alcohol serve as electron donors and carbon sources (Isaksen and Teske 1996). JTB16 (20 clones) cloned from both samples showed 87.7% homology with the 16S rDNA of sulfate-reducing dissimilatory bacterium *Desulfosarcina variabilis* (Devereux et al. 1989). JTB133 is related to the sulfate-reducing bacterium *Desulfobacterium* sp. BG33 (92.6% identity). JTB36 from sample JT2 is distantly related to *Nitrospina gracilis* (81.2% identity), a lithoautotrophic nitrite-oxidizing bacterium isolated from marine habitats belonging to the δ-Proteobacteria (Watson and Waterbury 1971; Teske et al. 1994). *Nitrospina* species can grow aerobically with nitrite as an electron donor or by oxidation of simple organic compounds. JTB38 is related to *Geobacter metallireducens* (84.1% identity). The large amount of δ-Proteobacteria related sequences in the two sediments demonstrates that the δ-Proteobacteria are abundant in the sediment of *Calyptogena* communities, and the sulfur-reducing bacteria may play an important role in sulfur circulation in these communities, reducing sulfate to H_2S, which is used by the sulfur oxidizing endosymbiotic bacteria of *Calyptogena.*

Four 16S rDNA sequences that clustered in the ε-subdivision of Proteobacteria were cloned from the two sediments. JTB129 and JTB360 are identical (92.4% and 92.5% homology, respectively) to those of the *Arcobacter*-related endosymbionts in a Vestimentiferan tube worm (Naganuma et al. 1997). JTB315 is related to the 16S rDNA of an *Alvinella pompejana* epibiont with 89.4% identity. The *Alvinella pompejana* epibiont 16S rDNA is one of the clones from the dorsal surface of *A. pompejana* that clustered in the ε-Proteobacteria (Haddad et al. 1995). JTB146 is related to *Thimicrospira denitrificans* (94.0% identity). The existence of ε-Proteobacteria related sequences in the sediment samples studied may indicate that a free -living group of ε-Proteobacteria is present in the sediments of *Calyptogena* communities.

The distribution of the bacterial sequences in this study demonstrates that certain groups of bacteria, phylogenetically related, are adapted to particular types of habitat. The δ- and ε-Proteobacteria are the abundant groups in the *Calyptogena* communities, which might play an important role in the metabolism and circulation of sulfur in this environment. Several sequences distantly related to cultivated bacteria indicate that many novel as yet uncultured or unculturable bacteria exist in these environments.

4.2.3 Phylogenetic Analysis of Archaeal 16S rDNA Sequences

Forty-five clones of archaeal 16S rDNA are similarly distributed in the two sediment samples and were analyzed together here. Eight sequences were selected from among all the clones by RFLP analysis and sequenced (Table 3). Large-scale phylogenetic trees of archaeal 16S rDNA sequences were constructed based on the sequenced archaeal clones and representative members of Archaea (Fig. 3). The trees were

Table 3. Distribution of archaeal 16S rDNA sequences cloned from the sediments of *Calyptogena* communities

Phylogenetic affiliation and sequenced clone no. (accession number)	No. of clones from sediment samples		
	JT1	JT2	Total
Crenarchaeota			
JTA173 (AB015273)		1	1
JTA168 (AB015274)	11	3	14
JTA167 (AB015275)	1		1
JTA153 (AB015276)	5	10	15
JTA47 (AB015277)	2	5	7
JTA266 (AB015278)	2	2	4
Total			42
Euryarchaeota			
JTA175 (AB015279)		1	1
JTA69 (AB015280)		2	2
Total			3

constructed on the basis of 27- to 958-bp aligned positions corresponding to archaeal 16S rDNA. The phylogenetic tree demonstrated that six sequences clustered in the domain of Crenarchaeota and two sequences in Euryarchaeota. JTA47, JTA153, JTA167, JTA168, and JTA266 are closely related to the unidentified archaeal marine clone SBAR5, clone SBAR12, and an unidentified archaeal clone Antarctic12 (96.4%-97.2% identity). JTA173 is distantly related to an unidentified Crenarchaeote (86.8%). Sequence JTA175 fell into the domain of Euryarchaeota, closely related to *Methanococcoides burtonii* (98.8%). Sequence JTA69 has no close identity to those of any Archaea yet cultured or cloned (identity values less than 78%); it branched deeply in the Euryarchaeota line of descent.

The diversity of archaeal rDNA clones recovered from *Calyptogena* communities in the Japan Trench in this study was great. Most sequences in this study clustering in Crenarchaeota showed high similarity with the marine group I archaeal rDNA from a variety of coastal and open sea sites in the North Pacific and Antarctic (represented by the sequence of the marine Archaea sp. SBAR5 in Fig. 3), peripherally related to hyperthermophilic Crenarchaeota (DeLong 1992; Fuhrman et al. 1992; 1993). This result implies that the marine Crenarchaeota are ubiquitous, active, and abundant in cold, aerobic marine water and deep-sea sediments worldwide. Moreover, marine Crenarchaeota, with their highly conserved sequences, might come from the same origin. Our results support the idea that the low-temperature marine Ar-

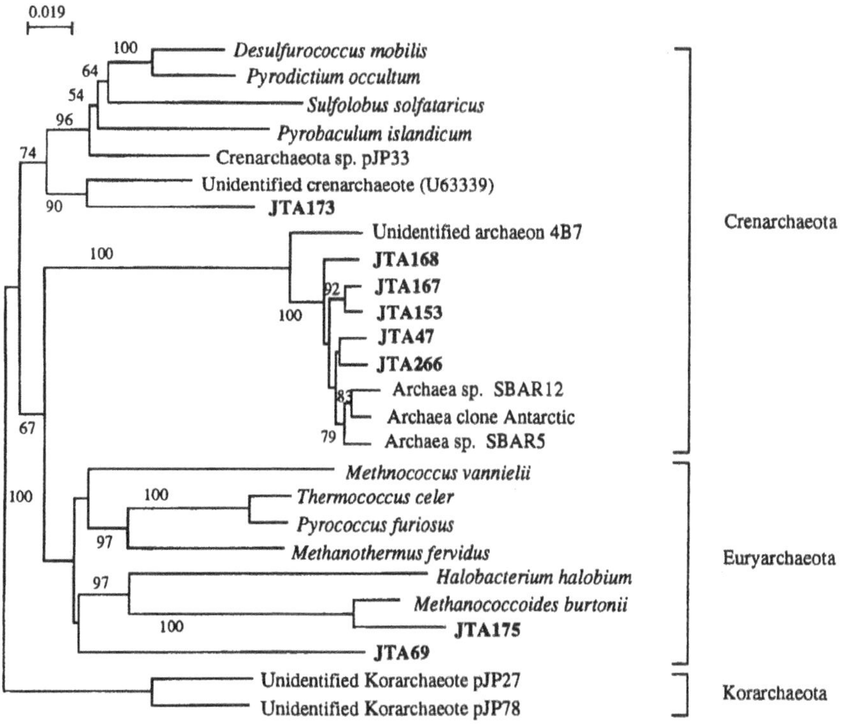

Fig. 3. Phylogenetic tree showing the relationship of archaeal 16S rDNA sequences amplified from sediments JT1 and JT2 to those of representatives of Archaea. The tree was constructed from a matrix of pairwise genetic distances by the maximum-parsimony algorithm and the neighbor-joining method. The Korarchaeota group was used as the outgroup. Bar: 0.019 substitutions per base position. Bootstrap values are shown for frequencies at or above a threshold of 50%

chaea are descendants of ancestral thermophiles, which is consistent with the theory of a high-temperature origin of life (Barns et al. 1996). *Methanococcoides burtonii* is a methylotrophic methanogenic bacterium isolated from the anoxic hypolimnion of Ace Lake, Antarctica. It utilizes methanol in addition to methylamines (Franzmann et al. 1992). JTA175 cloned in this study showed almost the same 16S rDNA sequence as *Methanococcoides burtonii,* indicating that *M. burtonii* exists in the sediments of *Calyptogena* communities. These communities contain high concentrations of methane and hydrogen sulfide (Masuzawa et al. 1992). The discovery of *M. burtonii* in the *Calyptogena* communities gives rise to the phenomenon that *M. burtonii* utilizing methanol contributes methane to the environment. This archaea, together with the δ-Proteobacteria, plays an important role in the circulation of inorganic compounds in *Calyptogena* communities (a proposed model of inorganic compound circulation is described later). It is of interest to understand the origin of *M. burtonii* in the area of *Calyptogena* communities, whether they became adapted to or originated from this environment.

The distantly related sequence (JTA173) and a sequence with no affiliation (JTA69) clustering in the Crenarchaeota and Euryarchaeota indicated the presence of novel archaea in the specific environment examined in this study. It seems likely that a good deal of microbial diversity remains to be discovered and described.

4.3 Microbial Diversity in the Area Inside and Outside of the *Calyptogena* Communities in the Nankai Trough

4.3.1 Phylogenetic Analysis of Cloned Bacterial 16S rDNA Sequences

Five sediment samples from inside and outside the area of *Calyptogena* communities were studied. Bacterial 16S rDNA were amplified from the purified DNAs from the sediments by PCR. The amplified fragments were cloned, and RFLP analyses were performed. Finally, nineteen 16S rDNA fragments that showed mutually distinct patterns were chosen, and these DNA sequences were determined.

The amplified 16S rRNA sequences selected by RFLP analysis were determined and compared with sequences in the DNA databases. The results suggested that 16 bacterial sequences belonged to the Proteobacteria, and 3 sequences belonged to gram-positive bacterial groups (Table 4). Phylogenetic trees constructed based on all the sequenced clones and representative members of bacteria (Fig. 4) indicated that five 16S rDNAs sequences (NKB1~5, 21 clones) were closely related to the Proteobacteria γ-subdivision, and that three of them (NKB1~3) were closely related to the genus *Pseudomonas*. Six 16S DNAs (NKB8–13, twenty clones) were closely related to the ε-Proteobacteria, and three (NKB14–16, eleven clones) were related to the δ-Proteobacteria. These sequences were found commonly in DNA from all of the sediment samples. However, sequences related to the α- and β-subdivisions were found only in DNA from the deeper sample 322.1 [322.1 (12 cm), inside the area of the *Calyptogena* community]. Gram-positive bacterial sequences were also found in DNA from sample 322.1.

Table 4. Distribution of bacterial 16S rDNA sequences cloned from the Nankai Trough sediment samples

16S rDNA sequence and its affiliation	No. of clones by depth from surface					
	Inside *Calyptogena* community, 322.1			Outside *Calyptogena* community, 322.2		
	0 cm	7 cm	12 cm	5 cm	15 cm	Total
Proteobacteria						
γ-Subdivision						
NKB1 [a]	2	3	4	4	0	13
NKB2	0	0	2	0	0	2
NKB3	0	0	0	1	0	1
NKB4	0	0	0	4	0	4
NKB5	0	0	0	0	1	1
Total						21
β-Subdivision						
NKB6	0	0	1	0	0	1
α-Subdivision						
NKB7	0	0	1	0	0	1
ε-Subdivision						
NKB8	3	5	1	1	1	11
NKB9	1	1	0	1	0	3
NKB10	1	0	0	0	0	1
NKB11	1	2	0	0	0	3
NKB12	0	0	0	1	0	1
NKB13	0	0	1	0	0	1
Total						20
δ-Subdivision						
NKB14	2	2	2	0	0	6
NKB15	0	3	0	1	0	4
NKB16	0	1	0	0	0	1
Total						11
Gram-positive bacteria						
NKB17	1	0	0	0	0	1
NKB18	0	1	0	0	0	1
NKB19	0	0	1	0	0	1
Total						3

[a] Accession numbers of the 16S rDNA sequences: NKB1 to NKB19 are AB013253 to AB013271, respectively.

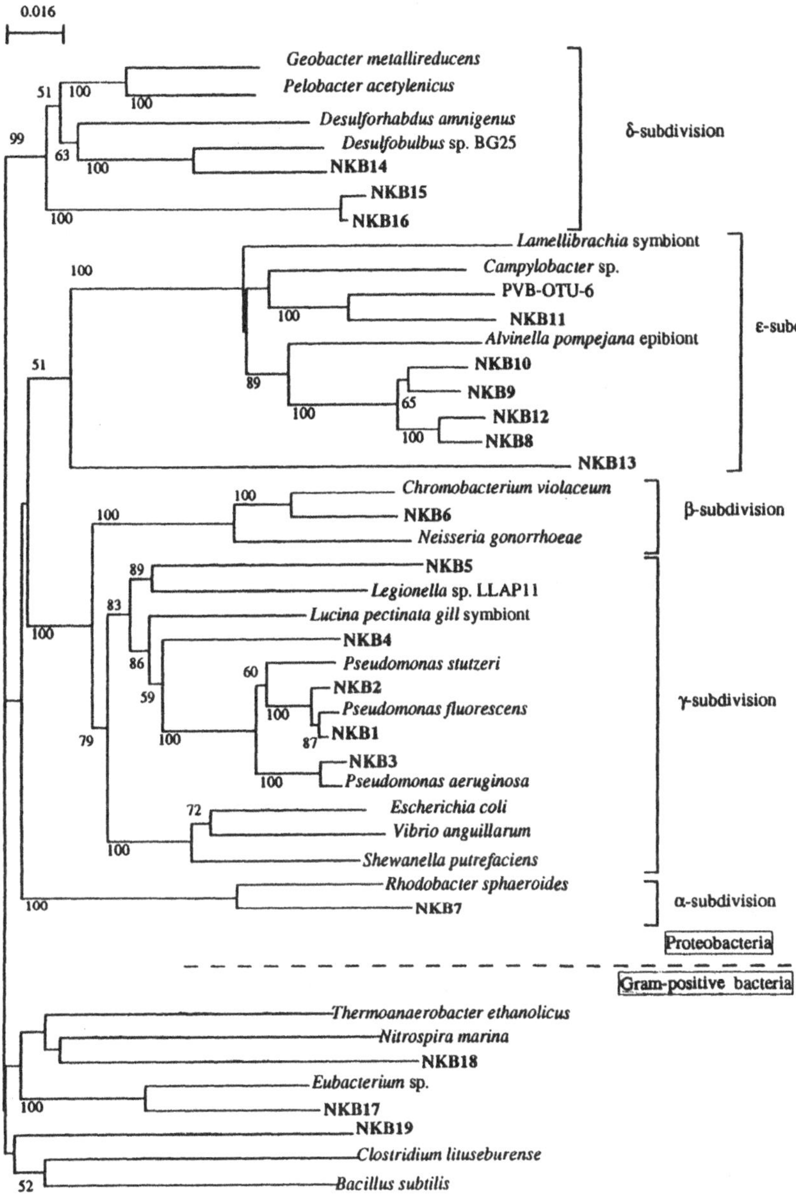

Fig. 4. Phylogenetic tree showing the relationship of 16S rDNA sequences cloned from the Nankai Trough sediment to those of representatives of bacteria. The tree was constructed from a matrix of pairwise genetic distances by the maximum-parsimony algorithm and the neighbor-joining method. Representatives of the gram-positive bacteria were used as the outgroup. Bar: 0.016 substitutions per base position. Bootstrap values are shown for frequencies at or above a threshold of 50%

4.3.2 Distribution of δ- and ε-Proteobacteria in the Sediments

Phylogenetic analyses demonstrated that there is greater microbial diversity in sediment 322.1 from inside the area of the *Calyptogena* community than in sediment 322.2 from outside of the bivalve community. More of the 16S rDNA clones (26 clones) were identified as belonging to the δ- and ε-Proteobacteria, which are known to include sulfate reducing bacteria and epibiont bacteria (Haddad et al. 1995, Naganuma et al. 1997), respectively, in the case of sediment 322.1 (0, 7, and 12 cm core deep). In contrast, only five clones belonging to δ- and ε-Proteobacteria were identified in sediment 322.2 (5 and 15 cm core deep), outside the *Calyptogena* community. It seems reasonable to find such strains inside the area of the *Calyptogena* community, as the environment of the *Calyptogena* community contained a high concentration of sulfate, hydrogen sulfide, and methane, and the sediment we used had a strong smell of hydrogen sulfide. Chemical composition analysis of sediment in the *Calyptogena* community indicated that sulfate reduction using methane is active within the sediment just underneath the living *Calyptogena* (Masuzawa et al. 1992). This result, together with the molecular analysis of the *Calyptogena* community in the Japan Trench, demonstrated that δ- and ε-Proteobacteria are abundant in the sediment of *Calyptogena* communities and give rise to evidence that microbial sulfate reduction takes place in the community by sulfate-reducing bacteria. The produced hydrogen sulfide was then used for the growth of sulfur-oxidizing endosymbionts in the *Calyptogena*. Hence, the δ- and ε-Proteobacteria may play an important role in the circulation of inorganic compounds in the *Calyptogena* communities.

4.3.3 Model of Sulfur Circulation in the Area of *Calyptogena* Communities

A schema of ecological and geochemical settings of cold seep *Calyptogena* communities based on the chemical composition was constructed by Masuzawa et al. (1992). It indicated that generation of hydrogen sulfide occurs through sulfate reduction in a process involving oxidation of methane. However, where such sulfate reduction occurs and where methane is generated were not made clear in their study. Our studies on molecular phylogenetic analysis of microorganisms in sediments from the area of the *Calyptogena* communities in the Japan Trench and Nankai Trough have expanded that schema. A model of inorganic compound circulation in *Calyptogena* communities is proposed from our study (Fig. 5). The abundance of δ- and ε-Proteobacteria-related sequences recovered from the sediment of the *Calyptogena* community in the Japan Trench and Nankai Trough serves as evidence that microbial sulfate reduction takes place in this community, as these groups include sulfate-reducing bacteria. The recovery of Euryarchaeota-related sequences in the *Calyptogena* community in the Japan Trench indicates that high concentrations of methane

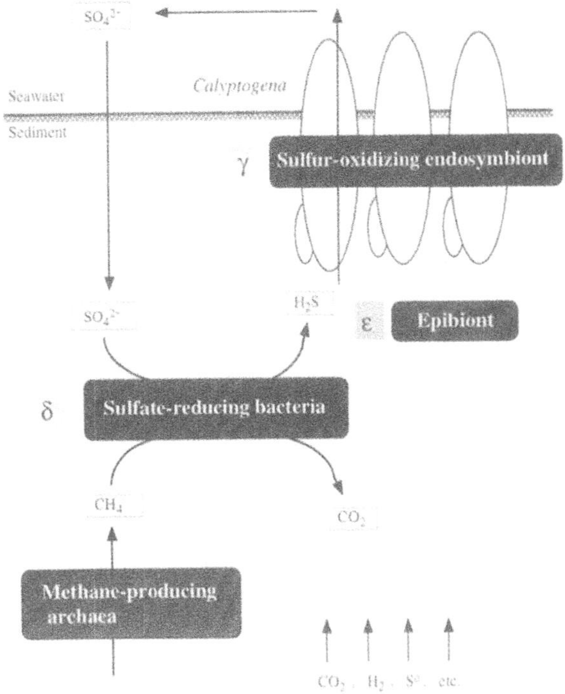

Fig. 5. Model of sulfur circulation within the microbial community and the community of *Calyptogena* clams in the cold-seep environment

in the sediment might be produced by methane-producing archaea. Crenarchaeota in the sediment may also play a role in the circulation of inorganic compounds through sulfur or sulfate reduction and sulfur oxidation.

4.4 Microbial Diversity in the Deep-Sea Sediments from Different Depths

4.4.1 Phylogenetic Analysis of all the 16S rDNA Clones

A total of 149 16S rDNA clones from the seven sediment samples obtained from different depths of the deep-sea were analyzed (Table 5). On the basis of the RFLP analysis, the 149 clones were classified as having 75 different sequences. Large-scale phylogenetic trees were constructed based on all of the sequenced clones and representative members of the major lines of bacterial descent. As shown in the phylogenetic trees, three sequences from sediments collected at depths of 1159, 3671, and 6379 m clustered within the δ-subdivision of Proteobacteria. Eight sequences (9 clones) from sediments collected at depths of 1159, 1521, 4031, and 6079 m fell in the α-subdivision of Proteobacteria. Two sequences from depths of 1159 and 6379 m fell into the β-subdivision of Proteobacteria. The sequences that clustered

Table 5. Distribution of 16S r DNA sequences cloned from sediments collected at different depths

Sample /depth (m)	Sequences (no.)	Clones (no.)	Distribution of clones among the bacterial groups [a]								
			Proteobacteria					G+	*Cytophaga*	Other	No affiliation
			δ	α	β	γ	ε				
BD1/1159	17	(23)	1	4 (5)	1	4 (9)	2	0	3	1	1
BD2/1521	16	(16)	0	1	0	2	2	1	3	1	6
BD3/2339	10	(13)	0	0	0	5 (8)	0	0	1	0	4
BD4/3671	7	(24)	1	0	0	3 (19)	1	0	0	1	1 (2)
BD5/4031	8	(18)	0	2	1	3 (13)	0	1	0	0	1
BD6/6482	5	(38)	0	0	0	5 (38)	0	0	0	0	0
BD7/6379	12	(17)	1 (2)	1	0	2	2 (3)	0	2 (4)	0	4 (5)
Total	75	(149)	3 (4)	8 (9)	2	24 (91)	7 (8)	2	9 (11)	3	17 (19)

Numbers in parentheses represent the number of the 16S rDNA clones.
[a] G+: gram-positive bacteria; Others; *Planctomyces*, *Actinomycetes*.

within the γ-subdivision of Proteobacteria were the dominant sequences found at all of the depths studied. The most common sequences were *Pseudomonas*-related sequences showing high similarity. Seven sequences from sediments collected at depths of 1159, 1521, 3671, and 6379 m clustered within the ε-Proteobacteria. Only two sequences from depths of 1521 and 4031 m fell into the gram-positive bacteria. There were also nine sequences from depths of 1159, 1521, and 6379 m that clustered in the *Cytophaga* group, and two from 1521 and 3671 m belonging to the actinomycete group. Seventeen sequences showed very low identity to known 16S rDNA sequences in the DNA databases, indicating that these sequences are from novel or unculturable bacteria in these environments.

4.4.2 Depth of 1159 Meters

The phylogenetic tree (Fig. 6) shows the relationship of the 17 sequences (a total of 23 clones) of 16S rDNAs cloned from sample BD1 at a depth of 1159 m with the representatives of the bacteria. These sequences clustered into seven major lineages of bacteria: the δ-, α-, β-, γ-, and ε-subdivisions of Proteobacteria, *Planctomyces*, and *Cytophaga*. FASTA searches revealed that sequence BD1-2 fell in the δ-Proteobacteria subdivision, which includes sulfate- and sulfur-reducing bacteria. It was found to show 89.2% homology with the 16S rDNA of *Desulfocapsa thiozymogene*, an anaerobic, sulfate-reducing bacterium isolated from a freshwater lake. This strain is able to grow by disproportionation of thiosulfate or sulfite to sulfate plus sulfide (Janssen et al. 1996). Sequences BD1-8, BD1-10, BD1-14, and BD1-17 are closely related to the α-Proteobacteria BD1-8 (total of two clones) shows 85.0% homology with the 16S DNA of *Pedomicrobium manganicum*; BD1-17 has 85.6% identity to that of *Rhodoapirillum fulvum*, a phototrophic purple nonsulfur bacterium. BD1-33 fell into the β-Proteobacteria and showed 89.1% homology with the 16S rDNA of *Alcaligenes xylosoxydans* and 93.3% homology with that of *Burkholderia* sp. BD1-1, BD1-3, BD1-7, and BD1-9 clustered within the γ-Proteobacteria. BD1-1 showed 86.5% homology with the 16S rDNA of *Stenotrophomonas maltophilia*. Sequence BD1-3 (two clones) was identified as having 99.3% homology with the 16S rDNA of a cultured *Pseudomonas* sp. BD1-7 showed 89.5% homology with the 16S rDNA of a chemoautotrophic bacterial symbiont of *Solemya velum*. This strain is widespread in marine habitats such as deep-sea hydrothermal vents and coastal sediments (Eisen et al. 1992). BD1-9 (a total of five clones) showed 89.0% homology with the 16S rDNA of *Vibrio vulnificus*. This organism is reported to be a common inhabitant of estuarine and marine waters; it can be isolated under a wide range of salinity and temperature conditions from oysters, clams, mussels, and fish, as well as from sediment and plankton (Aznar et al. 1994). Sequences BD1-4 and BD1-29 clustered within the ε-Proteobacteria. BD1-4 showed 87.9% homology with the 16S rDNA of a bacterial epibiont of *Alvinella pampejana*; and BD1-29 showed 93.9% homology with that of *Arcobacter nitrofigilis*. BD1-15 has a relatively low identity of 85.2% to the 16S rDNA of a thermophilic bacterium, *Rhodothermus marinus*, an aerobic heterotroph with optimum growth 65°-75°C. A number of *Rhodothermus marinus* strains have been isolated from submarine hot springs and vents. This bacterium belongs to the *Cytophaga* group (Anaresson and Fridjonsson 1994). Sequence BD1-27 also clustered in the *Cytophaga* group. One sequence, BD1-23, belonged to the *Planctomyces* group. BD1-5 shows very low identity to the known 16S rDNA sequences from the database.

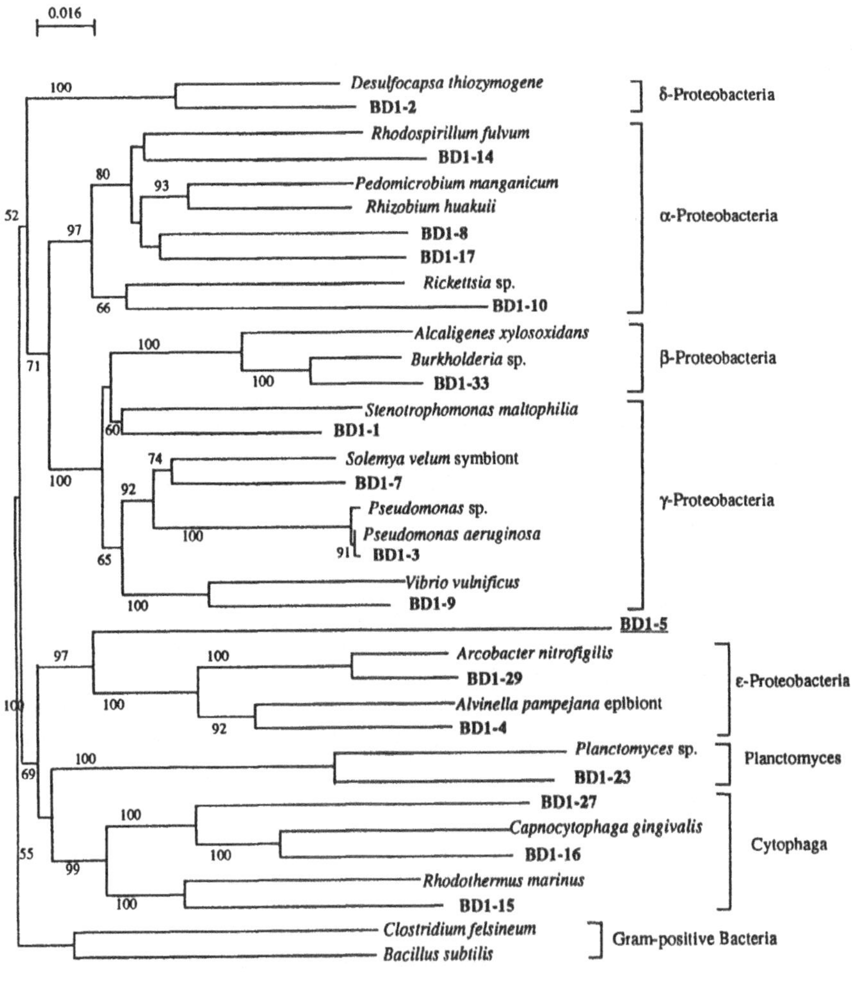

Fig. 6. Phylogenetic tree showing the relationship of 16S rDNA sequences from sediment BD1 to those of representatives of bacteria. The tree was constructed from a matrix of pairwise genetic distances by the maximum-parsimony algorithm and the neighbor-joining method. Representatives of the gram-positive bacteria were used as the outgroup. Bar: 0.016 substitution per base position. Bootstrap values are shown for frequencies at or above a threshold of 50%. An underline indicates that the sequence has no phylogenetic affiliation. These conditions are the same in Figures 7-12

4.4.3 Depth of 1521 Meters

Phylogenetic analysis demonstrated that sixteen 16S rDNA sequences from sample BD2, collected at a depth of 1521 m fell into six major lineages of bacteria: the α-, γ-, and ε-subdivisions of Proteobacteria, *Cytophaga*, actinomycetes, and the gram-positive bacteria (Fig. 7). Of the 16 sequences, one (BD2-8) was identical to the 16S rDNA of α-Proteobacteria, showing relatively low homology (81.6%) with that of *Rickettsia tsutsugamushi.* Sequences BD2-7 and BD2-13 clustered in the γ-Proteobacteria but with low identities to the strains in this subdivision, showing 87.7% and 88.0% similarity, respectively, to the 16S rDNA of the *Solemya velum* symbiont. In sample BD1, sequence BD1-7 also showed 89.5% homology with the 16S rDNA of the *Solemya velum* symbiont, but BD1-7 showed low similarity to BD2-7 (89.9%) and BD2-13 (90.2%). Two sequences (BD2-1 and BD2-5) with high similarity to the sequence BD1-4 clustered in the ε-subdivision of Proteobacteria, showing identities of 86.6% and 89.2%, respectively, to the 16S rDNA of the *Alvinella pompejana*

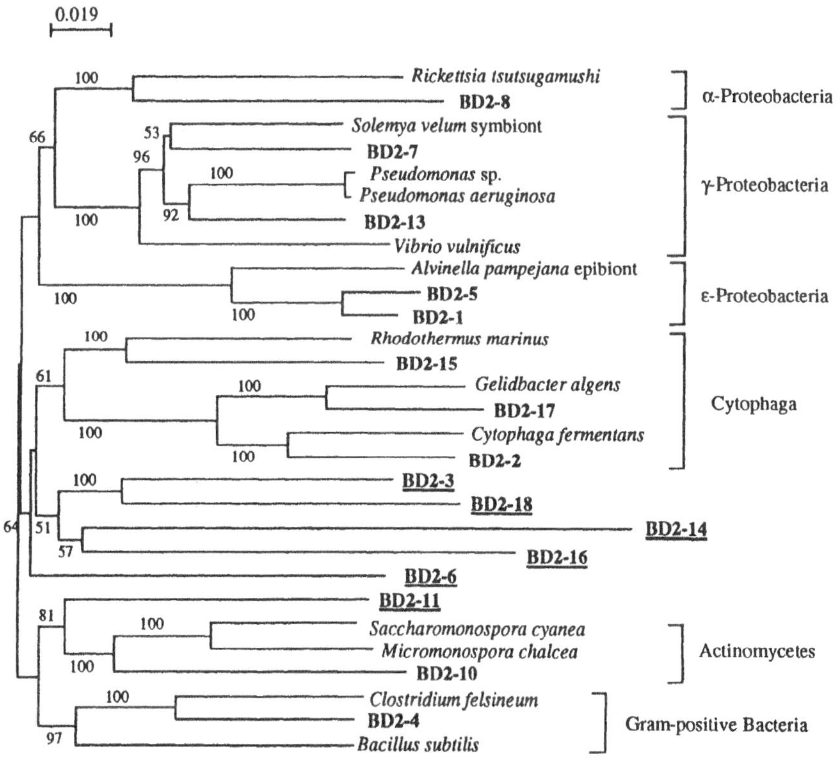

Fig. 7. Phylogenetic tree showing the relationship of 16S rDNA sequences amplified from sediment BD2 to those of representatives of bacteria. Representatives of the gram-positive bacteria were used as the outgroup. Bar: 0.019 substitution per base position

epibiont. Sequence BD2-2 showed 88.3% identity to that of *Cytophaga fermentans*; BD2-17 showed 90.8% identity to that of *Gelidbacter algens* in the *Cytophaga* group, which is a psychrophilic bacterium isolated from Antarctic marine habitats (Bowman et al. 1996). BD2-10 was found to belong to the actinomycetes. There was only one sequence (BD2-4) that fell into the gram-positive bacteria in the sample from this depth, showing 87.9% homology with the 16S rDNA of *Clostridium felsineum*. Of sixteen sequences, six (BD2-3, BD2-6, BD2-11, BD2-14, BD2-16, and BD2-18) were distinctly different from known bacterial sequences. The 1521 m depth sampling site was in Suruga Bay, and many unaffiliated sequences were amplified and cloned from the sample collected at this site. It may indicate that this site is a unique environment with many as yet uncultivated bacteria.

4.4.4 Depth of 2339 Meters

Phylogenetic analysis demonstrated that 10 sequences (13 clones) of 16S rDNA from sample BD3, collected at a depth of 2339 m, fell into two major lineages of bacteria: the γ-Proteobacteria and *Cytophaga* (Fig. 8). Sequences BD3-1, BD3-6, BD3-12, BD3-13, and BD3-15 clustered in the γ-subdivision of Proteobacteria, the most common group at every depth. BD3-1, BD3-13, and BD3-15 were found to be identical (with 91.0%-99.6% similarity) to 16S rDNA of members of the genus *Pseudomonas*. BD3-6 (two clones) showed 88.5% relatedness compared with the

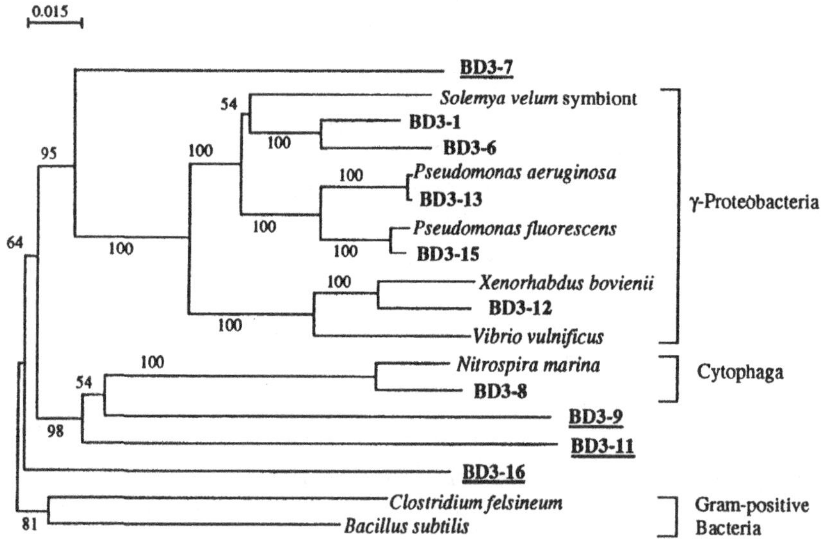

Fig. 8. Phylogenetic tree showing the relationship of 16S rDNA sequences amplified from sediment BD3 to those of representatives of bacteria. Representatives of the gram-positive bacteria were used as the outgroup. Bar: 0.015 substitution per base position

16S rDNA of the *Solemya velum* symbiont. BD3-12 showed 93.4% identity to that of *Xenorhabdus bovienii* and 91.2% to that of *Vibrio vulnificus*, while showing 89.4% identity to BD1-9 from sample BD1, which was found to have 89.0% identity to the 16S rDNA of *Vibrio vulnificus*. BD3-8 is related to that of *Nitrospira marina*, showing 94.5% identity. *Nitrospira* is one of the genera comprising the nitrifying bacteria; these obligate chemolithoautotrophs use carbon dioxide as the sole source of carbon and obtain energy from oxidation of ammonia to nitrite (Ehrich et al. 1995). BD3-7, BD3-9, BD3-11, and BD3-16 did not fall into any group of known bacterial sequences.

4.4.5 Depth of 3671 Meters

Seven 16S rDNA sequences (a total of 24 clones) from sample BD4, collected at a depth of 3671 m and analyzed phylogenetically, fell into four major lineages of bacteria: δ-, α-, and γ-Proteobacteria and the actinomycetes (Fig. 9). Sequences BD4-3 (a total of 13 clones) and BD4-16 (a total of 5 clones) were the dominant sequences in this sample and were related to the genus *Pseudomonas*. Sequence BD4-11 is identical to sequence BD1-9 (99.5%) from sample BD1, and it showed 89.0% similarity to the 16S rDNA of *Vibrio vulnificus*. Sequence BD4-12 was found to be identical to that of *Microbacterium arborescens* (98.4%), an actinomycete. BD4-10 and BD4-8 showed relatively low similarity to the 16S rDNA of δ- and ε-Proteobacteria, respectively. BD4-9 showed low identity to any of the known sequences.

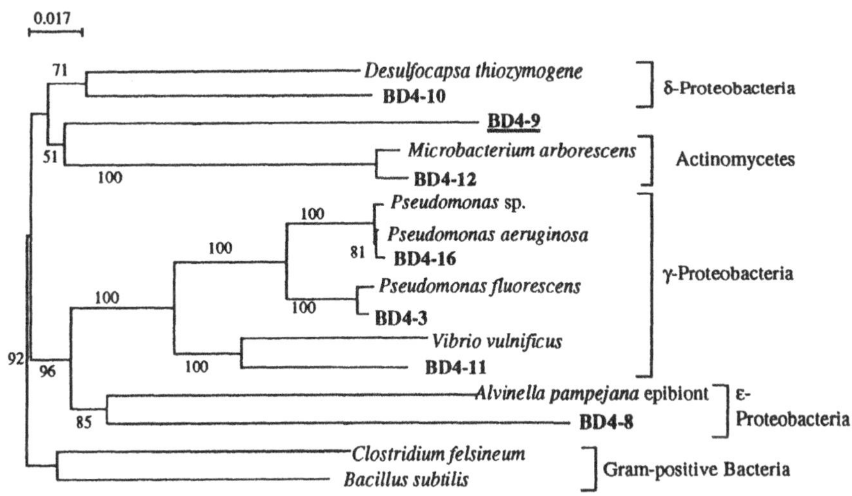

Fig. 9. Phylogenetic tree showing the relationship of 16S rDNA sequences amplified from sediment BD4 to those of representatives of bacteria. Representatives of the gram-positive bacteria were used as the outgroup. Bar: 0.017 substitution per base position

4.4.6 Depth of 4031 Meters

Eight 16S rDNA sequences (18 clones) from sample BD5, collected at a depth of 4031 m and analyzed phylogenetically, fell into four major lineages of bacteria: α-, β-, and γ-Proteobacteria and the gram-positive bacteria (Fig. 10). Two sequences were found to belong to the α-Proteobacteria: Sequence BD5-9 showed 97.8% similarity to the 16S rDNA of *Sphingomonas* sp.; and BD5-10 showed 97.5% similarity to that of *Methylobacterium* sp., a strictly aerobic, facultatively methylotrophic bacterium that is able to grow on one-carbon compounds more reduced than carbon dioxide as the sole carbon and energy source. Members of the genus *Methylobacterium* are distributed in a wide variety of natural habitats, including soil, dust, fresh water, and aquatic sediments (Hiraishi et al. 1995). BD5-11 showed 87.1% identity to *Alcaligenes xylosoxidans* in the β-Proteobacteria. Four sequences (a total of 13 clones) were found to be related to the γ-Proteobacteria: BD5-7(5 clones) and BD5-14 (6 clones) were identical to the genus *Pseudomonas*. BD5-16, which is identical to BD3-6 (99.4%), showed 88.3% homology with the *Solemya velum* symbiont. Sequence BD5-12 was found to be related to the gram-positive bacteria, showing 99.1% similarity to *Bacillus flavothermus*. BD5-13 showed very low identity to any of the known sequences.

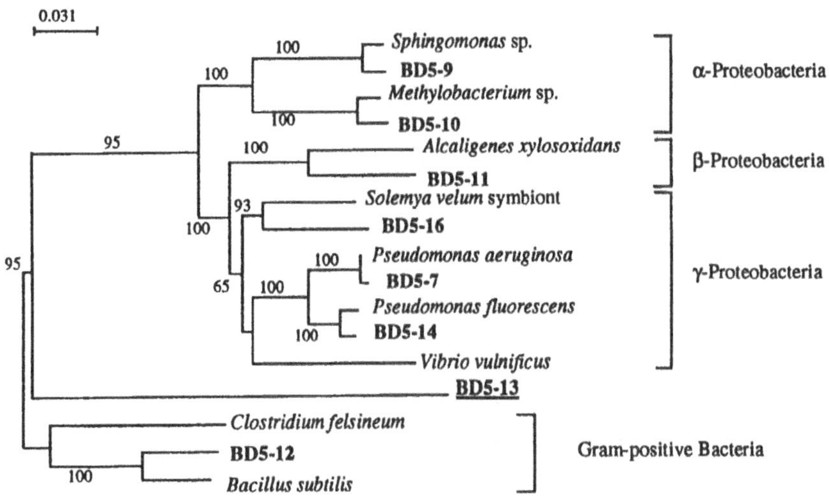

Fig. 10. Phylogenetic tree showing the relationship of 16S rDNA sequences amplified from sediment BD5 to those of representatives of bacteria. Representatives of the gram-positive bacteria were used as the outgroup. Bar: 0.031 substitution per base position

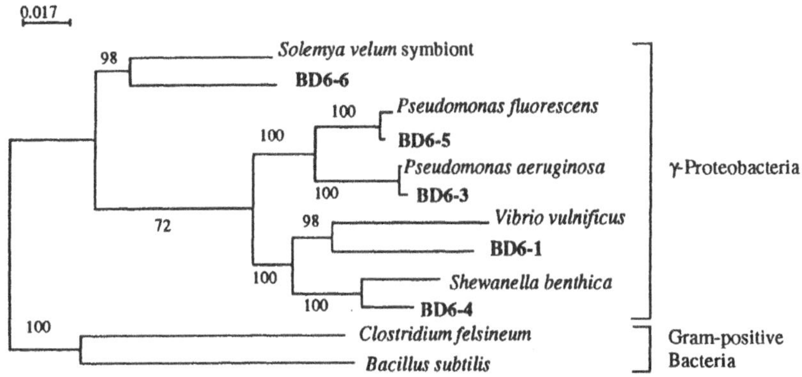

Fig. 11. Phylogenetic tree showing the relationship of 16S rDNA sequences amplified from sediment BD6 to those of representatives of bacteria. Representatives of the gram-positive bacteria were used as the outgroup. Bar: 0.017 substitution per base position

4.4.7 Depth of 6482 Meters

In total, 38 clones analyzed from sample BD6 collected at a depth of 6482 m, were classified as five different 16S rDNA sequences. Phylogenetic analysis showed that all of these sequences were related to the γ-Proteobacteria (Fig. 11). The genus *Pseudomonas* related sequences are the dominant ones in this depth as well and showed high similarity with the members of *Pseudomonas*. Sequence BD6-1(eight clones), which is identical to BD1-9 (99.8%) and BD4-11 (99.5%), showed 88.9% similarity to *Vibrio vulnificus*. BD6-6 showed 88.6% similarity to *Solemya velum* symbiont Sequence BD6-4 (three clones) showed high identity to the barophilic bacterium *Shewanella benthica*. Members of this group of bacteria have been isolated from deep-sea sediments and can grow better under high pressures (Kato et al. 1995, 1996). It seems reasonable to find sequences related to barophilic bacteria in the sediment at this depth.

4.4.8 Depth of 6379 Meters

Twelve 16S rDNA sequences (a total of 17 clones) from sample BD7, collected at a depth of 6379 m were phylogenetically clustered in five major lineages of bacteria: δ-, α-, γ-, and ε-Proteobacteria and the *Cytophaga* (Fig. 12). Sequence BD7-15 fell into the δ-Proteobacteria, showing 90.6% identity to the 16S rDNA of *Desulfocapsa thiozymogene*. BD7-3 fell into the α-Proteobacteria, showing low similarity to the 16S rDNA of *Rickettsia tsutsugamushi*. Two common sequences were found at this depth, as well as at other depths: BD7-8, which was found to be related to the 16S rDNA of the *Solemya velum* symbiont (87.8%), and BD7-5, a sequence that showed identity to that of *Ps. fluorescens* (98.9%). Two sequences (BD7-6 and BD7-9) fell

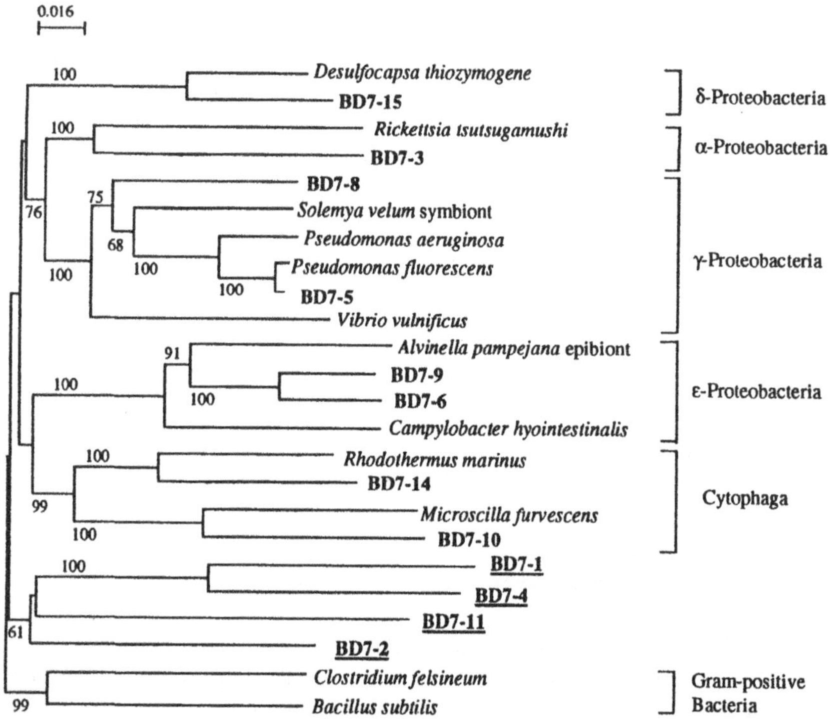

Fig. 12. Phylogenetic tree showing the relationship of 16S rDNA sequences amplified from sediment BD7 to those of representatives of bacteria. Representatives of the gram-positive bacteria were used as the outgroup. Bar: 0.016 substitution per base position

into the ε-Proteobacteria, showing 88.0% and 83.2% similarity, respectively, to the 16S rDNA of a *Alvinella pompejana* epibiont. BD7-14 (two clones) showed 85.5% identity to that of *Rhodothermus marinus* and 96.8% to BD1-15 in the *Cytophaga* group. Of the 12 sequences, 4 showed very low identity to any known bacterial sequences in the database.

4.4.9 Common and Particular Sequences Found at the Different Depths

Phylogenetic analysis of 16S rDNA sequences from sediments collected at various depths revealed a significant degree of microbial diversity. The 16S rDNA clone library is thought to approximate the distribution of populations of microorganisms in the sediments. The deep-sea sediments from seven depths were dominated by populations of bacterial sequences (43%) belonging to the genus *Pseudomonas*. Clones identical to 16S rDNA of members of the genus *Pseudomonas* were observed in all of the sediments examined. These clones were 90.3%–99.6% identical

Table 6. 16S rDNA sequences similarity matrix for *Solemya velum* symbiont

Clone	% Similarity to								
	S.v. s.	BD1-7	BD2-7	BD2-13	BD3-1	BD3-6	BD5-16	BD6-6	BD7-8
BD1-7	89.5	100							
BD2-7	87.7	89.9	100						
BD2-13	88.0	90.2	87.6	100					
BD3-1	89.6	88.0	87.9	88.3	100				
BD3-6	88.5	85.7	84.9	85.4	94.6	100			
BD5-16	88.3	85.7	84.7	85.2	94.5	99.4	100		
BD6-6	88.6	85.5	84.9	85.3	94.2	99.1	98.9	100	
BD7-8	87.8	87.3	85.5	86.9	88.5	86.3	86.3	86.4	100

S.v. s., *Solemya velum* symbiont.

to the 16S rDNA of the cultured *Pseudomonas.* sp. *P. fluorescens and P. aeruginosa.* In fact, many *Pseudomonas* strains have been isolated from the deep-sea. Lee et al. (1977) have analyzed some *Pseudomonas*-like marine bacteria growing on a single carbon source and examined their G+C content and the complement of some enzymes in intermediary metabolism. The strains tested were strict aerobes except those that could use denitrification as a means of anaerobic respiration. Nearly all of the strains were capable of growth in a minimal medium containing artificial seawater with glucose as the sole source of carbon and energy and ammonia as the sole source of nitrogen (Lee et al. 1977). A study of analysis of dissolved protein in seawater revealed that a 48-kDa protein is a major constituent at all stations; its N-terminal amino acid sequence was found to be homologous to the porin P, a *trans*-outmembrane channel protein of bacterium *P. aeruginosa* (Tanoue et al. 1995). A denitrifying bacterium, *Pseudomonas* sp., was isolated from the Mariana Trench. This strain showed greater tolerance to low temperature (4°C) and high hydrostatic pressure (50 MPa) than the denitrifiers obtained from the land (Tamegai et al. 1997). Our present results support the idea that *Pseudomonas* is one of the principle genera of bacteria present in the ocean (Fogg 1986). Some of the *Pseudomonas* are not likely to be the indigenous microflora of the deep-sea; for example, *Pseudomonas bathycetes* originating from a sediment sample from the Mariana Trench (Morita 1976) might be an intruder that entered the trench sediment with detritus and remained there in a preserved or low-activity state.

The second most common group of sequences found in the present study were 87.8%–89.5% identical to the 16S rDNA of a chemoautotrophic bacterium, the *Solemya velum* symbiont. In total, eight of such sequences were amplified and cloned from six sediment samples. The similarities between these sequences are shown in Table 6. They seem to be rather distantly related sequences (84.7%–94.6% similarity) except for three of them (BD3-6, BD5-16, and BD6-6), which are similar (99%). Chemoautotroph-invertebrate symbiosis appears to be widespread in marine habitats such as deep-sea hydrothermal vents and coastal sediments (Felbeck 1981; Fisher 1990). The bacterial symbionts utilize the energy released by the oxidation of reduced inorganic substrates, such as hydrogen sulfide, to fix carbon dioxide (Cavanaugh 1983, 1985). *Solemya velum* symbiont 16S rDNA was originally found in a gill sample from the bivalve *Solemya velum*, an Atlantic coast clam (Eisen et al. 1992). This sequence belongs to the chemoautotrophic cluster within the γ-subdivision of Proteobacteria. We extracted DNA from sediments, and the sediment symbiont-like 16S rDNA sequences are obviously different from the *Solemya velum* symbiont and were widely spread in the deep-sea sediments. These bacteria may therefore represent a source of chemoautotrophic productivity in the microbial community in these environments. The bacterial symbiont-related sequences might come from bacteria associated with the remains of dead clams or from free-living bacteria in the deep-sea sediment.

Common to three sediments samples were several 16S rDNA sequences related to the 16S rDNA of an *Alvinella pompejana* epibiont (Table 7). These epibiont-related sequences are similar to each other (89.2%–99.8% identity). *Alvinella pompejana*, a polychaete annelid that inhabits the walls of chimneys at active deep-sea hydrothermal vent sites, is thought to be one of the most thermotolerant metazoans known. A highly diverse and dense assemblage of microorganisms inhabits the dorsal surface of this worm (Haddad et al. 1995). The *Alvinella pompejana* epibiont 16S rDNA mentioned above is one of the clones obtained from a sample collected from the dorsal surface of *A. pompejana* and is related to the ε-Proteobacteria. The

Table 7. 16S rDNA sequences similarity matrix for *Alvinella pompejana* epibiont

Clone	% Similarity to					
	A. p. e.	BD1-4	BD2-1	BD2-5	BD7-9	BD7-6
BD1-4	87.9	100				
BD2-1	86.6	99.8	100			
BD2-5	89.2	95.7	95.7	100		
BD7-9	83.5	90.5	90.6	90.3	100	
BD7-6	88.0	97.1	97.2	97.4	91.9	100

A. p. e., *Alvinella pompejana* epibiont.

Table 8. 16S rDNA sequences similarity matrix for *Vibrio vulnificus*

Clone	% Similarity to				
	V. vulnificus	BD1-9	BD3-12	BD4-11	BD6-1
BD1-9	89.0	100			
BD3-12	91.2	89.4	100		
BD4-11	89.0	99.5	89.2	100	
BD6-1	88.9	99.8	89.3	99.5	100

existence of *Alvinella pompejana* epibiont-related 16S rDNA sequences in the sediment samples studied may indicate that there is a free living group of ε-Proteobacteria in the microbial community in these environments.

Four sequences from four different samples were found to be related to the 16S rRNA gene of *Vibrio vulnificus*, a naturally occurring estuarine bacterium (Table 8). It is abundant in a variety of habitats, such as oysters, clams, fish, and plankton; and it can also be isolated from sediments (DePaola et al. 1994). Our results demonstrate that *Vibrio vulnificus*-related bacteria also exist in the sediments far from an estuarine environment, even at a depth of more than 6000 m.

Cytophaga-related 16S rDNA sequences were also common in the sediments examined in this study. In total, eight sequences (10 clones) clustered within the *Cytophaga* group. Three sequences (four clones) from three samples were identical to the 16S rDNA of thermophilic bacterium *Rhodothermus marinus* in the *Cytophaga* group. A number of *Rhodothermus marinus* strains have been isolated from submarine hot springs and vents (Alfredsson et al. 1988; Nunes et al. 1992). Our results indicate that *Rhodothermus marinus*-like bacteria are also commonly distributed in cold deep-sea environments.

δ-Proteobacteria-related sequences were found in three samples: BD1, BD4, and BD7. BD1 and BD7 were collected from sites near manmade garbage and the *Calyptogena* community, respectively. These environments are obviously richer in nutrients than other sites and can be expected to show many types of bacteria. The δ-Proteobacteria were found largely distributed in the *Calyptogena* community as discussed above.

Only one sequence (BD6-4, three clones) amplified from the sample collected at a depth of 6482 m was found to be closely related to the 16S rDNA of *Shewanella benthica*. These bacteria are mostly barophilic bacteria; and for some of them the optimum pressure for growth is higher than 50 MPa. Many strains of *Shewanella benthica* and related strains have been isolated from the deep-sea at Suruga Bay (2485 m), the Ryukyu Trench (5110 m), the Japan Trench (6269 and 6356 m), the Izu-Bonin Trench (6499 m) and the Mariana Trench (10 898 m) (Kato et al. 1998b; Li et al. 1998). Theoretically, barophilic bacteria could be widespread in deep-sea high-pressure environments. We expected to obtain more clones related to barophilic *Shewanella* sp. in this study, but only one of these sequences was amplified from the sediments studied: the sample collected at a depth of 6482 m from the Japan Trench. A possible explanation for this phenomena is that barophilic *Shewanella* sp. is a minor microbial population in the deep-sea. The 16S rDNA analysis of sediment DNA is thought to identify most of the members of the bacterial community in the environment.

A large number of cloned sequences (23%) in this study showed low identity (below 80%) with known sequences in the DNA database. These sequences may represent as yet uncultivated microorganisms in the microbial communities in these sediments.

In conclusion, this study revealed a microbial diversity and distribution in deep-sea sediments collected at seven different depths and showed that many bacteria found in the deep-sea environment are similar to those living in soil, in shallow water, or in association with fish, clams and other organisms. Factors responsible for physical movement, such as wind, tide, or deposition of dead animals, could bring bacteria into the deep-sea environment. Some of these bacteria maybe not able to live under the high-pressure, low-nutrient conditions in such environments. The microorganisms indigenous to the deep-sea may represent a small portion of all living microorganisms because of the low amounts of nutrients in the deep-sea. The question remains as to whether the 16S rDNA sequences obtained as amplified products represent the proportion of 16S rDNA sequences present in the original sample. Although PCR is known to be somewhat biased in its amplification (Reysenbach et al. 1992), molecular phylogenetic methods using PCR and cloning of 16S rDNA are the most sensitive techniques available for describing the composition of complex microbial communities. Future studies on microbial diversity and ecology will tie the phylogenetic data together with data concerning specific microbial activities and sediment geochemistry.

4.5 Conclusions

Small-subunit ribosomal RNA genes (16S rDNA) were amplified by PCR from DNA extracted directly from the sediment samples. The phylogenetic analysis of the 16S rDNA sequences from three groups of sediment samples at different areas and depths revealed a significant degree of microbial diversity. Two sediment samples in group 1 from inside and beside the area of the *Calyptogena* communities in Japan Trench were studied (Li et al. 1999c). The results demonstrated that 33 different 16S rDNA sequences from the two samples analyzed fell into the similar phylogenetic affiliations: the α-, γ-, δ-, and ε-subdivisions of Proteobacteria, *Cytophaga,* and the gram-positive bacteria. Some of the 16S rDNA sequences were common to both samples. The δ- and ε-Proteobacteria-related sequences were abundant in both sediments. These sequences are mostly related to the sulfate- or sulfur-reducing bacteria and epibiont, respectively. Eight archaeal 16S rDNA sequences were cloned from the sediments. Most of the archaeal 16S rDNA sequences clustered in Crenarchaeota and showed high similarity to marine group I archaeal 16S rDNA. A *Methanococcoides burtonii*-related sequence obtained from the sediment clustered in the Euryarchaeota, indicating that *M. burtonii*-related strains in the area of *Calyptogena* communities may contribute to methane production in this environment.

Five sediments in group 2 from core samples inside and outside the area of *Calyptogena* communities in Nankai Trough were examined (Li et al. 1999a). At least 19 kinds of bacterial 16S rDNA related mostly to the Proteobacteria and a few gram-positive bacteria were identified. The results demonstrated that the microbial diversity is greater in the sediment inside the *Calyptogena* communities than in the sediments outside that area, especially the δ- and ε-Proteobacteria-related sequences. These results together with the results obtained from the *Calyptogena* communities in the Japan Trench give rise to evidence that microbial sulfate reduction takes place in the community by sulfate-reducing bacteria. The supply of methane by methanogenic archaea and the produced hydrogen sulfide may then be supplied for growth of the endosymbiont in the *Calyptogena*. A model of sulfur circulation within the microbial community and that of *Calyptogena* clams in the cold-seep environment was proposed from these studies (Fig. 5).

Seven sediment samples in group 3 from different depths of the deep-sea in the range of 1159–6482 m were examined (Li et al. 1999b). A total of 75 different 16S rDNA sequences (149 clones) analyzed clustered into the Proteobacteria, gram-positive bacteria, *Cytophaga*, *Planctomyces*, and actinomycetes groups; and many sequences showed no phylogenetic affiliation with known bacteria. The seven sediments from different depths are dominated by the population of 16S rDNA sequences (43%) belonging to the genus *Pseudomonas* in the γ-Proteobacteria subdivision. Clones identical to the genus *Pseudomonas* were observed in all of the sediments examined. The second group of common 16S rDNA sequences cloned from six sediment samples were related to that of a chemoautotrophic bacterium, the *Solemya velum* symbiont in the γ-Proteobacteria; they may represent a source of chemoautotrophic productivity for the microbial community in the environments. Five 16S rDNA sequences from three sediments were related to those of the *Alvinella pompejana* epibiont in the ε-Proteobacteria. *Vibrio vulnificus* related sequences were identified in four sediment samples, and *Cytophaga*-related sequences also were identified in four samples. Only one sequence was obtained that was closely related to the barophilic bacterium *Shewanella benthica*, which might be a minor population in the sediments. δ-Proteobacteria-related sequences were cloned from three samples. Two of the three sediments were from sites near manmade garbage and the *Calyptogena* community, respectively. These environments obviously would be richer in nutrients than other sites and would show more types of bacteria than normal areas. A large number of cloned sequences (23%) in this study showed low identity (below 80%) to known sequences. These sequences may represent communities of as-yet uncultivated microorganisms in the sediments.

The phylogenetic studies of the three groups of sediment demonstrated a distribution of populations of microorganisms in the sediments and functional specialized groups of bacteria.

References

Alfredsson GA, Kristjansson JK, Hjorleifsdottir S, Stetter KO (1988) *Rhodothermus marinus*, gen. nov., sp. nov., a thermophilic, halophilic bacterium from submarine hot springs in Iceland. J Gen Microbiol 134:299-306

Anaresson OS, Fridjonsson OH (1994) The sequence of the single 16S rRNA gene of the thermophilic eubacterium *Rhodothermus marinus* reveals a distant relationship to the group *containing flexbacter, Bacteroides*, and *Cytophaga* species J Bacterol 176:6165-6169

Aznar R, Ludwig W, Amann RI, Schleifer KH (1994) Sequence determination of rRNA genes of pathogenic *Vibrio* species and whole-cell identification of *Vibrio vulnificus* with rRNA-targeted oligonucleotide probes. Int J Syst Bacteriol 44:330-337

Barns SM, Delwiche CF, Palmer JD, Pace NR (1996) Perspective on archaeal diversity, thermophily and monophyly from environmental rDNA sequences. Proc Natl Acad Sci USA 93:9188-9193

Blackall LL, Rossetti S, Christensson C, Cunningham M, Hartman P, Hugenholtz P, Tandoi V (1997) The characterization and description of representatives of 'G' bacteria from activated sludge plants. Lett Appl Microbiol 25:63-69

Bowman JP, Cavanagh J, Austin JJ, Sanderson K (1996) Novel *Psychrobacter* species from Antarctic ornithogenic soils. Int J Syst Bacteriol 46:841-848

Cavanaugh CM (1983) Symbiotic chemoautotrophic bacteria in marine invertebrates from sulfide-rich habitats. Nature 302:58-61

Cavanaugh CM (1985) Symbiosis of chemoautotrophic bacteria and marine invertebrates from hydrothermal vents and reducing sediments. Bull Biol Soc Wash 6:373-388

DeLong EF (1992) Archaea in coastal marine environments. Proc Natl Acad Sci USA 89:5685-5689

DePaola A, Capers GM, Alexander D (1994) Densities of *Vibrio vulnificus* in the intestines of fish from the U.S. Gulf coast. Appl Environ Microbiol 60:984-988

Devereux R, Delaney M, Widdel F, Stahl DA (1989) Natural relationships among sulfate-reducing eubacteria. J Bacteriol 171:6689-6695

Distel DL, Felbeck H, Cavanaugh CM (1994) Evidence for phylogenetic congruence among sulfur-oxidizing chemoautotrophic bacterial endosymbionts and their bivalve hosts. J Mol Evol 38:533-542

Ehrich S, Behrens D, Lebedeva E, Ludwig W, Bock E (1995) A new obligately chemolithoautotrophic, nitrite-oxidizing bacterium, *Nitrospira moscoviensis* sp. nov. and its phylogenetic relationship. Arch Microbiol 164:16-23

Eisen JA, Smith SW, Cavanaugh CM (1992) Phylogenetic relationships of chemoautotrophic bacterial symbionts of *Solemya velum* say (Mollusca: Bivalvia) determined by 16S rRNA gene sequence analysis. J Bacteriol 174:3416-3421

Felbeck H (1981) Chemoautotrophic potential of the hydrothermal vent tube worm, Riftia pachyptila Jones (Vestimentifera). Science 213:336-338

Fisher C (1990) Chemoautotrophic and methanotrophic symbioses in marine invertebrates. Rev Aquat Sci 2:399-436

Fogg GE (1986) Picoplankton. Proc R Soc Lond Biol Sci 228:1-30

Franzmann PD, Springer N, Ludwig W, Demacario EC, Rohde M (1992) A methanogenic archaeon from Ace lake, Antarctica *Methanococcoides burtonii* new-species. Syst Appl Microbiol 15:573-581

Fuhrman JA, McCallum K, Davis AA (1992) Novel major archaebacterial group from marine plankton. Nature 356:148-149

Fuhrman JA, McCallum K, Davis AA (1993) Phylogenetic biodiversity of subsurface marine microbial communities from the Atlantic and Pacific oceans. Appl Environ Microbiol 59:1294-1302

Fujioka K, Murayama M (1992) Calyptogena community from landward slope of the Japan Trench and Mega shear. Proc JAMSTEC Symp Deep Sea Res 8:17-27

Haddad A, Camacho F, Durand P, Cary SC (1995) Phylogenetic characterization of the epibiotic bacteria associated with the hydrothermal vent polychaete *Alvinella pompejana*. Appl Environ Microbiol 61:1679-1687

Herrick JB, Madsen EL, Batt CA, Ghiorse WC (1993) Polymerase chain reaction amplification of naphthalene-catabolic and 16S rRNA gene sequences from indigenous sediment bacteria. Appl Environ Microbiol 59:687-694

Hiraishi A, Furuhata K, Matsumoto A, Koike KA, Fukuyama M, Tabuchi K (1995) Phenotypic and genetic diversity of chlorine-resistant *Methylobacterium* strains isolated from various environments. Appl Environ Microbiol 61:2099-2107

Isaksen MF, Teske A (1996) *Desulforhopalus vacuolatus* gen. nov., sp. nov., a new moderately psychrophilic sulfate-reducing bacterium with gas vacuoles isolated from a temperate estuary. Arch Microbiol 166:160-168

Jannasch HW, Taylor CD (1984) Deep-sea microbiology. Annu Rev Microbiol 38:487-514

Janssen PH, Schuhmann A, Bak F, Liesack W (1996) Disproportionation of inorganic sulfur compounds by the sulfate-reducing bacterium *Desulfocapsa thiozymogenes* gen. nov., sp. nov. Arch Microbiol 166:184-192

Kato C, Inoue A, Horikoshi K (1996) Isolating and characterizing deep-sea marine microorganisms. Trends Biotechnol 14: 6-12

Kato C, Inoue A, Horikoshi K (1998a) Characteristics of deep-sea microorganisms adapted to an extreme environment. In: Fingerman M, Nagabhushanam R, Thompson MF (eds) Marine biotechnology Vol 2: Environmental marine biotechnology. Oxford and IBH Publishing, New Delhi, pp 33-61

Kato C, Li L, Nogi Y, Nakamura Y, Tamaoka J, Horikoshi K (1998b) Extremely barophilic bacteria isolated from the Mariana Trench, Challenger Deep at a depth of 11,000 m. Appl Environ Microbiol 64:1510-1513

Kato C, Sato T, Horikoshi K (1995) Isolation and properties of barophilic and barotolerant bacteria from deep-sea mud samples. Biodiv Conserv 4:1-9

Kennicutt MC II, Brooks JM, Bridigare RR, Fay RR, Wade TL, McDonald SJ (1985) Vent-type taxa in a hydrocarbon seep region on the Louisiana slope. Nature 317:351-353

Kennicutt MC II, Brooks JM, Bridigare RR, McDonald SJ, Adkison DL (1989) An upper slope "cold" seep community: Northern California, Limnol. Oceanography 34:635-640

Kordick DL, Swaminathan B, Greene CE, Wilson KH, Whitney AM, O'Connor S, Hollis DG, Matar GM, Steigerwalt AG, Malcolm GB, Hayes PS, Hadfield TL, Breitschwerdt EB, Brenner DJ (1996) *Bartonella vinsonii* subsp. *berkhoffii* subsp. nov., isolated from dogs; *Bartonella vinsonii* subsp. *vinsonii*; and emended description of *Bartonella vinsonii*. Int J Syst Bacteriol 46:704-709

Kulm LD, Suess E, Moore JC, Carson B, Lewis BT, Ritger SD, Kadko DC, Thornburg TM, Embly RW, Rugh WD, Massoth GJ, Langseth MG, Cochrane GR, Scamman RL (1986) Oregon subduction zone: venting, fauna, and carbonates. Science 231:561-566

Lee JV, Gibson DM, Shewan JM (1977) A numerical taxonomic study of some *Pseudomonas*-like marine bacteria. J Gen Microbiol 98:439-451

Li L, Guenzennec J, Nichols P, Henry P, Yanagibayashi M, Kato C (1999a) Microbial diversity in Nankai Trough sediments at a depth of 3843 m. J Oceanography 55:635-642

Li L, Kato C, Nogi Y, Horikoshi K (1998) Distribution of the pressure-regulated operons in deep-sea bacteria. FEMS Microbiol Lett 159:159-166

Li L, Kato C, Horikoshi K (1999b) Bacterial diversity in deep-sea sediments from different depths. Biodiv Conserv 8:659-677

Li L, Kato C, Horikoshi K (1999c) Microbial diversity in sediments collected from the deepest cold-seep area, the Japan Trench at a depth of 6,400 m. Mar Biotechnol (in press)

Masuzawa T, Handa N, Kitagawa H, Kusakabe M (1992) Sulfate reduction using methane in sediments beneath a bathyal "cold seep" giant clam community off Hatsushima Island, SagamiBay, Japan. Earth Planet Sci Lett 110:39-50

Morita RY (1976) Survival of bacteria in cold and moderate hydrostatic pressure environments with special reference to psychrophilic and barophilic bacteria. In: Gray RG, Postgate JR (eds) The survival of vegetative microbes. Cambridge University Press, Cambridge. pp279-298

Naganuma T, Kato C, Hirayama H, Moriyama N, Hashimoto J, Horikoshi K (1997) Intracellular occurrence of ε-proteobacterial 16S rDNA sequences in the vestimentiferan trophosome. J Oceanography 53:193-197

Nunes OC, Donato MM, Da Costa MS (1992) Isolation and characterization of *Rhodothermus* strains from S. Miguel, Azores. Syst Appl Microbiol 15:92-97

Ohta S, Laubier L (1987) Deep biological communities in the subduction zone of Japan from bottom photographs taken during "Nautile" dives in the Kaiko project Earth Planet Sci Lett 83:329-342

Paull CK, Hecker B, Commeau R, Freeman-Lynde RP, Neumann C, Corso WP, Golubic S, Hook JE, Sikes E, Curray J (1984) Biological communities at the Florida escarpment resemble hydrothermal vent taxa. Science 226:965-967

Rasolofonirina RN (1989) Pathogenic bacteria encounted in diseases of alimentary origin. Arch Inst Pasteur Madagascar 56:81-88

Reysenbach AL, Giver LJ, Wickham GS, Pace NR (1992) Differential amplification of rDNA genes by polymerase chain reaction. Appl Environ Microbiol 58:3417-3418

Robb FT, Place AR, Sowers KR, Schreier HJ, Dassarma S, Fleischmann EM (1995) Archaea: a laboratory manual. Cold Spring Harbor Laboratory, Cold Spring Harbor, NY

Saitou N, Nei M (1987) The neighbor-joining method: a new method for reconstructing phylogenetic trees. Mol Biol Evol 4:406-425

Sibuet M, Juniper SK, Pautot G (1988) Cold-seep benthic communities in the Japan subduction zones: geological control of community development. J Mar Res 46:333-348

Tamegai H, Li L, Masui N, Kato C (1997) A denitrifying bacterium from the deep sea at 11000-m depth. Extremophiles 1:207-211

Tanoue E, Nishiyama S, Kamo M, Tsugita A (1995) Bacterial membranes: possible source of a major dissolved protein in seawater. Geochim Cosmochim Acta 59:2643-2648

Teske A, Alm E, Regan JM, Toze S, Rittmann BE, Stahl DA (1994) Evolutionary relationships among ammonia- and nitrite-oxidizing bacteria. J Bacteriol 176:6623-6630

Watson SW, Waterbury JB (1971) Characteristics of two marine nitrite-oxidizing bacteria, *Nitrospina gracilis* nov. gen. nov. spec. and *Nitrococcus mobilis* nov. gen. nov. sp. Arch Microbiol 77:203-230

2 Adaptation and Response of Extremophiles Toward Extreme Deep-Sea Conditions

5 Barophiles (Piezophiles)

CHIAKI KATO

The deep-sea is regarded as an extreme environment with high hydrostatic pressures [up to 110 megapascal (MPa)], predominantly low temperatures (1°-2°C) but with occasional regions of extremely high temperature (up to 375°C) at hydrothermal vents, darkness, and low nutrient availability. It is accepted that deep-sea microbiology as a definable field did not exist before the middle of this century and was paid little attention except for the efforts of Certes and Portier (Jannasch and Taylor 1984). Certes, during the Travaillier and Talisman Expeditions (1882-1883), examined sediment and water collected from depths to 5000 m and found bacteria in almost every sample. He noted that bacteria survived at great pressure and might live in a state of suspended animation (Certes 1884). In 1904 Portier used a sealed and autoclaved glass tube as a bacteriological sampling device and reported colony counts from various depths and locations (Richard 1907). In 1949 ZoBell and Johnson started work on the effect of hydrostatic pressure on microbial activities. The term "barophilic" was first used, defined today as optimal growth at pressure higher than 0.1 MPa or by a requirement of increased pressure for growth. Many microorganisms in the deep-sea are extremophiles, such as halophiles, thermophiles, psychrophiles, barophiles, or piezophiles [the term piezophile was proposed as a replacement to barophile as the Greek translations of the prefixes baro and piezo mean weight and pressure, respectively (Yayanos 1995)], and some of these microorganisms cannot survive in "moderate" environments. In this chapter, we focus on the isolation and taxonomy of microorganisms adapted to the deep-sea, and the molecular bases of their high pressure adaptations.

Key words. Deep-sea, High pressure, Barophilic bacteria, Barophily, Taxonomy, *Shewanella benthica, Shewanella violacea, Photobacterium profundum, Moritella japonica, Moritella yayanosii,* Gene expression, High pressure adaptation, Pressure-regulated operon, Respiratory chain, Cytochrome *c*, Quinol oxidase

5.1 Isolation and Taxonomy of Deep-Sea Barophilic (Piezophilic) Microorganisms

5.1.1 Isolation and Growth Properties

The development of manned and unmanned submersibles has supported the approaches used for investigating deep-sea environments and understanding the mechanisms of deep-sea bacterial adaptation. One of these is the manned submersible Shinkai 6500, which is operated by the Japan Marine Science and Technology Center (JAMSTEC). This vehicle can submerge to a depth of 6500 m (Takagawa et al. 1989). Recently, an unmanned submersible, Kaiko, was constructed and recorded a 10 911 m diving point in the Mariana Trench during a dive in March 1995 (Kyo et al. 1995).

Bacteria living in the deep-sea display several unusual features that allow them to thrive in their extreme environment. The first pure culture isolate of barophilic bacterium was reported in 1979 by Yayanos et al. The spirillum-like bacterium strain CNPT-3 had a rapid doubling rate at 50 MPa but did not grow into colonies at atmospheric pressure for several weeks. This bacterium has not lost any of its barophilic characteristics after at least 10 transfers, but it is maintained at 58 MPa between transfers. Numerous barophilic and barotolerant bacteria have since been isolated and characterized by the DEEPSTAR group at JAMSTEC, from deep-sea sediments at depths ranging from 2500 m to 11 000 m obtained by the sterilized mud samplers on the submersibles Shinkai 6500 and Kaiko (Kato et al. 1995a, 1996a, 1996b, 1998). Some of these strains are listed in Table 1.

Most of the isolated strains are not only barophilic or barotolerant but psychrophilic as well and cannot be cultured at temperatures above 20°C. Typical growth profiles of such bacteria as a function of several pressures and temperatures are shown in Fig. 1. The barophilic bacterium strain DB6705 was not able to grow at atmospheric pressure (0.1 MPa) at temperatures above 10°C, but it was able to grow

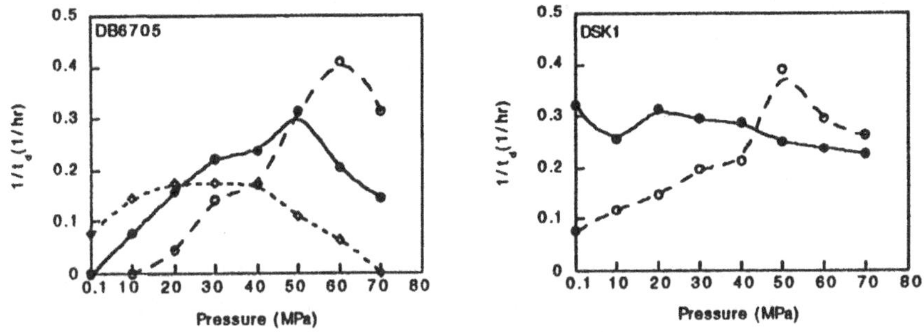

Fig. 1. Growth profiles of barophilic bacterium strain DB6705 and barotolerant bacterium strain DSK1 at various pressures and temperatures. *Open circles,* growth at 15°C; *filled circles,* growth at 10°C; *diamonds,* growth at 4°C. The growth rate is shown as $1/t_d$, where t_d represents the doubling time (in hours)

Table 1. High pressure-adapted bacterial strains isolated by the DEEPSTAR group

Bacterial strain	Properties	Source	Species	Reference
Extremely barophilic bacteria [a]				
DB21MT-2	Optimal growth at 70 MPa, 10°C No growth at less than 50 MPa	Mariana Trench, at 10 898 m depth	*Shewanella benthica*	Kato et al. 1998 Nogi and Kato 1999
DB21MT-5	Optimal growth at 80 MPa, 10°C No growth at less than 50 MPa	Mariana Trench, at 10 898 m depth	*Moritella yayanosii* [d]	Kato et al. 1998 Nogi and Kato 1999
Barophilic bacteria [b]				
DB5501	Optimal growth at 50 MPa, 15°C	Suruga Bay, at 2485 m depth	*Shewanella benthica*	Kato et al. 1995a Li et al. 1998 Nogi et al. 1998b
DB6101	Optimal growth at 50 MPa, 15°C	Ryukyu Trench, at 5110 m depth	*Shewanella benthica*	Kato et al. 1995a Li et al. 1998 Nogi et al. 1998b
DB6705	Optimal growth at 60 MPa, 15°C No growth at 0.1 MPa	Japan Trench, at 6356 m depth	*Shewanella benthica*	Kato et al. 1995a Li et al. 1998 Nogi et al. 1998b
DB6906	Optimal growth at 60 MPa, 15°C No growth at 0.1 MPa	Japan Trench, at 6,269 m depth	*Shewanella benthica*	Kato et al. 1995a Li et al. 1998 Nogi et al. 1998b
DB172F	Optimal growth at 70 MPa, 10°C No growth at 0.1 MPa	Izu-Bonin Trench, at 6499 m depth	*Shewanella benthica*	Kato et al. 1996b Li et al. 1998 Nogi et al. 1998b
DB172R	Optimal growth at 60 MPa, 10°C No growth at 0.1 MPa	Izu-Bonin Trench, at 6499 m depth	*Shewanella benthica*	Kato et al. 1996b Li et al. 1998 Nogi et al. 1998b

Table 1. continued

Bacterial strain	Properties	Source	Species	Reference
Moderately barophilic bacteria [c]				
DSS12	Optimal growth at 30 MPa, 8°C	Ryukyu Trench, at 5110 m depth	*Shewanella violacea* [d]	Kato et al. 1995a Nogi et al. 1998b
DSJ4	Optimal growth at 10 MPa, 10°C	Ryukyu Trench, at 5110 m depth	*Photobacterium profundum* [d]	Nogi et al. 1998c
DSK1	Optimal growth at 50 MPa, 15°C	Japan Trench, at 6356 m depth	*Moritella japonica* [d]	Kato et al. 1995a Nogi et al. 1998c
Barotolerant bacteria				
DSK25	Optimal growth at 0.1 MPa, 35°C	Japan Trench, at 6500 m depth	*Sporosarcina* sp.	Kato et al. 1995c

[a] Extremely barophilic bacteria are defined as those unable to growth at pressures below 50 MPa but able to grow well at 100 MPa.
[b] Barophilic bacteria are defined as those displaying optimal growth at a pressure of more than 40 MPa.
[c] Moderately barophilic bacteria are defined as displaying optimal growth at a pressure of less than 40 MPa and able to grow well at atmospheric pressure.
[d] Novel deep-sea species reported by DEEPSTAR group.

better at higher temperatures than lower temperatures when incubated at higher pressures (> 50 MPa). The growth of the barotolerant bacterial strain DSK1 was also better at a higher temperature (15°C) than at a lower temperature (10°C) at pressures above 50 MPa. These results illustrate the generalization that deep-sea bacteria show their strongest barophilic responses near their upper temperature limits for growth (typically about 15°C).

The effects of pressure and temperature on cell growth for those deep-sea barophilic strains isolated by Yayanos and the DEEPSTAR group are similar, in that all strains become more barophilic at higher temperatures (Yayanos 1986; Kato et al. 1995a). These studies indicate that all barophilic isolates are obligately barophilic

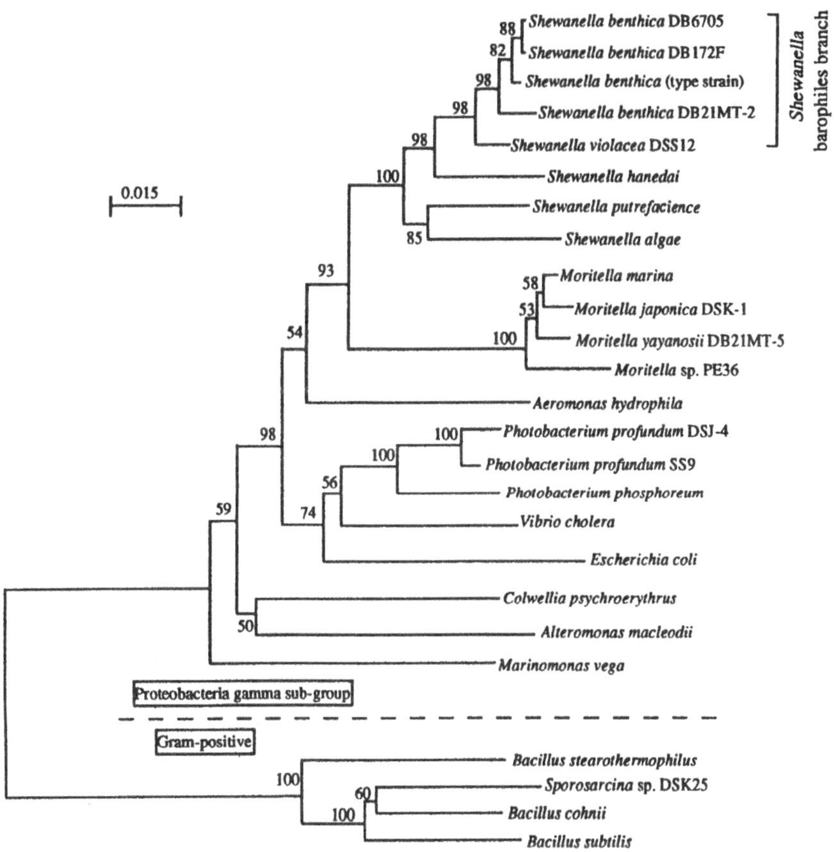

Fig. 2. Phylogenetic tree showing the relationships between isolated deep-sea barophilic bacteria within the γ-subgroup of Proteobacteria and genus *Bacillus* as determined by comparing 16S rDNA sequences using the neighbor joining method. Bar represents the average number of nucleotide substitutions per site. Bootstrap values (%) are shown for frequencies above the threshold of 50%

above the temperature that permits growth at atmospheric pressure. This means that the upper temperature limit for growth can be extended by high pressure. Likewise, barophilic bacteria reproduce more rapidly at a lower temperature (e.g., 2°C) when the pressure is less than that at its capture depth. It also appears to be true as a general rule that the pressure needed for the maximal rate of reproduction at 2°C may reflect the true habitat depth of an isolate (Yayanos et al. 1982). The reproduction rate of barophiles at pressures near the pressure of the depth of capture increased with increasing temperature over an interval of 6°-10°C. Why do barophilic cells grow best at temperatures that on first thought might never be encountered in the cold deep-sea? There are three possible explanations. One is that this behavior is simply a consequence of adaptation to high pressure and low temperature. Second, the response may have been inherited by the bacteria from ancestors who lived in a warmer environment (Deming et al. 1984). A third explanation is that bacteria of the cold deep-sea do periodically encounter warmer temperatures.

5.1.2 Taxonomy

Many deep-sea barophilic bacteria have been shown to belong to the γ-Proteobacteria through comparison of 5S and 16S rDNA sequences. The G+C content of chromosomal DNA from *Vibrio* spp. that belong to the γ-Proteobacteria is 40%-50% (Kreig and Holt 1984) and is considered to be typical of this subgroup. The G+C content of chromosomal DNA from the isolated barophiles in our laboratory was found to be similar, at 40%-46%. As a result of a taxonomic study based on its 5S rDNA sequences, Deming et al. (1988) reported that the obligate barophilic bacterium *Cowellia hadaliensis* belongs to the Proteobacteria γ-subgroup. DeLong et al. (1997) have also documented the existence of barophilic and psychrophilic deep-sea bacteria that belong to this subgroup, as indicated by their 16S rDNA sequence. It is interesting to note that the 16S rDNA sequences of the barophilic strains DB6906, DB172F, and DB172R and the psychrophilic and moderately barophilic strain DSS12 (Kato et al. 1995a) show the highest homology of all, indicating that these strains are closely related (Fig. 2) (Li et al. 1998). The barophilic bacteria reported by Liesack et al. (1991) are also included in the same branch of the γ-Proteobacteria, as are the strains isolated by the DEEPSTAR group (Kato et al. 1995a). These data suggest that most deep-sea high-pressure adapted (barophilic or piezophilic) bacteria that can be readily cultured belong to the Proteobacteria γ-subgroup and may not be widely distributed within the domain Bacteria.

DeLong et al. (1997) reported that 11 cultivated psychrophilic and barophilic deep-sea bacteria are affiliated with one of five genera within the γ-subgroup: *Shewanella*, *Photobacterium*, *Colwellia*, *Moritella*, and an unidentified genus. The only deep-sea barophilic species of two of these genera were reported to be *S. benthica* in the genus *Shewanella* (Deming et al. 1984; MacDonell and Colwell 1985) and *C. hadaliensis* in the genus *Colwellia* (Deming et al. 1988) prior to the reports from the DEEPSTAR group. They have identified four new barophilic species within those genera based on chromosomal DNA–DNA hybridization studies and several other taxonomic properties. The percent values of DNA–DNA relatedness between the deep-sea barophilic strains are shown in Table 2.

Both previously described and new species of bacteria have been identified among barophilic bacterial isolates. *Photobacterium profundum*, a new species, was identified in the moderately barophilic strains DSJ4 and SS9 (Nogi et al. 1998c). *P. profundum* strain SS9 has been extensively studied with regard to the molecular mech-

Table 2. Levels of homology for chromosomal DNAs betweenbarophilic isolates and reference species

Genus *Photobacterium*

Strain		% Homology with DNA from						
	G+C%	SS9	DSJ4	*P. ang*	*P. dam*	*P. his*	*P. lei*	*P. pho*
SS9	41.5	100	78	27	23	24	17	19
DSJ4	42.0	78	100	28	19	28	18	20
P. ang	39.6	24	21	100	25	30	34	24
P. dam	41.8	20	20	28	100	66	19	19
P. his	41.8	29	16	22	65	100	17	16
P. lei	39.7	20	20	46	23	28	100	25
P. pho	39.1	23	21	43	23	37	24	100

SS9, DSJ4, *P. profundum strains* SS9, DSJ4; *P. ang*, *P. angustum*; *P. dam*, *P. damselae*; *P. his*, *P. histaminum*; *P. lei*, *P. leiognathi*; *P. Pho*, *P. phosphoreum*; G+C%,

Genus *Shewanella*

Strain		% Homology with DNA from							
	G+C%	DSS12	DB172F	DB6705	DB21MT-2	*S. bent*	*S. peal*	*S. hane*	*S. put*
DSS12	47.0	100	33	35	40	31	21	12	13
DB172F	48.3	39	100	87	77	86	ND	14	9
DB172R	48.5	28	ND	87	ND	82	ND	14	8
DB5501	49.9	35	98	81	ND	99	ND	9	9
DB6101	49.3	34	100	97	ND	81	ND	12	10
DB6705	48.0	36	79	100	84	87	23	29	12
DB6906	48.0	37	88	75	ND	83	ND	11	10
DB21MT-2	48.7	48	79	78	100	81	ND	20	21
PT-99	47.0	46	80	100	ND	82	ND	16	12
S. bent	47.9	41	79	88	80	100	20	15	7
S. peal	47.0	21	ND	21	ND	20	100	26	ND
S. hane	45.4	15	5	10	10	8	22	100	7
S. alg	51.2	4	7	7	ND	8	ND	6	7
S. put	46.5	13	6	9	9	8	ND	1	100

DSS12, *S. violacea*; DB-series and PT-99, *S. benthica strains*; *S. bent*, *S. benthica*; *S. peal*, *S. pealeana*; *S. hane*, *S. hanedai*; *S. alg*, *S. algae*; *S. put*, *S. putrefaciens*.
ND, not determined.

Table 2. continued

Genus *Moritella*

Strain		% Homology with DNA from						
	G+C%	DSK1	DB21MT-5	*M. mar*	*S. put*	*S. hane*	*S. bent*	*S. viol*
DSK1	45.0	100	22	17	16	23	4	18
DB21MT-5	44.6	41	100	32	10	ND	8	ND
M.mar	42.5	33	28	100	9	12	8	11
S.put	46.5	10	7	9	100	1	8	13
S.hane	45.4	14	ND	9	7	100	8	15
S.bent	47.9	14	12	11	7	15	100	41
S.viol	47.0	18	ND	10	13	12	31	100

DSK1, *M. japonica*; DB21MT-5, *M. Yayanosii*; *M. mar*, *M. marina*; *S. viol*, *S. violacea*.

anisms for pressure regulation (Bartlett et al. 1989; 1996; Welch and Bartlett 1998). *P. profundum* is the only species to express barophily within the genus *Photobacterium* or to produce the long-chain polyunsaturated fatty acid (PUFA) eicosapentaenoic acid (EPA). No other species of *Photobacterium* has been reported to produce EPA (Nogi et al. 1998c).

The moderately barophilic strain DSS12 isolated from the Ryukyu Trench at a depth of 5110 m was identified to be *Shewanella violacea* (Nogi et al. 1998b), a novel barophilic species belonging within the *Shewanella* barophile branch (Li et al. 1998). Other *Shewanella* barophilic strains [PT-99 (DeLong and Yayanos 1986); DB5501, DB6101, DB6705, DB6906 (Kato et al. 1995a); DB172F, DB172R (Kato et al. 1996b); and DB21MT-2 (Kato et al. 1998)] were identified as all belonging to the same species of *S. benthica* (Li et al. 1998; Nogi et al. 1998b; Nogi and Kato 1999). The barophilic and psychrophilic *Shewanella* strains, including *S. violacea* and *S. benthica*, also produce EPA (DeLong et al. 1997; Kato et al. 1998; Nogi et al. 1998b); thus the occurrence of this PUFA is a property shared by many deep-sea bacteria. *Shewanella violacea* strain DSS12 has been well studied at JAMSTEC, particularly with respect to its molecular mechanisms of adaptation to high pressure. This strain is moderately barophilic, with a fairly constant doubling between 0.1and 70 MPa, whereas the doubling times of most of the barophilic *S. benthica* strains change substantially with increasing pressure (Kato et al. 1995a, 1996a, 1996b, 1998). Because there are few differences in the growth characteristics of strain DSS12 under different pressure conditions, this strain is a convenient deep-sea bacterium for use in studies on the mechanisms of adaptation to high pressure environments. Studies using this strain include analyses of the pressure regulation of gene expression (Kato et al. 1997a, 1997b) and of the role of *d*-type cytochromes in the growth of cells under high pressure (Kato et al. 1996d; Tamegai et al. 1998). The molecular mechanisms of gene expression have been analyzed focusing on a cloned pressure-regulated promoter; and more detailed studies are in progress, as we describe in the

Table 3. Identified deep-sea barophilic bacterial species

Genus and species	Strain	Properties	Reference
Colwellia			
C. hadaliensis	BNL-1	Extremely barophilic	Deming et al. 1988
Photobacterium			
P. profundum	SS9, DSJ4	Moderately barophilic	Nogi et al. 1998c
Shewanella			
S. benthica	PT-99, DB-series	Moderately, obligately, and extremely barophilics	Kato et al. 1998 Deming et al. 1984 Li et al. 1998 Macdonell and Colwell 1985 Nogi et al. 1998b
S. violacea	DSS12	Moderately barophilic	Nogi et al. 1998b
Moritella			
M. japonica	DSK1	Moderately barophilic	Nogi et al. 1998a
M. yayanosii	DB21MT-5	Extremely barophilic	Kato et al. 1998 Nogi and Kato 1999

Boldface type indicates the four newly identified species.

following section (Nakasone et al. 1998).

Strain DSK1, which is a moderately barophilic bacterium isolated from the Japan Trench, was identified as *Moritella japonica.* This is the first barophilic species identified in the genus *Moritella* (Nogi et al. 1998a). The type strain of the genus *Moritella* is *M. marina,* previously known as *Vibrio marinus* (Urakawa et al. 1998), which is one of the most common psychrophilic organisms isolated from marine environments (Colwell and Morita 1964). *M. marina* is closely related to the genus *Shewanella* on the basis of 16S rDNA data and is not a barophilic bacterium (Nogi et al. 1998a). The extremely barophilic bacterium strain DB21MT-5 isolated from the world deepest Mariana Trench, Challenger Deep, at a depth of 10 898 m (Kato et al. 1998) was identified as a *Moritella* species and designated *M. yayanosii* (Nogi and Kato 1999). The optimal pressure for growth of *M. yayanosii* strain DB21MT-5 is 80 MPa; this strain is not able to grow at pressures less than 50 MPa but is able to grow well at higher pressures, even as high as 100 MPa (Kato et al. 1998). Production of the long-chain PUFA docosahexaenoic acid (DHA) is one of the characteristic properties of the genus *Moritella* (DeLong et al. 1997; Kato et al. 1998; Nogi et al. 1998a). DeLong and Yayanos (1985, 1986) reported that the fatty acid composition of the barophilic strain changed as a function of pressure, and in general greater amounts of PUFAs were synthesized at the higher growth pressures. Approximately 70% of the membrane lipids in *M. yayanosii* is PUFA, which is consistent with its adaptation to high pressures. From the above studies, four more identified barophilic species could be added to the list of deep-sea barophiles, as shown in Table 3.

5.2. Molecular Mechanisms of the Adaptation in Deep-Sea Microorganisms

Before discussing pressure signal transduction and adaptation in deep-sea bacteria, it is useful to consider the effects of hydrostatic pressure on the responses of biological processes in the broader context of mechanostimulation, which is present in all organisms (Erdos et al. 1991; Blount et al. 1996). Pressure differentials across cells exert a variety of effects on cells and tissues. In mammals, differences in blood pressure of less than 1% of atmospheric pressure can lead to changes in cardiac gene expression, presumably as part of a homeostatic mechanism to maintain blood pressure within a narrow range (Izumo et al. 1988). Turgor pressure must be sensed and regulated; and in microorganisms with large cell walls, such as gram-positive bacteria and certain species of algae, it can be as high as 2 MPa (Whatmore and Reed 1990; Bisson and Kirst 1995). Bioluminescence in dinoflagellates can be greatly increased by relatively small pressure changes associated with laminar fluid shear on the order of 1 dyn cm^{-2} (Gooch and Vidaver 1980; Latz et al. 1994).

Sensing pressure not as a function of space, as in the examples above, but as a function of time exists in marine organisms other than deep-sea bacteria. Many fishes have evolved physiological responses to changes in hydrostatic pressure that influence the concentration of gases within a swim bladder and hence the buoyancy (Palster and Sheid 1992). Planktonic crustaceans can display responses to pressure changes of less than 1 kPa (10^{-2} atm) (Fraser and MacDonald 1994). Hence the ability to respond to pressure changes is not unique to deep-sea bacteria, although the capacity to respond to large changes in pressure in an apparently adaptive manner probably is restricted to those bacteria and higher organisms inhabiting high pressure environments.

5.2.1 Pressure Regulation in Microorganisms Outside of the Genus *Shewanella*

Pressure-inducible proteins have been observed in *Methanococcus thermolithotrophicus*, *Rhodotorula rubra*, and *Escherichia coli*, representing all three domains of life (Jaenicke et al. 1988; Gross et al. 1993; Welch et al. 1993). Even microorganisms that do not come from high-pressure environments may respond to high pressure by inducing the synthesis of certain stress proteins. An overlapping response of organisms to high pressure and high temperature has been observed. The steady-state levels of the heat shock protein GroES appear to be elevated during growth of *E. coli* at 40 MPa (Gross et al. 1993).

High-pressure effects on the *lac* and *tac* promoters of *E. coli* have also been studied, using promoters fused to the chloramphenicol acetyltransferase (CAT) gene (Kato et al. 1994; Sato et al. 1995). Plasmid-encoded *lac*- or *tac*-CAT activity in *E. coli* cells grown at high pressure was increased 80- to 90-fold compared with the level expressed at atmospheric pressure. Quantitative primer extension analysis indicated that growth at 30 and 50 MPa induced high CAT transcript levels in *E. coli* carrying the *lac*-CAT and *tac*-CAT transcriptional fusion plasmids (Sato et al. 1995)(see 10.3.1).

Another major effect of pressure is on the membrane constituents of bacteria. Elevated pressure represses the synthesis of several outer membrane proteins in *E. coli*, including the OmpC, OmpF, and LamB proteins (Nakashima et al. 1995; Sato et al. 1996). One outer membrane protein, OmpH, in the moderately barophilic bacterium *Photobacterium profundum* strain SS9 has been found to be induced 10- to 100-fold in mid-log cells grown at the optimal pressure 28 MPa compared with cells grown at atmospheric pressure, and the *ompH* regulatory promoter was confirmed (Bartlett et al. 1989). A second OMP of strain SS9, designated OmpL, is repressed at elevated pressures; and a third OMP, designated OmpI, is induced at pressures beyond the optimal pressure of strain SS9 and is the most abundant OMP at 40 MPa (Bartlett et al. 1993; Chi and Bartlett 1993; Bartlett and Chi 1994). It has also been discovered that certain species of marine bacteria, particularly those from the cold deep-sea, can produce long chain and branched PUFAs in their membranes (DeLong and Yayanos 1985; 1986; Wirsen et al. 1987; Kamimura et al. 1993). Increasing the extent of fatty acid unsaturation can maintain the membrane in a functional liquid crystalline state at high pressures, low temperatures, or both (MacDonald and Cossins 1985).

5.2.2 Pressure Regulated Operons in *Shewanella* sp.

A promoter, activated by growth at high pressure, was cloned from the barophilic *Shewanella benthica* strain DB6705 into *Escherichia coli* (Kato et al. 1995b). Gene expression initiating from this promoter was induced by high pressure in both the barophilic strain DB6705 and in *E. coli* transformants harboring this promoter. Downstream from this promoter, two open reading frames (ORF1 and ORF2) were identified as one operon, designated a pressure-regulated operon (Kato et al. 1996c). The structure of the promoter sequence was similar to that of the promoter of *ompH* from *Photobacterium profundum* strain SS9 (Bartlett and Welch 1995). The highly conserved pressure-regulated operon from the moderately barophilic bacterium strain DSS12 was also cloned and sequenced (Kato et al. 1997a). Its sequence is almost identical to the operon from barophilic strain DB6705 (Fig. 3).

The molecular details of pressure regulation in the region upstream of the pressure-regulated operon (ORF1 and ORF2) from *Shewanella violacea* strain DSS12 have been analyzed. Five transcription initiation sites have been identified by primer extension analysis in this region, and transcription starting from each of these sites is increased by an elevated pressure (Kato et al. 1997a). Nucleotide sequence analysis of this region indicated the existence of two potential regulatory regions (regions A and B) as follows.

1. Similarity between a consensus sequence for σ^{54} binding and part of the region designated region A was detected (homology of more than 90%) when the region flanking the 5' portion of the second transcription initiation site was com-

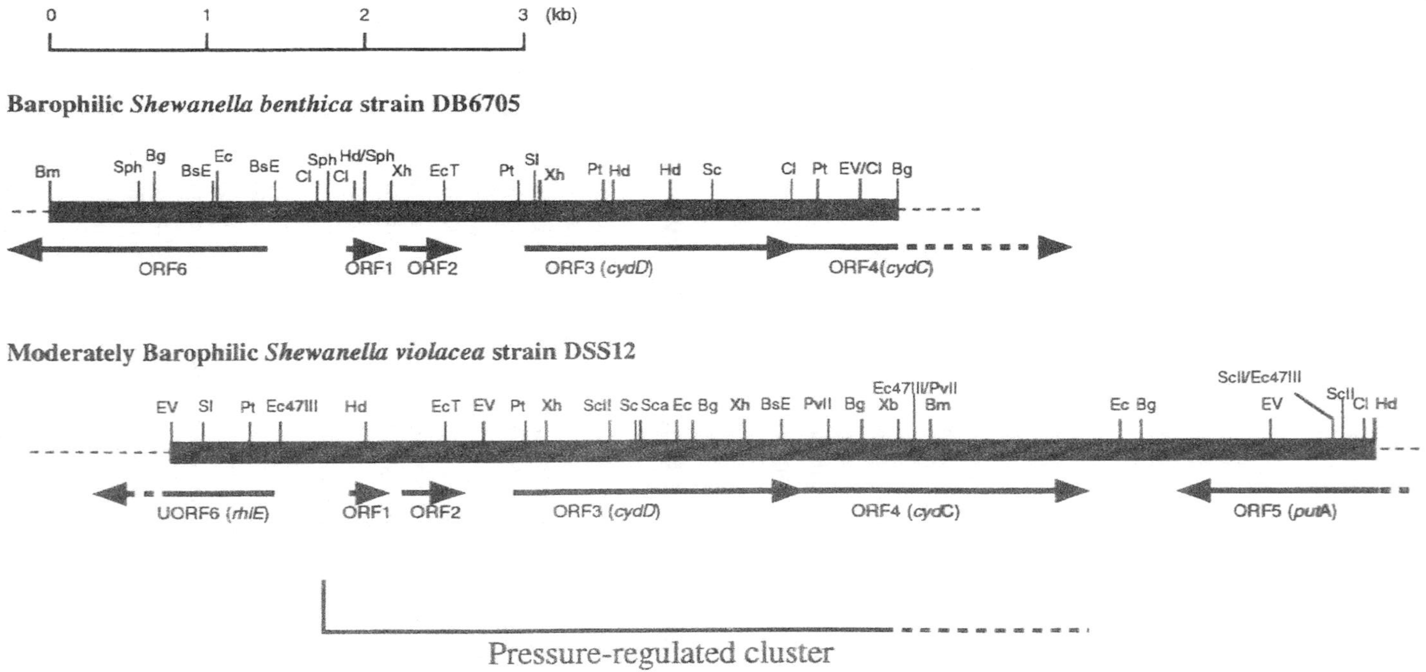

Fig. 3. Physical and genetic map of the DNA locus containing the two pressure-regulated operons and flanking ORFs from *Shewanella* strains DB6705 and DSS12. *Arrows* indicate open reading frames. Restriction endonuclease sites are indicated as follows; Bg, *Bgl*II; Bm, *Bam*HI; BsE, *Bst*EII; Cl, *Cla*I; Ec, *Eco*RI; EcT, *Eco*T22; Ec47III, *Eco*47III; EV, *Eco*RV; Hd, *Hin*dIII; Pt, *Pst*I; PvII, *Pvu*II; Sc, *Sac*I; ScII, *Sac*II; Sca, *Sca*I; Sl, *Sal*I; Sph, *Sph*I; Xb, *Xba*I; Xh, *Xho*I

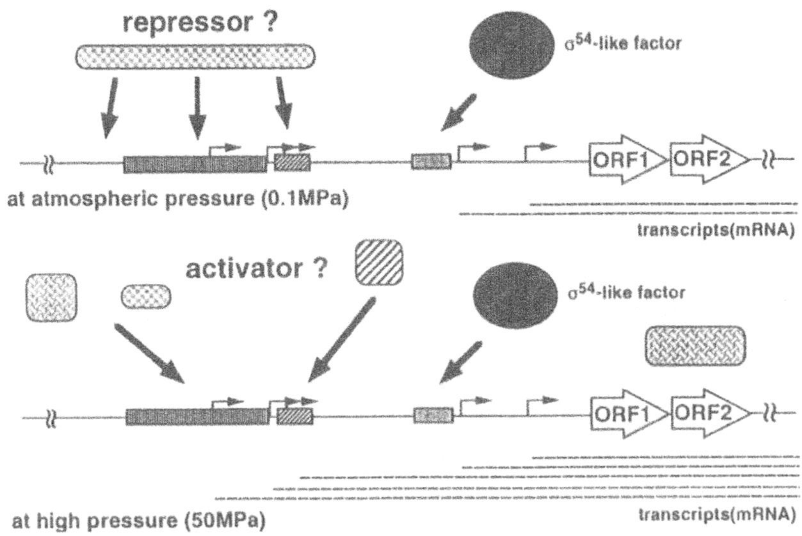

Fig. 4. Summary of binding of *trans*-acting factors to the region upstream of the pressure-regulated operon of the deep-sea barophilic *Shewanella violacea* strain DSS12 under atmospheric pressure (0.1 MPa) and high pressure (50 MPa) conditions. *Vertical shading*, 13 copy octamer repeat sequence; *diagonal hatching*, palindrome structure, *dotted shading*, σ^{54} consensus sequence

pared with several consensus sequences of *Escherichia coli* σ^{54} factors. However, typical promoter consensus sequences were not found in the flanking regions of other transcription initiation sites. These findings suggest that this strain not only may have new types of promoter sequences but perhaps also binding sites for previously unknown types of σ factors or other regulatory proteins.

2. A unique octamer motif, AAGGTAAG, was found to be tandemly repeated 13 times in the region designated region B containing the third to fifth transcription initiation sites just upstream of region A. A palindromic sequence, AGTTAAAGATTAAACT, was found in a region downstream of the tandemly repeated sequence.

These regions may have a role in regulating expression of the operon, and it seems possible that some pressure-regulated DNA-binding proteins might bind there under high pressure conditions.

To analyze whether such DNA-binding proteins exist in strain DSS12, an electrophoretic mobility shift assay (EMSA) was performed (Nakasone et al. 1998). Interactions of factors binding to the upstream region under both atmospheric and high-pressure (50 MPa) conditions considered from the results of EMSA are summarized in Fig. 4. They indicated that a σ^{54}-like factor recognizes region A, and other unknown factors recognize region B. Different shift patterns of protein–DNA complexes were observed using cell lysates prepared from cells of strain DSS12 cultured at 0.1 or 50 MPa. A factor present in cells grown at 0.1 MPa binds to region B, and a factor in cells grown at 50 MPa binds particularly to the tandem octamer motif (AAGGTAAG; region B). Because transcripts from region B are not detectable in cells grown at atmospheric pressure (0.1 MPa) (Kato et al. 1997a), the factors binding to region B might function as a transcriptional repressor. In contrast, in cells grown at high pressure (50 MPa), all transcripts are detectable (Kato et al. 1997a), suggesting that the factors expressed under these conditions might act as transcriptional activators. These results indicate that the deep-sea barophilic strain DSS12 expresses different DNA-binding factors under different pressure conditions.

5.2.3 Respiratory Mechanisms Controlled by Pressure in Piezophilic *Shewanella*

Downstream from the pressure-regulated operon of piezophilic *Shewanella* sp., another pressure-regulated operon was discovered whose first ORF was designated ORF3 and whose gene expression was enhanced by high pressure (Kato et al. 1997a). These two operons represented a pressure-regulated gene cluster (Fig. 3). According to the results of transcriptional analyses, the pressure-regulated operon is expressed at elevated pressure; and at 70 MPa the transcripts are present in largest amount. Based on the deduced amino acid sequence and the results of heterologous complementation studies in *Escherichia coli*, ORF3 appears to encode the CydD protein (Kato et al. 1996c). In *E. coli*, CydD is required for assembly of the cyto-

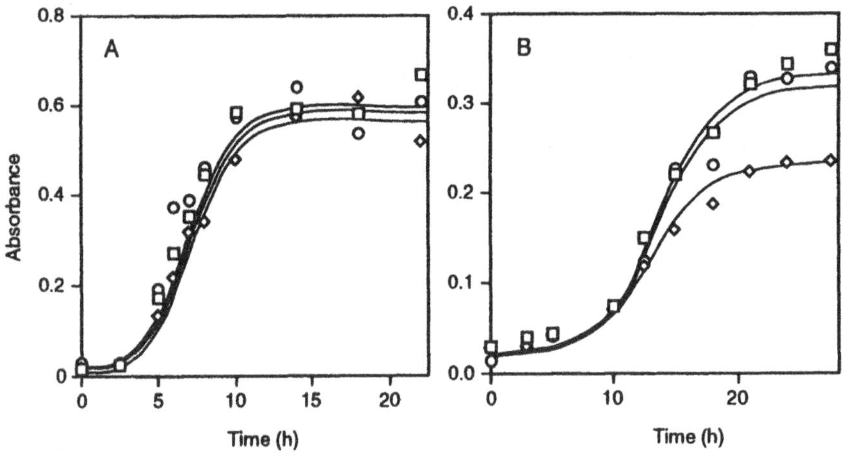

Fig. 5. Growth profiles of the *E. coli* strains, B178 (wild type, *circles*), JD518 (*cydD*$^-$, *diamonds*), and JD518 carrying the plasmid pAI-16 including ORF3 of strain DSS12 (*squares*), grown at 30°C in LB broth with oxygenated Fluorinert in the pressure vessels. The absorbance of the culture was determined at 660 nm. **A** Cells grown under 0.1 MPa **B** Cells grown under 30 MPa

chrome *bd* complex, one of the components of the aerobic respiratory chain (Poole et al. 1994). *E. coli cydD* mutants display increased sensitivity to high pressure but can be converted to cells that display wild type levels of high pressure sensitivity if the DSS12 ORF3 gene is introduced into the cells on a plasmid, as shown in Fig. 5. The cytochrome *bd* protein complex of strain DSS12 was observed spectrophotometrically only in the case of cells grown under high pressure conditions (Tamegai et al. 1998). It seems likely that the pressure regulation of this respiratory system in strain DSS12 plays a significant role in cell growth under high-pressure conditions (Kato et al. 1997b).

To study the pressure-regulated respiratory system in piezophilic *Shewanella* sp. in more detail, *c*-type cytochromes and a terminal quinol oxidase were purified and characterized from *S. benthica* strain DB172F (Qureshi et al. 1998a, 1998b). Two kinds of cytochrome *c* from the membrane and the cytoplasm were purified and named cytochrome *c*-551 and cytochrome *c*-552, respectively (Qureshi et al. 1998a). Cytochrome *c*-551 was found to contain 44.2 nmol of heme *c*/mg of protein and cytochrome *c*-552 was found to contain 31.3 nmol of heme *c*/mg of protein. The CO difference spectrum of cytochrome *c*-551 showed a peak at 413.7 nm and troughs at 423.2, 522, and 552 nm, which showed that this cytochrome combined with CO. Cytochrome *c*-551 was found to consist of two subunits with molecular masses of 29.1 and 14.7 kDa, and each subunit contains a heme *c* molecule. Cytochrome *c*-552 was found to consist also of two subunits with molecular masses of 16.9 and

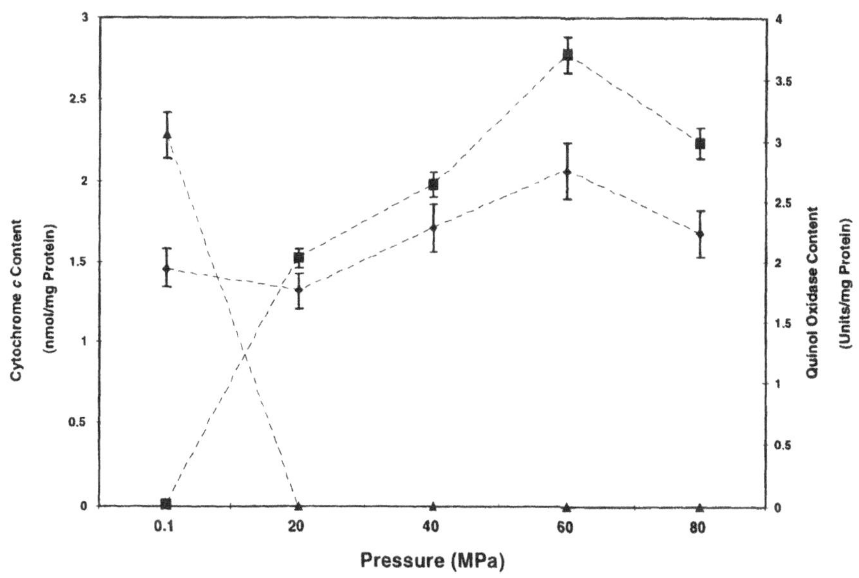

Fig. 6. Expression of respiratory components in piezopilic *Shewanella benthica* strain DB172F grown under elevated pressure conditions. *Diamonds*, membrane-bound *c*-type cytochrome; *squares*, quinol oxidase; *triangles*, cytoplasmic *c*-type cytochrome

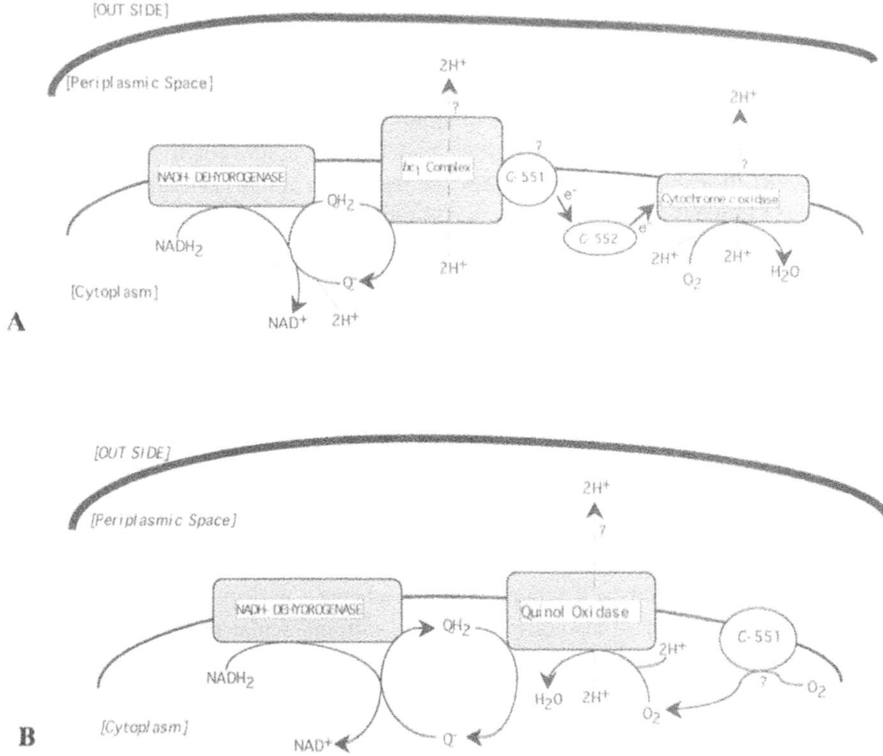

Fig. 7. Proposed electron transport system for piezophilic *Shewanella benthica* strain DB172F grown at 0.1 MPa **A** and 60 MPa **B**. Q^-, quinone; QH_2, quinol; C-551, cytochrome *c*-551; C-552, cytochrome *c*-552

14.7 kDa, and it contains only one heme *c*. Cytochrome *c*-551 is constantly expressed during growth under either 0.1 or 60.0 MPa, whereas cytochrome *c*-552 is expressed only during growth at 0.1 MPa (Qureshi et al. 1998a) (Fig. 6).

A novel membrane bound *ccb*-type quinol oxidase, from cells grown at 60 MPa, was purified to an electrophoretically homogeneous state from the same strain (Qureshi et al. 1998b). The purified enzyme complex was found to consist of four subunits with molecular masses of 98, 66, 18.5, and 15 kDa; the complex contained 0.96 mol of protoheme and 1.95 mol of covalently bound heme *c* per mole of enzyme. Only the protoheme in the enzyme reacted with CO and CN^-, and the catalytic activity of the enzyme was 50% inhibited by 4 μM CN^-. The isoelectric point of the native enzyme complex was determined to be 5.0. This enzyme was found to be specifically induced only under elevated hydrostatic pressure conditions, and it is expressed at high levels in cells grown at 60 MPa (Fig. 6). In contrast, the membranes of cells grown at atmospheric pressure (0.1 MPa) exhibit high levels of cytochrome *c* oxidase activity and *N,N,N',N'*-tetramethyl-*p*-phenylenediamine (TMPDH$_2$)-oxidase activity (Qureshi et al. 1998b).

These interesting observations suggest that the external growth pressure significantly alters the respiratory chain components, and that two respiratory chains regulated in response to pressure in the deep-sea *S. benthica* strain DB172F are present. A possible model of the respiratory system regulated by pressure is shown in Fig. 7.

Proposed Respiratory System Functioning at 0.1 MPa

In a proposed respiratory system functioning at 0.1 MPa (Fig. 7A), three respiratory chain enzyme complexes are present; NADH-dehydrogenase (also called complex I), bc_1 complex (also called complex III), and terminal cytochrome *c* oxidase (also called complex IV). Apparently, this is the typical respiratory chain found in mitochondria and mesophilic bacteria. In this model the flux of respiratory electrons is from complex I to complex IV. Pairs of high-energy electrons derived from the metabolism of glucose are stored in the form of NADH and FADH$_2$. Complex I oxidizes NADH$_2$ to NAD, and the two electrons released are transferred to quinone Q (oxidized form), which is then converted to the reduced state, quinol QH$_2$. Complex III then transfers the two electrons from quinol to membrane-bound cytochrome *c*-551. The soluble cytochrome *c*-552 then takes up the two electrons from *c*-551 and transfers them to the terminal oxidase. The terminal oxidase reduces oxygen and pumps protons into the periplasmic space. The proton pump function of complex III is driven by the flow of electrons. Under normal physiological conditions, the donor of electrons to terminal oxidase is *c*-552. We suppose that *c*-551 might perform this function to involve with the complex III as an electron acceptor, but further studies are needed in this regard.

Proposed Respiratory System Functioning at 60 MPa

At high pressure the terminal oxidase enzyme in strain DB172F is quinol oxidase (Fig. 7B). The membrane-bound cytochrome *c*-551 is thought to have the ability to bind oxygen and supply it to the terminal quinol oxidase. This function is suggested by the finding that cytochrome *c*-551 is capable of binding with ligands such as carbon monoxide and cyanide. Hence, under elevated pressure conditions the respiratory chain seems to be more compact and a "short-cut" compared to that under normal atmospheric pressure conditions. This may intimately relate to the bacterial piezophilic property. Being a new field of study, this area needs further exploration regarding the novel bioenergetic characteristics of the deep-sea bacteria.

References

Bartlett D, Chi E (1994) Genetic characterization of *ompH* mutants in the deep-sea bacterium *Photobacterium* species strain SS9. Arch Microbiol 162:323-328

Bartlett DH, Welch TJ (1995) *ompH* gene expression is regulated by multiple environmental cues in addition to high pressure in the deep-sea bacterium *Photobacterium* species strain SS9. J Bacteriol 177:1008-1016

Bartlett DH, Chi E, Welch TJ (1996) High pressure sensing and adaptation in the deep-sea bacterium *Photobacterium* species strain SS9. High pressure bioscience and biotechnology. In: Hayashi R, Balny C (eds) Elsevier Science BV, The Netherlands, pp29-36

Bartlett DH, Chi E, Wright WE (1993) Sequence of the *ompH* gene from the deep-sea bacterium *Photobacterium* SS9. Gene 131:125-128

Bartlett D, Wright M, Yayanos AA, Silverman M (1989) Isolation of a gene regulated by hydrostatic pressure in a deep-sea bacterium. Nature 342:572-574

Bisson MA, Kirst GO (1995) Pressure regulation in *algae*. Naturwissenschaften 82:461-471

Blount P, Sukharev SI, Moe PC, Schreder MJ, Guy HR, Kung C (1996) Membrane topology and multimeric structure of a mechanosensitive channel protein of *Escherichia coli*. EMBO J 15:4798-4805

Certes A (1884) C R Acad Sci Paris 98:690-693

Chi E, Bartlett DH (1993) Use of a reporter gene to follow high pressure signal transduction in the deep-sea bacterium *Photobacterium* SS9. J Bacteriol 175:7533-7540

Colwell RR, Morita RY (1964) Reisolation and emendation of description of *Vibrio marinus* (Russell) Ford. J Bacteriol 88:831-837

DeLong EF, Yayanos AA (1985) Adaptation of the membrane lipids of a deep-sea bacterium to changes in hydrostatic pressure. Science 228:1101-1103

DeLong EF, Yayanos AA (1986) Biochemical function and ecological significance of novel bacterial lipids in deep-sea prokaryotes. Appl Environ Microbiol 51:730-737

DeLong EF, Franks DG, Yayanos AA (1997) Evolutionary relationship of cultivated psychrophilic and barophilic deep-sea bacteria. Appl Environ Microbiol 63:2105-2108

Deming JW, Hada H, Colwell RR, Luehrsen KR, Fox GE (1984) The ribonucleotide sequence of 5S rRNA from two strains of deep-sea barophilic bacteria. J Gen Microbiol 130:1911-1920

Deming JW, Somers LK, Straube WL, Swartz DG, Macdonell MT (1988) Isolation of an obligately barophilic bacterium and description of a new genus, *Colwellia* gen. nov. Syst Appl Microbiol 10:152-160

Erdos T, Butler-Browne GS, Rappaport L (1991) Mechanogenetic regulation of transcription. Biochimie 73:1219-1231

Fraser PJ, MacDonald AG (1994) Crab hydrostatic pressure sensors. Nature 371:383-384

Gooch VD, Vidaver W (1980) Kinetic analysis of the influence of hydrostatic pressure on bioluminescence of *Gonyaulax polyhedra*. Photochem Photobiol 31:397-402

Gross M, Lehle K, Jaenicke R, Nierhaus KH (1993) Pressure-induced dissociation of ribosomes and elongation cycle intermediates: stablilizing conditions and identification of the most sensitive functional state. Eur J Biochem 218:463-468

Izumo S, Nadal-Ginard B, Mahdavi V (1988) Protooncogene induction and reprogramming of cardiac gene expression produced by pressure overload. Proc Natl Acad USA 85:339-343

Jaenicke R, Berndardt G, Ludemann HD, Stetter KO (1988) Pressure-induced alteration in the protein pattern of the thermophilic archaebacterium *Methanococcus thermolithotrophicus*. Appl Environ Microbiol 54:2375-2380

Jannasch HW, Taylor CD (1984) Deep-sea microbiology. Annu Rev Microbiol 38:487-514

Kamimura K, Fuse H, Takimura O, Yamaoka Y (1993) Effects of growth pressure and temperature on fatty acid composition of a barotolerant deep-sea bacterium. Appl Environ Microbiol 59:924-926

Kato C, Ikegami A, Smorawinska M, Usami R, Horikoshi K (1997a) Structure of genes in a pressure-regulated operon and adjacent regions from a barotolerant bacterium strain DSS12. J Mar Biotechnol 5:210-218

Kato C, Inoue A, Horikoshi K (1996a) Isolating and characterizing deep-sea marine microorganisms. Trends Biotechnol 14:6-12

Kato C, Li L, Nakamura Y, Nogi Y, Tamaoka J, Horikoshi K (1998) Extremely barophilic bacteria isolated from the Mariana Trench, Challenger Deep, at a depth of 11,000 meters. Appl Environ Microbiol 64:1510-1513

Kato C, Li L, Tamegai H, Smorawinska M, Horikoshi K (1997b) A pressure-regulated gene cluster in deep-sea adapted bacteria with reference to its distribution. Recent Res Dev Agric Biol Chem 1:25-32

Kato C, Masui N, Horikoshi K (1996b) Properties of obligately barophilic bacteria isolated from a sample of deep-sea sediment from the Izu-Bonin Trench. J Mar Biotechnol 4:96-99

Kato C, Sato T, Horikoshi K (1995a) Isolation and properties of barophilic and barotolerant bacteria from deep-sea mud samples. Biodiv Conserv 4:1-9

Kato C, Sato T, Smorawinska M, Horikoshi K (1994) High pressure conditions stimulate expression of chloramphenicol acetyltransferase regulated by the *lac* promoter in *Escherichia coli*. FEMS Microbiol Lett 122:91-96

Kato C, Smorawinska M, Sato T, Horikoshi K (1995b) Cloning and expression in *Escherichia coli* of a pressure-regulated promoter region from a barophilic bacterium, strain DB6705. J Mar Biotechnol 2:125-129

Kato C, Smorawinska M, Sato T, Horikoshi K (1996c) Analysis of a pressure-regulated operon from the barophilic bacterium strain DB6705. Biosci Biotechnol Biochem 60:166-168

Kato C, Suzuki S, Hata S, Ito T, Horikoshi K (1995c) The properties of a protease activated by high pressure from *Sporosarcina* sp. strain DSK25 isolated from deep-sea sediment. JAMSTEC R 32:7-13

Kato C, Tamegai H, Ikegami A, Usami R, Horikoshi K (1996d) Open reading frame 3 of the barotolerant bacterium strain DSS12 is complementary with *cydD* in *Escherichia coli*: *cydD* functions are required for cell stability at high pressure. J Biochem 120:301-305

Kreig NR, Holt JG (1984) Bergey's manual of systematic bacteriology, vol 1. Williams & Wilkins, Baltimore

Kyo M, Miyazaki E, Tsukioka S, Ochi H, Amitani Y, Tsuchiya T, Aoki T, Takagawa S (1995) The sea trial of "KAIKO," the full ocean depth research ROV. Oceans'95 3:1991-1996

Latz MI, Case JF, Gran RL (1994) Excitation of bioluminescence by laminar fluid shear associated with simple Couette flow. Limnol Oceanogr 39:1424-1439

Li L, Kato C, Nogi Y, Horikoshi K (1998) Distribution of the pressure-regulated operons in deep-sea bacteria. FEMS Microbiol Lett 159:159-166

Liesack W, Weyland H, Stackebrandt E (1991) Potential risks of gene amplification by PCR as determined by 16S rDNA analysis of a mixed culture of strict barophilic bacteria. Microb Ecol 21:191-198

MacDonald AG, Cossins AR (1985) The theory of homeoviscous adaptation of membranes applied to deep-sea animals. Soc Exp Biol Symp 39:301-322

MacDonell MT, Colwell RR (1985) Phylogeny of the *Vibrionaceae* and recommendation for two new genera, *Listonella* and *Shewanella*. Syst Appl Microbiol 6:171-182

Nakashima K, Horikoshi K, Mizuno T (1995) Effect of hydrostatic pressure on the synthesis of outer membrane proteins in *Escherichia coli*. Biosci Biotechnol Biochem 59:130-132

Nakasone K, Ikegami A, Kato C, Usami R, Horikoshi K (1998) Mechanisms of gene expression controlled by pressure in deep-sea microorganisms. Extremophiles 2:149-154

Nogi Y, Kato C (1999) Taxonomic studies of extremely barophilic bacteria isolated from the Mariana Trench and *Moritella yayanosii* sp. nov., a new barophilic bacterial species. Extremophiles 3:71-77

Nogi Y, Kato C, Horikoshi K (1998a) *Moritella japonica* sp. nov., a novel barophilic bacterium isolated from a Japan Trench sediment. J Gen Appl Microbiol 44:289-295

Nogi Y, Kato C, Horikoshi K (1998b) Taxonomic studies of deep-sea barophilic *Shewanella* species and *Shewanella violacea* sp. nov., a new barophilic bacterial species. Arch Microbiol 170:331-338.

Nogi Y, Masui N, Kato C (1998c) *Photobacterium profundum* sp. nov., a new, moderately barophilic bacterial species isolated from a deep-sea sediment. Extremophiles 2:1-7

Pelster B, Scheid P (1992) Countercurrent concentration and gas secretion in the fish swim bladder. Physiol Zool 65:1-16

Poole RK, Gibson F, Wu G (1994) The *cydD* gene product, componentof a heterodimeric ABC transporter, is required for assembly of periplasmeric cytochrome *c* and of cytochrome *bd* in *Escherichia coli*. FEMS Microbiol Lett 117:217-224

Qureshi MH, Kato C, Horikoshi K (1998a) Purification of two pressure-regulated *c*-type cytochromes from a deep-sea barophilic bacterium, *Shewanella* sp. strain DB-172F. FEMS Microbiol Lett 161:301-309

Qureshi MH, Kato C, Horikoshi K (1998b) Purification of a ccb-type quinol oxidase specifically induced in a deep-sea barophilic bacterium, *Shewanella* sp. strain DB-172F. Extremophiles 2:93-99

Richard J (1907) L'Oceanographie. Vuibert & Nony, Paris

Sato T, Kato C, Horikoshi K (1995) The effect on high pressure on gene expression by the *lac* and *tac* promoters in *Escherichia coli*. J Mar Biotechnol 3:89-92

Sato T, Nakamura Y, Nakashima K, Kato C, Horikoshi K (1996) High pressure represses expression of the *malB* operon in *Escherichia coli*. FEMS Microbiol Lett 135:111-116

Takagawa S, Takahashi K, Sano T, Mori Y, Nakanishi T, Kyo M (1989) 6500 m Deep manned research submersible "Shinkai 6500" system. Oceans'89 3:741-746

Tamegai H, Kato C, Horikoshi K (1998) Pressure-regulated respiratory system in barotolerant bacterium, *Shewanella* sp. strain DSS12. J Biochem Mol Biol Biophys 1:213-220

Urakawa H, Kita-Tsukamoto K, Steven SE, Ohwada K, Colwell RR (1998) A proposal to transfer *Vibrio marinus* (Russell 1891) to a new genus *Moritella* gen. nov. as *Moritella marina* comb. nov. FEMS Microbiol Lett 165:373-378

Welch TJ, Bartlett DH (1998) Identification of a regulatory protein required for pressure-responsive gene expression in the deep-sea bacterium *Photobacterium* species strain SS9. Mol Microbiol 27:977-985

Welch TJ, Farewell A, Neidhardt FC, Bartlett DH (1993) Stress response in *Escherichia coli* induced by elevated hydrostatic pressure. J Bacteriol 175:7170-7177

Whatmore AM, Reed RH (1990) Determination of turgor pressure in *Bacillus subtilis*: a possible role for K^+ in turgor regulation. J Gen Microbiol 136:2521-2526

Wirsen CO, Jannasch HW, Wakeham SG, Canuel EA (1987) Membrane lipids of a psychrophilic and barophilic deep-sea bacterium. Curr Microbiol 14:319-322

Yayanos AA (1986) Evolutional and ecological implications of the properties of deep-sea barophilic bacteria. Proc Natl Acad Sci USA 83:9542-9546

Yayanos AA (1995) Microbiology to 10,500 meters in the deep sea. Annu Rev Microbiol

49:777-805

Yayanos AA, Dietz AS, Boxtel RV (1979) Isolation of a deep-sea barophilic bacterium and some of its growth characteristics. Science 205:808-810

Yayanos AA, Dietz AS, Boxtel RV (1982) Dependence of reproduction rate on pressure as a hallmark of deep-sea bacteria. Appl Environ Microbiol 44:1356-1361

ZoBell CE, Johnson FH (1949) The influence of hydrostatic pressure on the growth and viability of terrestrial and marine bacteria. J Bacteriol 57:179-189

6 Thermophiles

Juan M. González

Among the extremophiles, one important group of microorganisms are those able to grow at exceptionally high temperatures. This chapter is mostly about extreme thermophiles, a group of thermophiles whose optimum growing temperatures are above 80°C. They are the so called hyperthermophiles, which can grow at temperatures up to the boiling point of water (100°C or 212°F).

Most hyperthermophiles so far described have been isolated from areas with geothermal activity. Both terrestrial and marine hydrothermal areas have been the source of hyperthermophiles (Brock 1978; Stetter 1986; Zeihkus et al. 1980). The discovery of submarine hydrothermal vents in deep-sea environments (Corliss et al. 1979) revealed the existence of rich and diverse biotic communities fueled by reduced elements, chemicals, and gases leached from the vents (Jannasch and Mottl 1985; Jones et al. 1983; Turncliffe 1991). These deep-sea volcanic systems are characterized by sharp temperature and chemical gradients. Cold deep-sea water is usually close to 0°C; but at the vents, owing the high pressure existing at the depths of these vent environments, water exists in its liquid form at temperatures of more than 400°C. Thus, these environments offer unique opportunities for finding unique microorganisms able to thrive under extreme conditions of temperature and pressure (Baross and Hoffman 1985; Jannasch and Mottl 1985; Stetter 1986).

Life on Earth is today understood as three clearly defined domains: bacteria, archaea, and eukarya. Hyperthermophiles in both the bacteria and archaea domains have been described. One characteristic of hyperthermophiles is that usually they branch at the deepest known positions of phylogenetic trees (Fig. 1). Thus, hyperthermophiles may provide information on the characteristics and physiology of the

Key words. Hydrothermal vent, Okinawa Trough, Izu-Bonin Trough, Mariana Trough, Hyperthermophiles, Archaea, *Pyrococcus*, *Thermococcus*, *Thermoaerobacter*, Hyperthermophilic bacteria, Mariana Trench

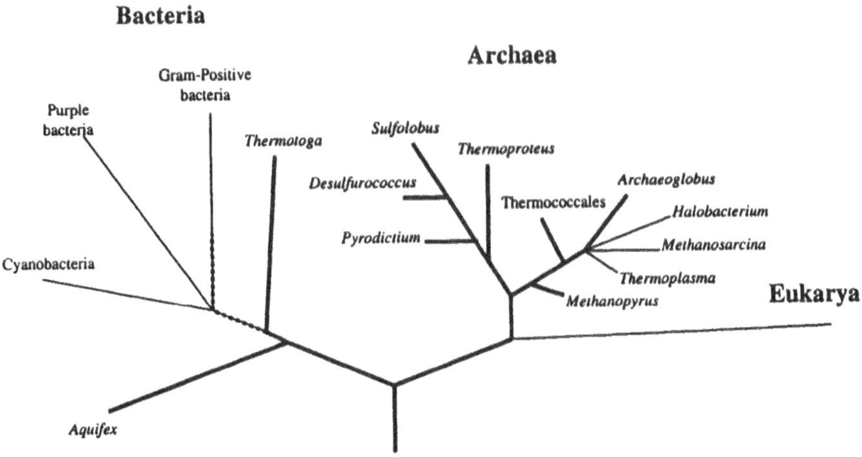

Fig. 1. Phylogenetic position of hyperthermophiles (*thick lines*) in a 16S-ribosomal-DNA based phylogenetic tree. The three domains (archaea, bacteria, and eukarya) are shown together with some of the most representative hyperthermophilic species. The position of the Order Thermococcales (archaea) and the extreme thermophiles belonging to the gram-positive bacteria (*dashed thick line*) are shown

most ancestral cells (Woese 1987; Pace 1991), which according to current theories on the origin of life on Earth probably originated in a hot environment much like the hydrothermal vents.

Growth at high temperatures well above the extremes that rapidly destroy critical cellular components (e.g., DNA, RNA, and enzymes) in mesophilic organisms implies the presence of novel and profound adaptations. These novel and often surprising physiological properties made these organisms a focus of interest in both microbiological and biochemical research (Stetter 1995). For instance, most industrial enzymes are currently utilized at temperatures above 50°C (Horikoshi 1991), which gives an idea of the tremendous potential for thermophiles and their commercial interest.

During the first phase of the DEEPSTAR project, an effort has been made to retrieve novel hyperthermophiles from deep-sea environments and to study some of their characteristics and physiological properties. This research aimed to better understand the diversity and physiology of deep-sea organisms in hot environments.

6.1 Isolation and Description of Novel Thermophiles

Most of the species newly described by the DEEPSTAR group belong to the Order Thermococcales which currently includes two genera: *Thermococcus* and *Pyrococ-*

cus (Zillig 1992). The optimum growth temperature appears to be the simplest and most obvious characteristic differentiating these genera. *Thermococcus* species usually show optimum growth temperatures at about 80°-85°C and *Pyrococcus* species at 95°-100°C. The Thermococcales are the most diverse group of hyperthermophiles known today. They are heterotrophic microorganisms that usually grow optimally on complex substrates such as yeast extract or peptone.

A novel extreme thermophile described by the DEEPSTAR group belonging to the bacteria domain is the new genus *Thermoaerobacter*, which is related to the gram-positive bacteria and has been isolated from the deepest seafloor area in our planet.

Following is a description of the isolation and characteristics of the thermophiles retrieved during the first phase of the DEEPSTAR project: *Thermococcus profundus* (Kobayashi et al. 1994), *T. peptonophilus* (González et al. 1994, 1995), strain DT1331 (Kwak et al. 1995), *T. siculi* (Grote et al. 1999), *Pyrococcus horikoshii* (González et al. 1998), and *Thermaerobacter marianensis* (Takai et al. 1999).

6.1.1 *Thermococcus profundus*

Hydrothermal vents from the Iheya Small Ridge at the Mid-Okinawa Trough (27° 33'N, 126° 58'E) were sampled. Hydrothermal fluids were at about 100°C. These vents were located at a depth of 1395 m and were accessed using the manned submersible Shinkai 2000. Hydrothermal sediment particles containing elemental sulfur were inoculated in anaerobic culture medium (Kobayashi et al. 1994). One isolate, designated DT5432, was obtained by repeated serial dilutions of preliminary enrichments. Based on the results of physiological and genetic analyses, strain DT5432 was considered a novel species and named *Thermococcus profundus* (Kobayashi et

A B

Fig. 2. Electron micrographs of *Thermococcus profundus*. **A** platinum-shadowed cell showing a polar tuft of flagella. Bar 0.5 μm. **B** thin sectioned cell. Bar 0.5 μm. (From Kobayashi et al. 1994, with permission)

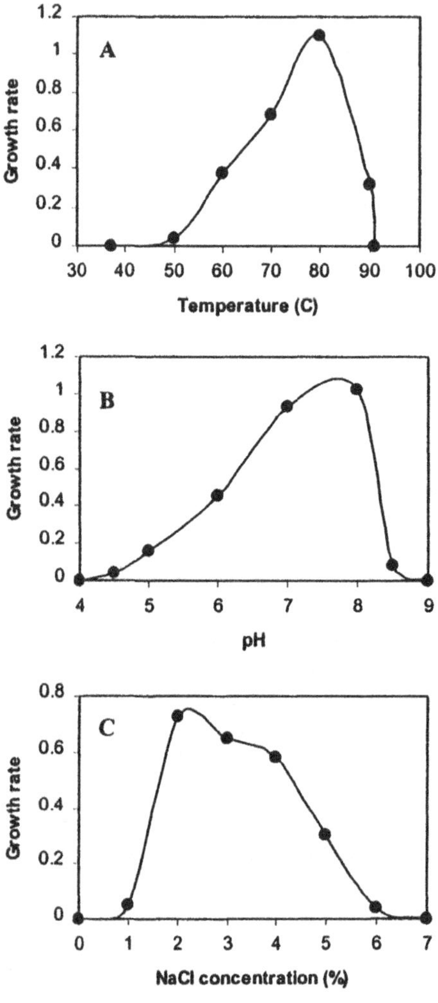

Fig. 3. Effect of temperature (**A**), pH (**B**), and NaCl concentration (**C**) on *Thermococcus profundus* growth. (From Kobayashi et al. 1994, with permission)

al. 1994); its physiological characteristics are described below.

Cells of *T. profundus* were cocci 1-2 μm in diameter showing a polar tuft of flagella (Fig. 2). Cells were surrounded by a single layer of low electron density and a thickness of 40-45 nm (Fig. 2). The *T. profundus* cell wall shows differences from other *Thermococcus* species.

Thermococcus profundus grew at 50°–90°C with an optimum at 80°C (Fig. 3). The pH range for growth was 4.5–8.5; its optimum pH was 7.5 (Fig. 3). Growth of *T. profundus* was observed in media containing 1%-6% NaCl; the optimum NaCl, concentration was 2% (Fig. 3). Cells lysed in the absence of NaCl, suggesting their dependence on the salt concentration in their culture medium, as for most marine microorganisms. *T. profundus* under optimum growth conditions showed a generation time of 50 min.

The *T. profundus* cells grew heterotrophically using as substrates complex proteinaceous substrates such as yeast extract, peptone, tryptone, casein, and gelatin. It was also able to utilize pyruvate, maltose, and starch as carbon sources. *T. profundus* required elemental sulfur for growth producing H_2S. Cells of *T. profundus* were strict anaerobes.

Sensitivity to several antibiotics was tested. *T. profundus* was resistant to ampicillin, chloramphenicol, kanamycin, streptomycin, and tetracycline, a pattern of resistance typical of archaea (Böck and Kandler, 1985). *T. profundus* was inhibited by rifampicin (100 μg/ml).

DNA analysis resulted in a G+C content of 52.5 mol%. The 16S ribosomal DNA gene was amplified by the polymerase chain reaction (PCR) and sequenced. Phylogenetic analysis of this DNA sequence placed *T. profundus* as a distinct *Thermococcus* species (see below). As well, DNA–DNA hybridization also pointed to significant differences between *T. profundus* and other species of the genus *Thermococcus*. At the time the description of *T. profundus* was published, only *T. celer* (the type species in the genus) (Zillig et al. 1983), *T. stetteri* (Miroshnichenko et al. 1989), and *T. litoralis* (Neuer et al. 1990) had been described.

6.1.2 *Thermococcus peptonophilus*

Several strains of archaeal hyperthermophiles were obtained from deep-sea hydrothermal vents located on the western Pacific Ocean (González et al. 1994). Among these strains the best characterized are OG1 and SM2. OG1 strain was obtained from the Izu-Bonin forearc area (28°34.2'N, 140°38.9'E) at hydrothermal vents sampled by the Shinkai 2000 at a depth of 1380 m. Thermal fluids collected at this location were at about 280°-300°C. SM2 strain was obtained from hydrothermal vents sampled by the Shinkai 6500 at the Southern Marina Trough (13°23.7'N, 143°55.2'E) at a depth of about 1450 m. Hydrothermal fluids at this site were at 90°-100°C. Strains OG1 and SM2 were classified as novel *Thermococcus* species, *Thermococcus peptonophilus* (González et al. 1995); the type strain was OG1.

Both OG1 and SM2 strains were cocci 0.7-2.0 μm in diameter with a polar tuft of flagella (Fig. 4). The cell envelopes of OG1 and SM2 were different: The OG1 cell envelope had two layers. The interior cell envelope, about 20-35 nm thick, was uniform and showed low electron-density; the OG1 exterior cell envelope was irregular, electron-dense, and 10–35 nm thick (Fig. 4). The cell envelope of SM2 strain showed a single low-electron-density layer with a thickness of 10-15 nm. The cell envelopes of strains OG1 and SM2 were different from those of other *Thermococcus* species. In fact, the diversity found in the cell envelopes among the early described species of *Thermococcus* appeared to be related to species diversity, as every *Thermococcus* species reported by the year 1995 showed different cell envelopes.

Growth of *T. peptonophilus*, strains OG1 and SM2, was observed at 60°–100°C (Fig. 5). The optimum growing temperature for OG1 strain 85°C and for SM2 strain 90°C. Growth of *T. peptonophilus* was observed between 1% and 5% NaCl. Maximum growth rates were observed at 3% NaCl (Fig. 5). *T. peptonophilus*, strains OG1 and SM2 grew optimally at pH 6. Growth was detected between pH 4 and 8 (Fig. 5).

T. peptonophilus growth was inhibited by the antibiotic rifampicin. OG1 and SM2 showed maximum growth rates in media containing 0.2% tryptone plus yeast extract (0.1% each), although at lower rates (Fig. 5) growth was also observed over a large range of nutrient concentrations, from 0.005% to 5.0%. These results suggest that the nutrient concentration can be a limiting factor in hyperthermophiles' growth because of a lack or an excess of nutrients. Generation times for *T. peptonophilus* under optimum conditions were as low as 25 min. This result is in the same range of growth rates obtained for *Escherichia coli*, and it shows that *T. peptonophilus* is the

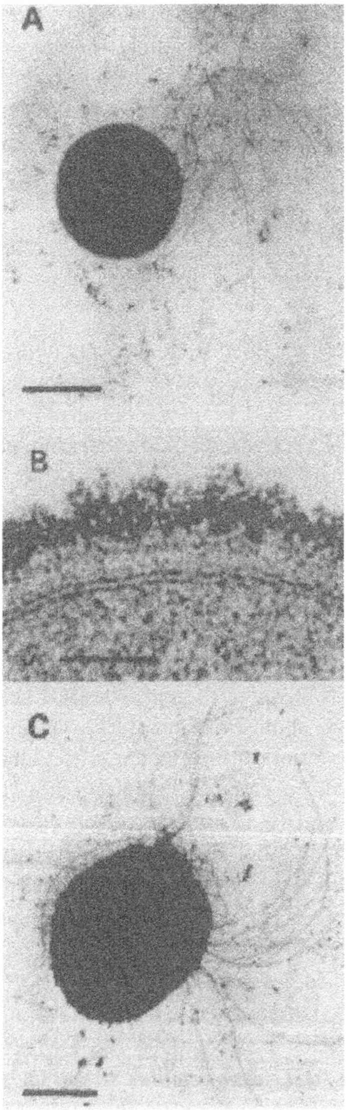

Fig. 4. Electron micrographs of *Thermococcus peptonophilus* cells. **A** platinum-shadowed cell of strain OG1 showing its flagella. Bar 0.5 μm. **B** thin section of a cell from strain OG1 showing the cell envelope structures. Bar 0.05 μm. **C** platinum-shadowed cell of strain SM2 showing its flagella. Bar 0.5 μm. (From González et al. 1995, with permission)

fastest growing hyperthermophile so far described.

Thermococcus peptonophilus is a heterotrophic, strict anaerobe. Most *Thermococcus* species grow optimally on complex proteinaceous substrates such as tryptone, yeast extract, casein, peptone, or similar extracts. *T. peptonophilus* is able to grow only on complex substrates. *T. peptonophilus* was unable to grow on mixtures of 21 amino acids even in the presence of vitamin mixtures. No autotrophic growth occurred with CO_2 as the carbon source. Elemental sulfur was not required for growth, but poor growth yields were obtained in its absence. The presence of elemental sulfur enhanced rate and yield *of T. peptonophilus* more than 10-fold growth.

T. peptonophilus showed a G+C content of 52 mol% which is in the range of most *Thermococcus* species with the exception of *T. litoralis* (Neuer et al. 1990). Sequencing the 16S ribosomal DNA and its phylogenetic analysis showed that *T. peptonophilus* was phylogenetically related to the *Thermococcus* species DNA–DNA hybridization experiments resulted in a low homology between strains OG1 or SM2 and other described *Thermococcus* species, which confirms that OG1 and SM2 strains represented a novel *Thermococcus* species. In addition, DNA–DNA hybridization results also showed about 90% similarity between strains OG1 and SM2, suggesting that they belong to the same species.

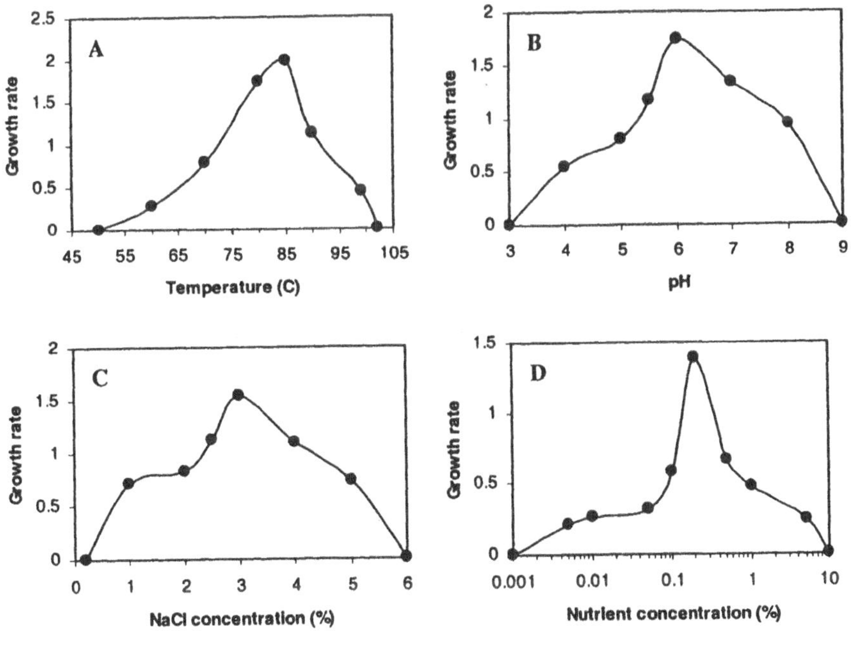

Fig. 5. Growing characteristics of *Thermococcus peptonophilus* strain OG1. **A** temperature. **B** pH. **C** NaCl concentration. **D** nutrient concentration as 50% yeast extract and 50% tryptone. (From González et al. 1995, with permission)

6.1.3 *Thermococcus* sp. Strain DT1331

Samples from a black smoker chimney (depth 760 m) at the Minami-ensei Knoll (28° 23'N, 127° 38'E) were collected using the unmanned submersible Dolphin 3K (Kwak et al. 1995). At this site the hydrothermal vents diffused from the chimney at about 100°C. The grayish sediment particles erupting from the vent were collected. One of those isolated was designated DT1331 and was further characterized.

DT1331 cells were cocci of 0.5-1.0 μm in diameter. No flagella were found by electron microscopy, so it is assumed they were nonmotile cells. Strain DT1331 had

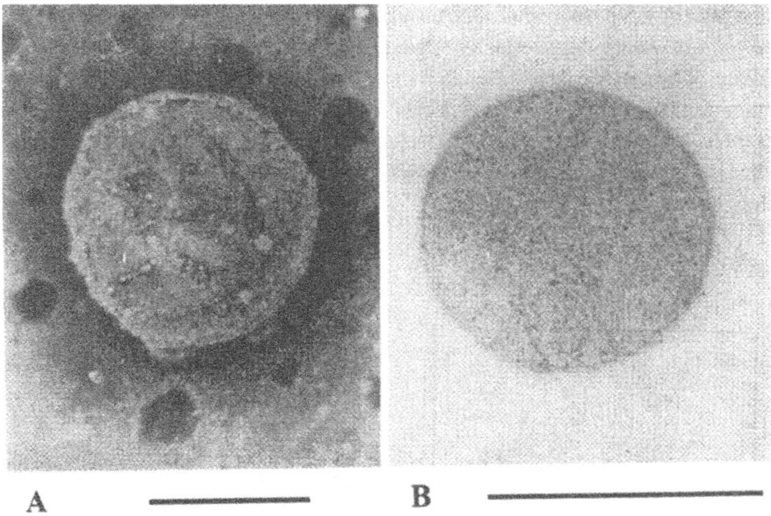

Fig. 6. Electron micrographs of cells from the *Thermococcus* sp. strain DT1331. **A** phosphotungstic acid stained cell. **B** thin section of a DT1331 cell. Bars 1.0 μm. (From Kwak et al. 1995, with permission)

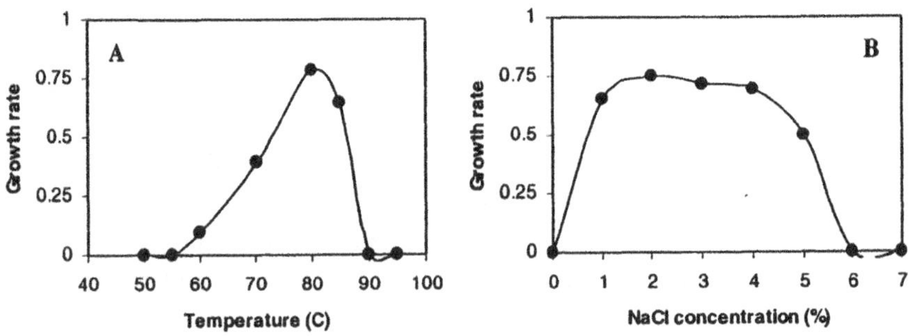

Fig. 7. Effect of temperature (**A**) and NaCl concentration (**B**) on growth of *Thermococcus* sp. strain DT1331. (From Kwak et al. 1995, with permission)

a single-layer cell envelope (Fig. 6).

Strain DT1331 grew between 55° and 93°C with an optimum at 80°C (Fig. 7). The optimum pH was 6.0, and DT1331 grew in the pH range 4.5–8.5. DT1331 required 1%–5% NaCl for growth. The optimum NaCl concentration was around 2%–4% NaCl (Fig. 7). The cells lysed at NaCl concentrations lower than 1%. The generation time of strain DT1331 under optimum growth conditions was 50 min.

Strain DT1331 grows on complex proteinaceous substrates. Growth was also observed on casamino acids with supplemented tryptophan. This finding suggests that tryptophan is required for growth and that strain DT1331 can grow on single amino acid mixtures. However, there was no growth of strain DT1331 on other substrates. Strain DT1331 obligately required elemental sulfur for growth, producing H_2S. Strain DT1331 was resistant to the following antibiotics: ampicillin, chloramphenicol, erythromycin, kanamycin, streptomycin, and tetracycline; it was inhibited by rifampicin (100 μg/ml).

The G+C content in strain DT1331 was 52.3 mol%. Amplification of the 16S ribosomal DNA from strain DT1331 by PCR followed by digestion with several restriction enzymes (Kwak et al. 1995) showed patterns identical with those of *T. celer*. Thus, strain DT1331 should be classified as a *Thermococcus*. Sequencing of its 16S ribosomal DNA gene and DNA–DNA hybridization experiments should be performed to define strain DT1331 as a new *Thermococcus* species.

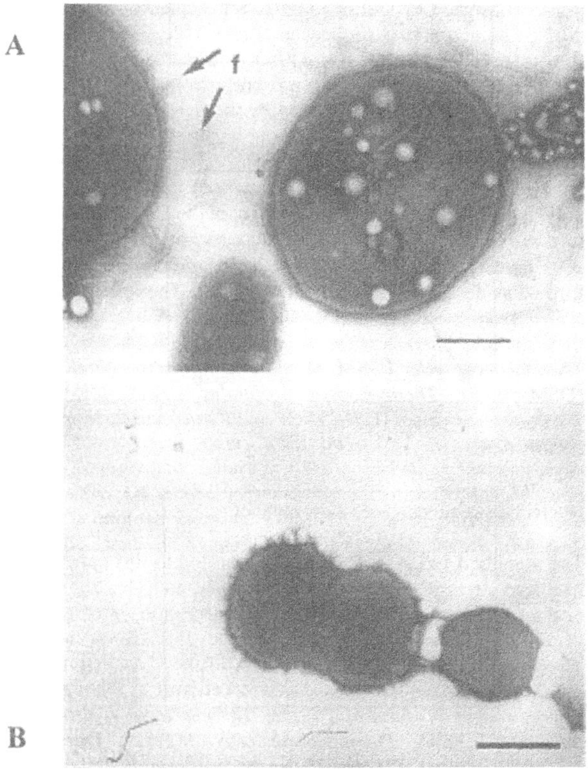

Fig. 8. Transmission electron micrographs of *Thermococcus siculi* strain RG-20. A negative-stained cells with phosphotungstic acid. *f*, fimbriae. Bars 0.5 μm (**A**) and 1 μm (**B**). (From Grote et al. 1999, with permission)

6.1.4 *Thermococcus siculi*

Samples of fluids were collected from a hydrothermal vent at 1394 m depth at the Mid-Okinawa Trough (27° 32.7'N, 126° 58.5'E) employing the manned submersible Shinkai 2000 (Grote et al. 1999). An isolate, strain RG-20, was obtained and it was classified as a new *Thermococcus* species named *T. siculi* (Grote et al. 1998).

Thermococcus siculi cells are nonmotile cocci 0.8–2.0 μm in diameter. Electron micrographs (Fig. 8) revealed the presence of fimbriae. *T. siculi* grew between 50° and 93°C, with an optimum temperature for growth at 85°C (Fig. 9). The pH range for growth was 5.0–9.0 and the optimum pH was 7.0 (Fig. 9). Growth was observed at a NaCl concentration of 1%–4%; the optimum was 2% NaCl. Cells lysed in medium with less than 1% NaCl. Under optimum growth conditions *T. siculi* showed a generation time of 130 min, which is the longest reported among the Thermococ-

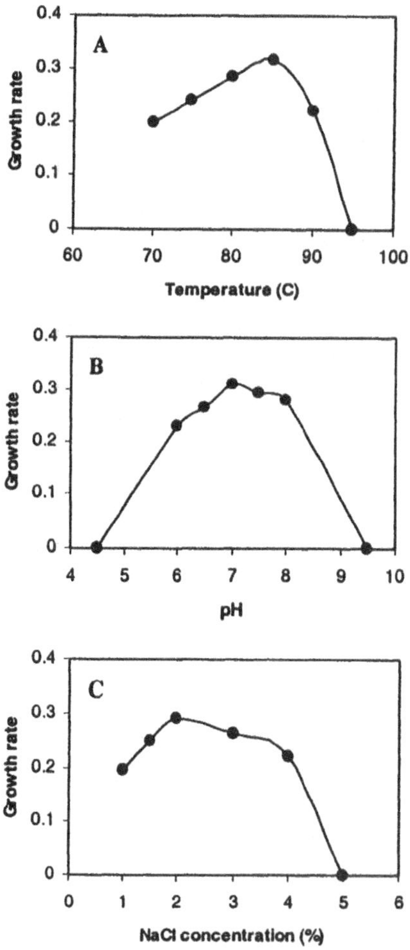

Fig. 9. Growth of *Thermococcus siculi* as affected by temperature (**A**), pH (**B**), and NaCl concentration (**C**). (From Grote et al. 1999, with permission)

Table 1. Physiological characteristics of the described species of the genus *Thermococcus* and the *Thermococcus* sp. strain DT1331

Species	G+C content (mol%)	Temperature (°C)		NaCl conc. (g/l)		pH		Carbon source	S° effect	Doubling time (min)	Rifampicin resistance
		Range	Optimum	Range	Optimum	Range	Optimum				
T. celer	57	≤ 75-93	88	> 10 to > 40	38-40	n/a	5.8	P, Su, AA	E	50	+
T. litoralis	38	65-95	88	18-65	25	6.2-8.5	7.2	P, M, Pyr, AA	E	n/a	–
T. stetteri	50	60-85	75	10-40	25	5.7-7.2	6.5	P, S, M, Pc, AA	R	70	–
T. profundus	52.2	50-90	80	10-60	20	4.4-8.5	7.5	P, S, M, Pyr, AA	R	50	–
T. peptonophilus	52	60-100	85	10-50	30	4.0-8.0	6.0	P	E	25	–
T. alcaliphilus	42.4	54-91	85	10-60	20-30	6.5-10.5	9.0	P	E	80	n/a
T. chitinophagus	46.5	60-93	85	8-80	20	3.5-9.0	6.7	P, Ch	E	75	n/a
T. fumicolans	54-55	73-103	85	6-40	13-26	4.5-9.5	8.0	P, M, Pyr, AA	E	~57	–
T. hydrothermalis	58	55-100	85	20-80	30-40	3.5-9.5	6.0	P, M, Pyr, C, AA	E	~79	+
T. zilligii	46.2	55-85	75-80	≤ 11.6	2.9	5.4-9.2	7.4	P	R	n/a	+
T. gorgonarius	50.6	68-95	80-88	10-50	20-35	5.8-8.5	6.5-7.2	P	R	n/a	n/a
T. pacificus	53.3	70-95	80-88	10-60	20-35	6.0-8.0	6.5	P, S, M	R	n/a	n/a
T. acidaminovorans	49	56-93	85	10-60	20-30	5.0-9.5	9.0	P, AA	E	120	n/a
T. siculi	55.8	50-93	85	10-40	20	5.0-9.0	7.0	P	E	130	–
Strain DT1331	52.3	55-93	80	10-50	20	4.5-8.5	6.0	P, AA	R	50	–

Thermococcus strains from the DEEPSTAR project are shown in boldface.
P, peptides; Su, sucrose; M, maltose; S, starch; Pyr, pyruvate; C, cellobiose; Pc, pectin, Ch, chitin; AA, amino acids; n/a, not available.

cales.

Growth of *T. siculi* was observed in the presence of complex proteinaceous substrates but was unable to use any other compound as a carbon source (Grote et al. 1999). Elemental sulfur could be substituted by cystine (10 g/l) resulting in similar growth responses by *T. siculi*. *T. siculi* is a heterotrophic, strict anaerobe. Rifampicin (150 µg/ml) inhibited *T. siculi* growth. Analysis of the lipid composition by thin-layer chromatography (TLC) revealed the presence of archaeol and caldarchaeol in *T. siculi* core lipids. Glycolipids and aminolipids were not detected by TLC.

DNA analysis resulted in a G+C content for *T. siculi* of 55.8 mol%. Comparison of the 16S ribosomal DNA gene sequence with described *Thermococcus* species clearly revealed that strain RG-20 belongs to the genus *Thermococcus* (see below). DNA–DNA hybridization showed low genetic homology between *T. siculi* and other *Thermococcus* species, which unequivocally suggested that strain RG-20, *T. siculi*, should be classified as a distinct species. Table 1 shows the characteristics of the *Thermococcus* species so far described.

6.1.5 *Pyrococcus horikoshii*

A novel hyperthermophile was isolated from hydrothermal samples obtained at the Okinawa Trough vents (27° 33'N, 126° 56'E) in the northeastern Pacific Ocean at a depth of 1395 m. The manned submersible Shinkai 2000 was used for sample col-

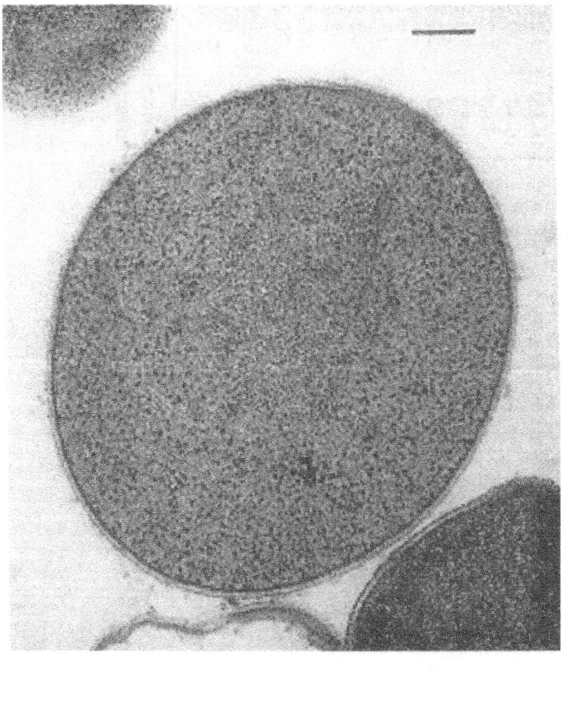

Fig. 10. Electron micrograph of a thin section of *Pyrococcus horikoshii*. Bar, 0.1 um. (From González et al. 1998, with permission)

lection. The isolate was classified as a novel species belonging to the Thermococcales, *Pyrococcus horikoshii* (González et al. 1998). *P. horikoshii* was named in honor of Dr. K. Horikoshi.

The cells were slightly irregular cocci with a diameter of 0.8-2.0 μm depending on growth conditions. During the exponential phase of growth, these cells are mostly observed in pairs, probably dividing cells; they are mostly seen as single cells during the early stationary phase. Electron microscopy revealed the presence of a polar tuft of flagella. Thin sections showed a complex cell envelope formed by a cytoplasmic membrane and an external electron-dense layer enclosing a periplasmic space (Fig. 10).

The growth of *P. horikoshii* under different conditions of temperature, NaCl concentration and pH are shown in Fig. 11. Growth was observed from less than 80°C up to 102°C. The optimum growth temperature was 98°C. Although growth did not

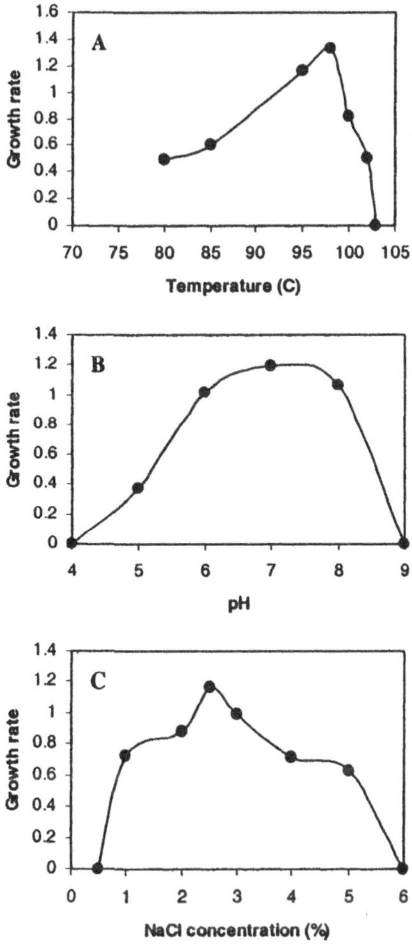

Fig. 11. Growth of *Pyrococcus horikoshii*. Effect of temperature (**A**), pH (**B**), and NaCl concentration (**C**). (From González et al. 1998, with permission)

occur at temperatures higher than 102°C, *P. horikoshii* showed prolonged survival at higher temperatures (e.g., 105°C). *P. horikoshii* grows optimally at pH 6–8, with pH 7 allowing the fastest growth. No growth was observed below pH 5 or at pH 8.5 or higher. NaCl concentrations of 1%–5% allowed growth; a sharp optimum was observed at 2.4% NaCl.

Pyrococcus horikoshii cells are heterotrophic, strict anaerobes. Growth was observed on complex proteinaceous substrates such as yeast extract, peptone, tryptone, beef extract, and nonhydrolyzed casein. No autotrophic growth was seen. None of the substrates tested stimulated growth or supported growth as the sole carbon source, although utilization of mannose has been experimentally proved after analysis of its genome sequence (Kawarabayashi et al. 1998). *P. horikoshii* failed to grow in mineral medium amended with casamino acids. Because casamino acids (Difco) lack the amino acids tryptophan, glutamine, and asparagine (Trp, Gln, and Asn) due to acid

Table 2. Comparisons among *Pyrococcus* species: *P. furiosus, P. abyssi, P. endeavori*, and *P. horikoshii*

Parameter	*P. furiosus*	*P. abyssi*	*P. endeavori*	*P. horikoshii*
Morphology	Slightly irregular cocci	Slightly irregular cocci	Slightly irregular cocci	Slightly irregular cocci
Growth conditions				
Temperature (°C)				
Optimum	100	96	98	98
Range	70-103	67-102	80-110	< 80-102
pH				
Optimum	7	6.8	7	7
Range	5-9	4.0-8.5	4-8	5-8
Salt concentr. (%)				
Optimum	2	3	2.8	2.4
Range	0.5-5	0.7-5	n/a	1-5
Substrates				
Complex substrates (i.e., yeast extract, peptone)	Yes	Yes	Yes	Yes
Casamino acids	Very weak	Yes	Yes	No
21-Amino-acid mixture	Yes	Yes	Yes	Yes
Other carbon sources	Maltose, starch, pyruvate, cellobiose	Maltose, starch, pyruvate		Mannose
Trp requirement	No	No	n/a	Yes
Growth with S°	Enhanced	Enhanced	Enhanced	Enhanced
Doubling time (min)	37	33	60	32
G+C content (mol%)	38	45	55	44
Genome sequenced	In progress	In progress	No	Yes
Origin of type strain	Vulcano Island, Italy	North Fiji Basin	Juan DeFuca Ridge	Okinawa Trough
Depth (m)	0	2000	2200	1395

Modified from González et al. 1998.

hydrolysis during production (Moore and Stein 1963), the medium with casamino acids was supplemented with these three amino acids. This supplemented casamino acids medium supported growth, but *P. horikoshii* was unable to grow in the casamino acids medium supplemented only with Gln and Asn. This confirms the Trp requirement by *P. horikoshii.* Other described *Pyrococcus* species do not require Trp in their culture medium (Table 2). Elemental sulfur greatly enhanced the growth of *P. horikoshii*.

Like other archaea, *P. horikoshii* has ether lipid membranes. The core lipids were separated by TLC and showed that the membrane contains predominantly tetraether lipids, with a small proportion of diether lipids (Fig. 12). This characteristic shows that *P. horikoshii* is related to *P. abyssi* (Erauso et al. 1993).

It was found that osmotic shock of *P. horikoshii* cells produced the release of high alkaline phosphatase activity; about 75% of the total alkaline phosphatase activity in untreated cells (after cell breakage) was released during osmotic shock. Alkaline phosphatase showed high temperature resistance with a half-life of about 1 h (57 min) at 97°C; this is the most thermostable alkaline phosphatase reported to date and the second report of this enzyme in the archaea domain. These results suggest the presence of alkaline phosphatase activity in a periplasmic space (Neu and Heppel 1965), as envisioned by electron microscopy (Fig. 10). The characteristics of the cell envelope in *P. horikoshii* suggest the existence of significant differences with the cell envelope described in *P. abyssi* (Erauso et al. 1993) and resembles more the one reported in *P. furiosus* (Fiala and Stetter 1986).

DNA analyses showed a G+C content in *P. horikoshii* of 42 mol%. Alignment of 16S ribosomal DNA sequences from *P. horikoshii* and related species allowed the classification of this isolate as a member of the genus *Pyrococcus* (see below). DNA–

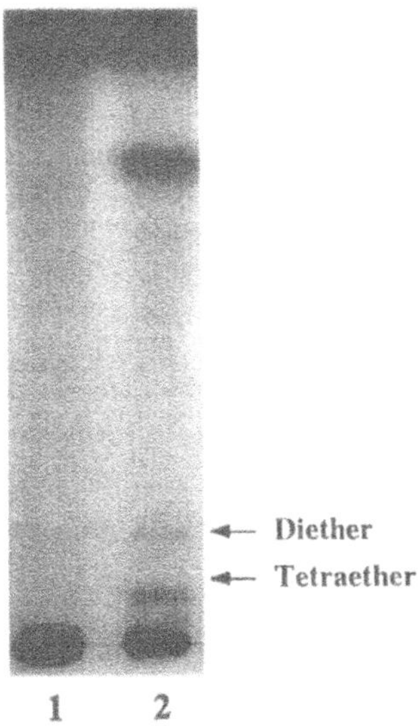

Fig. 12. Diether and tetraether lipids in extracts from purified membrane fractions of *Pyrococcus horikoshii* resolved by thin layer chromatography. (From González et al. 1998, with permission)

DNA hybridization showed relatively low homology between *P. horikoshii* and other *Pyrococcus* species. *P. furiosus* showed the highest homology to *P. horikoshii*, but this level of homology is lower than the recommended values for phylogenetic definition of a species (< 70%) (Wayne et al. 1987). This genetic evidence, together with the above morphological and nutritional characteristics, indicates that *P. horikoshii* is a distinct species belonging to the Thermococcales and establishes it as a novel species of the genus *Pyrococcus*.

6.1.6 Phylogenetic Analysis of Hyperthermophilic Archaea Isolated During the DEEPSTAR Project

Ribosomal RNAs are ancient molecules found in all organisms; thus they are conserved sequences that can provide excellent evolutionary information. Comparisons of the ribosomal DNA gene sequences can be used to determine the phylogenetic relationships between organisms (Fig. 1). In this study, the results and characteristics of several hyperthermophiles isolated during the DEEPSTAR project has been presented. The phylogenetic relationships between these isolates and related organ-

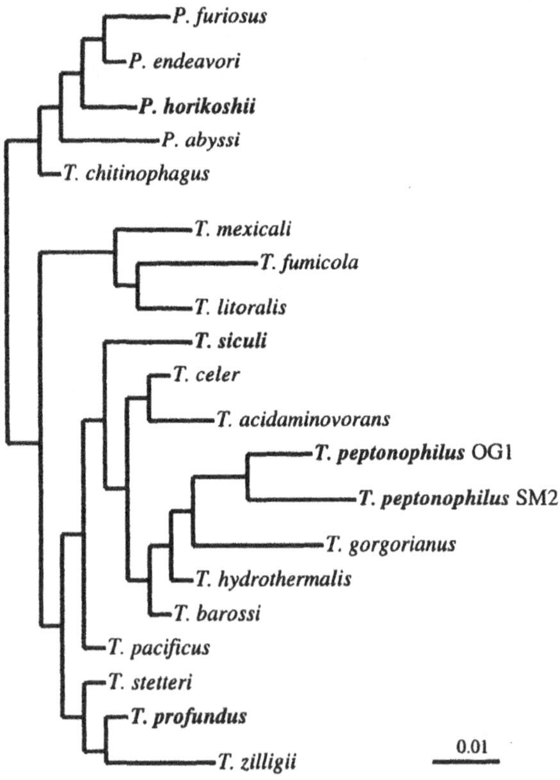

Fig. 13. Unrooted phylogenetic tree based on 16S ribosomal-DNA gene sequences showing the position of *Thermococcus* and *Pyrococcus* species. Bar represents the expected number of changes (0.01) per sequence position. The species described in the text are shown in boldface

isms of the order Thermococcales are presented in Fig. 13, showing the described species of *Pyrococcus* and *Thermococcus*. The two genus occupy separate branches with the exception of *Thermococcus chitinophagus,* which clusters with *Pyrococcus* species. This phylogeny highlights the great diversity existing within the Thermococcales.

6.1.7 *Thermoaerobacter marianensis*: An Extremely Thermophilic Bacterium

A mud sample was collected from the world's deepest seafloor, the Mariana Trench Challenger Deep (11°21.111'N, 142°25.949'E) at a depth of 10 897 m, using the unmanned submersible Kaiko (Takami et al. 1997). Isolate 7p75a was obtained (Takai et al. 1999).

Cells of the isolate 7p75a were thin, straight or slightly curved, nonmotile, rods 2-7 μm long and 0.3-0.6 μm wide during the exponential phase of growth (Fig. 14).

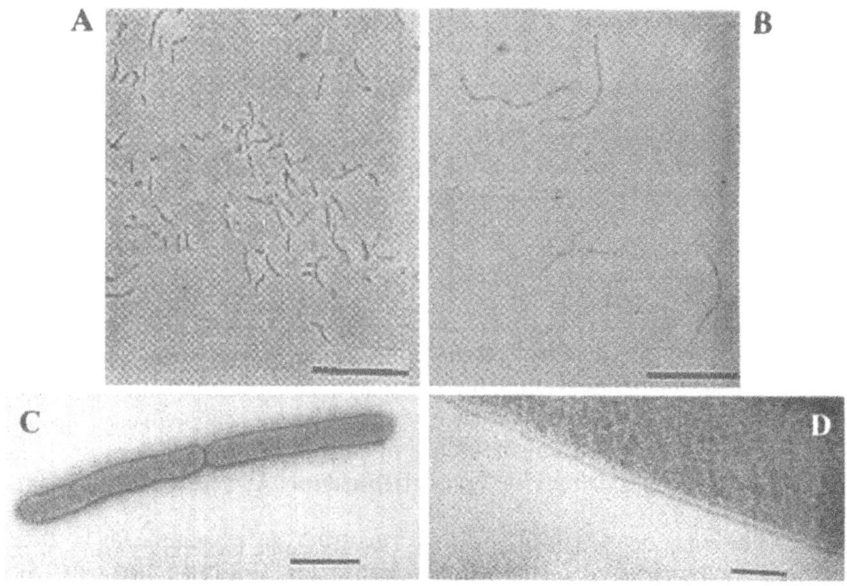

Fig. 14. Light and electron micrographs of *Thermoaerobacter marianensis*. **A** Phase-contrast micrograph of exponentially growing cells. Bar 10 μm. **B** Phase-contrast micrograph of cells in stationary phase of growth. Bar 10 μm. **C** Electron micrograph of a pair of cells negatively stained. Bar 1 μm. **D** Electron micrograph showing a thin section of an exponentially growing cell. Bar 0.5 μm. (From Takai et al. 1999, with permission)

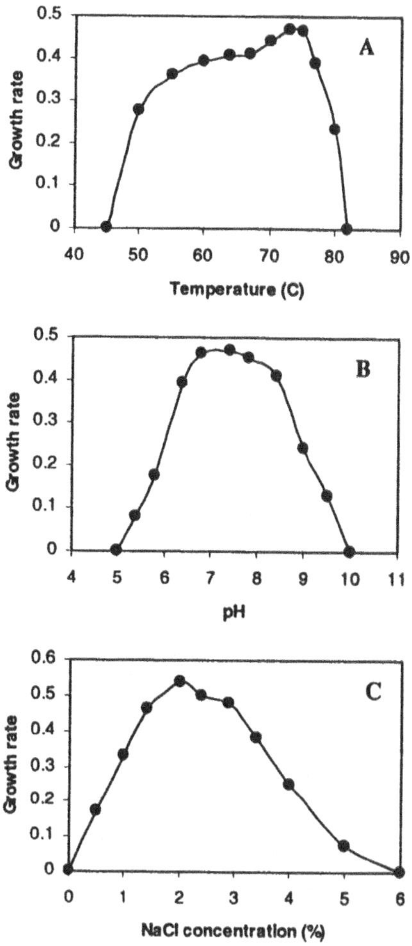

Fig. 15. Effect of temperature (**A**), pH (**B**), and NaCl concentration (**C**) on *Thermaerobacter marianensis* growth. (From Takai et al. 1999, with permission)

Cells showed variable Gram stain results, although they stained weakly positive during the exponential phase. During the stationary phase, cells become longer (>10 µm). Thin sections showed the presence of a thin layer of peptidoglycan (Fig. 14), which is consistent with a variable results of Gram staining. Formation of spores was not observed, and spore-related genes (*ssp* and *spo*0A) were not found by PCR techniques.

Cells of strain 7p75a are strict aerobic, heterotrophic bacteria growing in the range of 50°-80°C; the optimum growth temperature is 74°-76°C (Fig. 15). The optimum pH for growth is 7.5, although 7p75a cells grow in a pH range of 5.5–9.5 (Fig. 15). Strain 7p75a requires salts for growth; it grows at 0.5%–5.0% salt concentration, with an optimum at 2.0% (Fig. 15). Isolate 7p75a was sensitive to hydrostatic pressure; under optimum conditions, it grows with a generation time of about 90 min. In addition to peptone and yeast extract, strain 7p75a was able to utilize a large

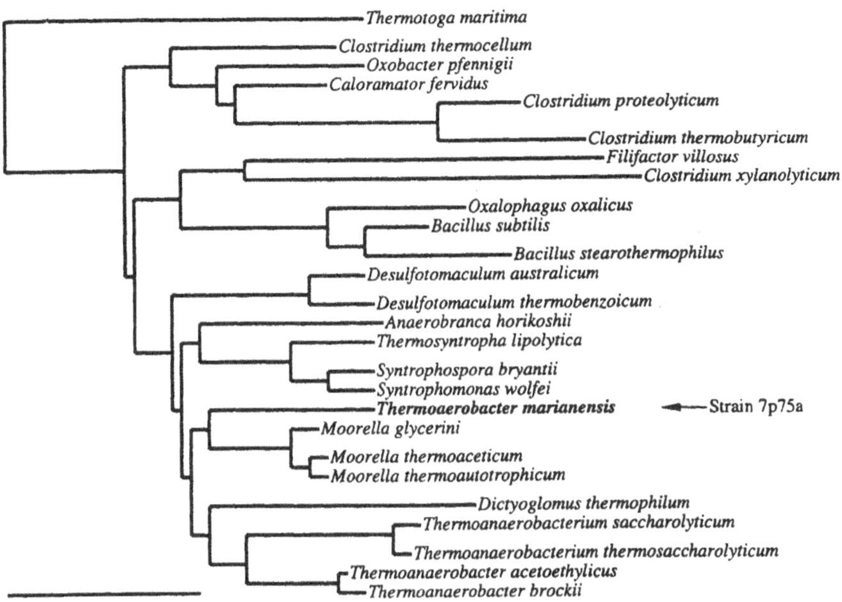

Fig. 16. Unrooted phylogenetic tree based on 16S ribosomal-DNA gene sequences showing the phylogenetic position of *Thermoaerobacter marianensis* among other gram-positive bacteria. (From Takai et al. 1999, with permission)

number of sugars, organic acids, and amino acids as its sole carbon and energy sources (Table 3). Strain 7p75a appears to be sensitive to the concentration of potential substrates for growth; high concentrations of some substrates (Table 3) inhibited growth, whereas the increasing concentrations of other substrates (Table 3) resulted in higher growth rates and yields. The major cellular fatty acids in strain 7p75a were 52.3% iso-$C_{17:0}$, 27.6% $C_{14:1}$, 9.3% anteiso-$C_{17:0}$, 4.3% iso- $C_{15:0}$, 3.5% anteiso- $C_{15:0}$, 2.1% $C_{16:0}$, and 0.9% $C_{18:0}$ acids.

The DNA analyses of strain 7p75a revealed a high G+C content (72.5 mol%). The 16S ribosomal DNA sequence was determined, and a phylogenetic tree based on these sequences was constructed (Fig. 16). These results show that strain 7p75a is a high G+C content, non-spore-forming, aerobic isolate phylogenetically close to low G+C, spore-forming, anaerobic bacterial species. Fig. 16 also shows that strain 7p75a occupies an independent branch among these bacterial species. Based on the physiological and genetic analyses mentioned above, strain 7p75a was classified as a representative of a new bacterial genus *Thermoaerobacter*, and its type species is *Thermoaerobacter marianensis* (Takai et al. 1999).

Table 3. Substrate utilization by *Thermoaerobacter marianensis* at two concentrations (0.2 and 0.02%)

Substrate	Concentration		Growth enhancement at low concentrations
	0.2%	0.02%	
Peptone	++	++	
Yeast extract+peptone	+++	++	
Yeast extract	++	++	
Lignin	–	–	
Cellulose	+	+	
Starch	++	++	
Xylan	++	++	
Chitin	++	++	
Casein	+	++	y
Casamino acids	+	++	y
Maltotriose	++	++	
Cellobiose	++	++	
Maltose	++	+	
Lactose	++	++	
Trehalose	++	++	
Saccharose	++	++	
Glucose	+	++	y
Galactose	+	++	y
Fructose	–	–	
Xylose	–	++	y
Mannitol	++	+	
Inositol	++	+	
Tartrate	++	++	
Succinate	++	++	
Propionate	++	++	
2-Aminobutyric acid	++	++	
Malate	++	++	
Lactate	++	++	
Pyruvate	++	++	
Acetate	++	++	
Glycerol	++	++	
Glycine	+	+	
Alanine	+	+	
Leucine	+	+	
Isoleucine	+	++	y

Table 3. continued

Substrate	Concentration		Growth enhancement at low concentrations
	0.2%	0.02%	
Valine	+	++	y
Cysteine	++	++	
Methionine	+	+	
Proline	++	++	
Phenylalanine	+	+	
Tryptophane	–	–	
Tyrosine	–	–	
Serine	+	++	y
Threonine	+	++	y
Asparagine	++	++	
Glutamine	++	++	
Aspartate	++	++	
Glutamate	++	++	
Lysine	++	++	
Arginine	++	++	
Histidine	++	++	

Modified from Takai et al. 1999, with permission.
+++, > 5×10^8 cells/ml; ++, 8×10^7 to 5×10^8 cells/ml; +, 3×10^7 to 8×10^7 cells/ml; –, no growth; y, growth enhancement at low concentrations (0.02%).

6.2 Some Physiological Characteristics of Hyperthermophiles

6.2.1 Glutamate Dehydrogenase

Glutamate dehydrogenase (GDH) is a key enzyme in nitrogen metabolism and may function in ammonia assimilation (amination) or glutamate catabolism (deamination). GDH from a hyperthermophile (i.e., *Pyrococcus furiosus*) was first described by Consalvi et al. (1991). Consalvi et al. (1991) observed that GDH was abundant in *P. furiosus* cells. The marked abundance of GDH raised questions concerning the physiological role of this enzyme in hyperthermophiles. Although preliminary studies suggested that the role of GDH in hyperthermophiles was mainly biosynthetic (ammonia incorporation) (Consalvi et al. 1991; Ohshima and Nishida 1993), others proposed a critical role of GDH in glutamate catabolism (Klump et al. 1992; Robb et al. 1992).

Kobayashi et al. (1995) carried out an interesting study on the characterization of GDH from the hyperthermophilic archaeon *Thermococcus profundus* and on the involvement of this enzyme in carbohydrate metabolism and consequently in alanine production. González et al. (1998) reported the GDH gene sequence from the hyperthermophile *Pyrococcus horikoshii* and the phylogenetic relationships of this gene in several hyperthermophiles belonging to the Thermococcales.

Role of Glutamate Dehydrogenase in *T. Profundus*

There are two known routes of ammonia incorporation into the cells. One of them is, as mentioned above, the reductive assimilation of ammonia by means of GDH. Another important means of incorporating ammonia is via the enzyme glutamine synthetase (GS), which produces glutamine from ammonia and glutamate, and glutamate synthase (GOGAT), which converts glutamine to glutamate; glutamate in turn is used to form other amino acids by transferring the amino moiety using specific transaminases (Brock and Madigan 1991). Results from *T. profundus* did not exhibit detectable GS activity. GS activity detection was performed over a range of ammonium sulfate concentrations (0%–1%), at different growth temperatures, pH levels, and carbon sources. Thus, ammonia assimilation does not seem to be carried out by

Table 4. GDH purification from *T. profundus*

Step	Protein (mg)	Activity (U)	Specific activity (U/mg)	Recovery (%)	Purification (-fold)
Crude extract	29.1	242	8.3	100	1.0
Matrex gel red A	3.1	193	62.2	80	7.5
Superdex HR200	2.5	183	74.5	76	9.0

From Kobayashi et al 1995, with permission.
GDH, glutamate dehydrogenase.

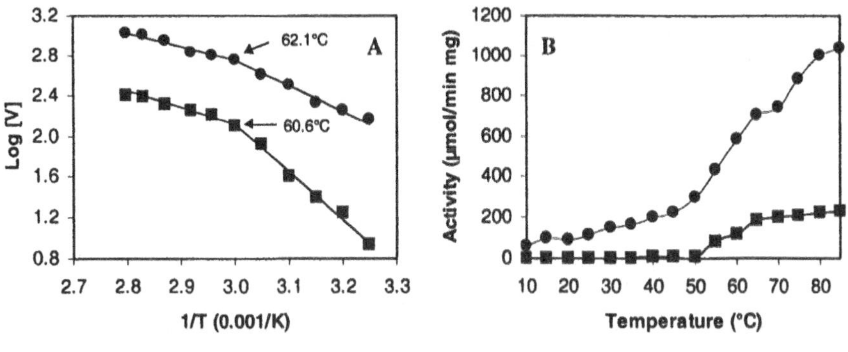

Fig. 17. Effects of temperature on glutamate dehydrogenase activity from *Thermococcus profundus*. Effects on both amination (*circles*) and deamination (*squares*) reactions are presented. **A** Arrhenius plot of these reactions showing the transition temperatures at 62.1°C (amination) and 60.6°C (deamination). **B** Temperature dependence of both reactions. (From Kobayashi et al. 1995, with permission)

GS in this organism. Because GS activity levels are undetectable in *T. profundus*, the main pathway for ammonia incorporation appears to be the use of the enzyme GDH, as described above.

The enzyme GDH from *T. profundus* was purified as summarized in Table 4. The molecular weight of the native GDH from *T. profundus* was estimated to be 263 000 daltons, and it was composed of six identical monomers of molecular weight 43 000 each (Kobayashi et al. 1995). GDH optimum pH values were 7.0 and 7.2 for the process of deamination and amination, respectively. This enzyme was extremely stable across a broad pH range, as no loss of activity was detected within a range of pH 4–10 after incubation at 70°C for 60 min. GDH activity (in both reactions, amination and deamination) gradually increased with temperatures up to 85°C (Fig. 17). The enzyme showed a discontinuity in Arrhenius plot with a transition temperature at around 61°–62°C (Fig. 17). The calculated values of activation energy were 87.1 KJ/mol (amination) and 45.3 KJ/mol (deamination) at the lower temperature range, and in the higher range of temperatures the energy decreased to 27.5 KJ/mol (amination) and 24.6 KJ/mol (deamination). The GDH from *T. profundus* was stable up to 1 h at all temperatures tested (50°–100°C).

Analyses of the kinetic constants of *T. profundus* GDH with respect to L-glutamate, 2-ketoglutarate, ammonia, NADPH, $NADP^+$, NAD^+, and NADH are shown in Table 5. The enzyme showed negative cooperativity for 2-ketoglutarate and ammonia based on estimates of their Hill coefficients of 0.7 and 0.45, respectively. *T. profundus* GDH shows 94 and 3.2 times higher affinity for 2-ketoglutarate and ammonia, respectively, than for glutamate (Table 5). These results suggest that the reaction catalyzed by GDH may be strongly biased toward amination of 2-ketoglutarate, that is, toward glutamate formation. This suggests a predominant biosynthetic role of *T. profundus* GDH. No activity was detected with the coenzyme NAD^+, indicating that *T. profundus* GDH should be classified as EC 1.4.1.4 (NADP-dependent) rather than EC 1.4.1.3 (NAD, NADP-dependent).

Table 5. Apparent K_m and k_{cat} for *T. profundus* GDH

Substance	Concentration range (mM)	K_m (mM)	kcat [a]
2-Ketoglutarate	0.1-1.0	0.20	4.7
	1.0-4.0	0.87	1.8
NH_3	0.22-1.30	1.6	0.16
	1.6-50.0	22	0.096
NADPH	0.01-0.10	0.035	59
NADH	0.01-1.00	0.076	3.2
L-Glutamate	0.9-25.0	6.8	0.05
$NADP^+$	0.06-0.80	0.077	3.8

From Kobayashi et al. 1995, with permission.

[a] The k_{cat} values were calculated as V_{max}/K_m, where V_{max} was the maximum velocity (mmol/min) per milligram of GDH under each condition.

Results of this study showed that cells of *T. profundus* have a large amount of GDH, which accounts for more than 10% of the total soluble protein. Similar large contents of GDH has also been found in other hyperthermophiles, such as *P. furiosus*, *P. woesei*, *P. endeavori*, and *T. zilligii* (Consalvi et al. 1991; DiRuggiero et al. 1993; Hudson et al. 1993; Ohshima and Nishida 1993). This extraordinarily high content of GDH among the Thermococcales is an additional evidence of its critical role in the physiology of these hyperthermophiles. As a result of the large amount of GDH in *T. profundus* cells and the finding that its activity is clearly biased toward glutamate formation, one would expect an accumulation of glutamate or other amino acids. As shown in Fig. 18, *T. profundus* excreted alanine into the culture medium during growth in a medium with pyruvate (1.5 mM Ala at 24 h) or maltose (0.7 mM). Excretion of other amino acids was not detected. Moreover, Kobayashi et al. (1995) observed that *T. profundus* growth was not affected by the absence of ammonia, although the production of alanine decreased significantly (0.6 mM at 24 h in pyruvate-based medium) when the medium was not amended with ammonia. These results suggest that ammonia is being incorporated into amino acids and finally transferred, producing alanine. Ammonia was also produced during *T. profundus* growth (Fig. 18), although alanine appears to be the preferred form for eliminating excess nitrogen by *T. profundus*.

One might still ask about the origin of alanine in *T. profundus* cultures. In general, there are two enzymes involved in L-alanine formation, alanine dehydrogenase (ADH), and alanine aminotransferase (AAT). No ADH activity was detected in *T. profundus*, but high AAT activity was found. Affinity analyses for alanine and pyru-

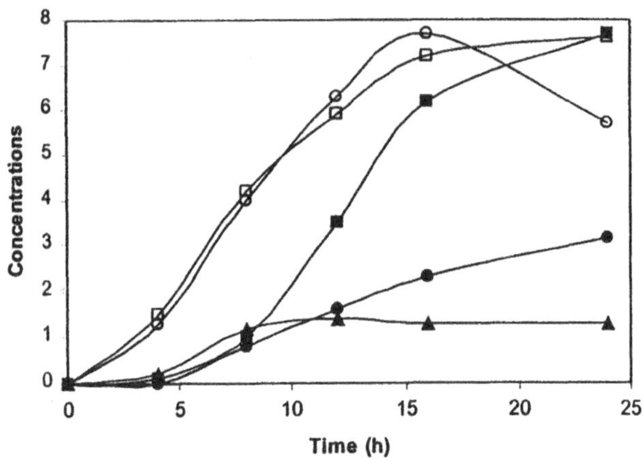

Fig. 18. Time course of alanine excretion from *Thermococcus profundus* in the presence (*closed squares*) and absence (*closed circles*) of ammonia. Cell growth in the presence (*open squares*) and absence (*open circles*) of ammonia is also indicated as total amount of protein per milliliter of culture. Production of ammonia (*closed triangles*) in the ammonia-free medium is also shown. Protein concentration are in microgram per milliliter; alanine and ammonia concentrations are expressed as 10× mllimoles. (From Kobayashi et al. 1995, with permission)

vate formation resulted in 0.87 and 0.12 U/mg protein, respectively. This indicates that AAT activity in *T. profundus* is biased toward alanine formation. Thus, transference of the amino group from glutamate to pyruvate by AAT seems the responsible step in alanine formation. Similar findings have also been reported for the hyperthermophile *Pyrococcus furiosus* (Kengen and Stams 1994), and alanine was suggested as a reduced end product of fermentation.

A major role of GDH and AAT in *T. profundus* would be in the control of the ammonia concentration inside the cells. In addition, GDH shows other major functions in *T. profundus* cells. For instance, GDH would be responsible for glutamate formation from ammonia during growth on pyruvate or maltose. Glutamate would be used for synthesis of other amino acids by using specific aminotransferases; for example, high aspartate aminotransferase has been found in *T. profundus* (Kobayashi et al. 1995). During growth on peptides, GDH would function to deamine amino acids to allow their use as energy sources and synthesis of sugars. Whenever necessary, AAT would serve as a route for removing excess of nitrogen from the cells as alanine; the amino group would be transferred from glutamate.

Glutamate Dehydrogenase Gene Sequences from *T. Profundus* and *P. horikoshii*

At present, the GDH gene sequence of several hyperthermophilic microorganisms are available. Among the Thermococcales, the GDH gene has been sequenced from several species. The DEEPSTAR project has participated with the sequences from two of its novel isolates: *T. profundus* (Higuchi et al. 1997) and *P. horikoshii* (González et al. 1998), the accession numbers of which are D87814 and AF035935, respectively.

A phylogenetic tree constructed from the GDH sequences of several organisms is shown in Fig. 19. 16S-rDNA- and GDH-based phylogenies (Fig. 13 and 19, respectively) differentiate between the described species of the Order Thermococcales. These results together with conclusions from the previous section, suggest that ammonia assimilation by GDH would be characteristic in ancient microorganisms (Kobayashi et al. 1995) as supported by the position of these hyperthermophiles at the deepest branches of ribosomal DNA-based phylogenies (Fig. 1).

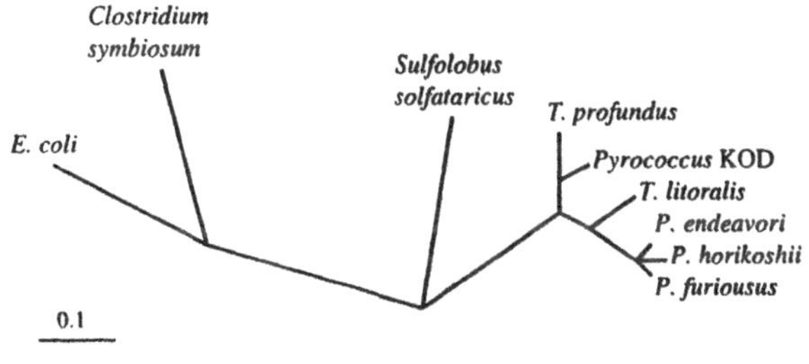

Fig. 19. Unrooted phylogenetic tree based on glutamate dehydrogenase (GDH) gene sequences showing the relationship among the known sequences of this gene in *Thermococcus* and *Pyrococcus* species

6.2.2 Protease Activity in *T. peptonophilus*

Among the Thermococcales, growth depends basically on peptides (Schönheit and Schäfer 1995; Dirmeier et al. 1998). All of them grow on peptide-containing substrates, although most can utilize some other substrates (i.e., maltose, pyruvate, sucrose, amino acids) (Schönheit and Schäfer 1995; Dirmeier et al. 1998). Interestingly, *Thermococcus peptonophilus* (González et al. 1995) is able to grow only on peptides. *T. peptonophilus* was unable to grow on mixtures of 21 amino acids even in the presence of vitamin mixtures. These properties make *T. peptonophilus* a good candidate for research on extracellular proteolytic activity, as *T. peptonophilus* growth depends on its extracellular proteolytic activity for nutrient uptake. In this study (González et al. 1996a), the characteristics of the extracellularly bound proteolytic activity produced by the extremely thermophilic archaeon *Thermococcus peptonophilus* were examined.

High proteolytic activity was found in cell suspensions of *T. peptonophilus* and was detectable at temperatures higher than 60°C and above 120°C. Proteolytic activity was measured by hydrolysis of casein. More than 50% of this proteolytic activity could be recovered in one sodium dodecyl sulfate (SDS) extraction (cells did not

Table 6. Distribution of proteolytic activity in fractions and extracts from *T. peptonophilus* (OG1 strain)

Fraction	Percent of total
Total (cell suspension)	100
Supernatant (1st extraction)	57.6
Supernatant (2nd extraction)	24.9
Supernatant (3rd extraction)	13.9
Sonicated cell suspension	100.6

From González et al. 1996, with permission.

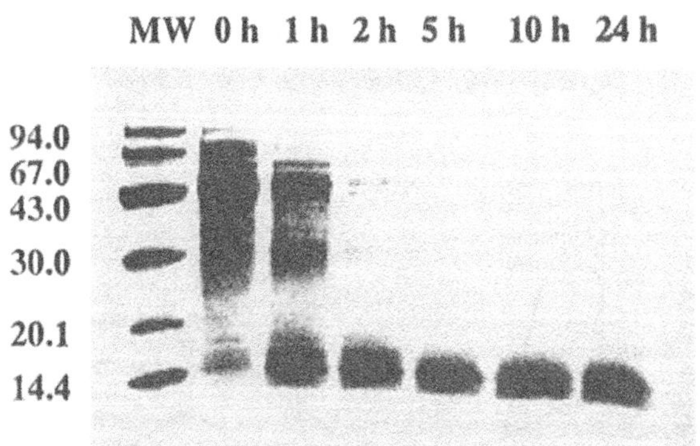

Fig. 20. Sodium dodecyl sulfate-polyacrylamide gel electrophoresis (SDS-PAGE) showing the protein fractions present over time in the SDS extract from *Thermococcus peptonophilus* during self-digestion at 90°C in the presence of 0.5% SDS. Molecular mass markers (*MW*) are indicated in kilodaltons. (From González et al. 1996b, with permission)

lyse at this point); more than 95% of the activity could be extracted after three SDS extractions (Table 6). These results suggest that most of this highly thermostable proteolytic activity was extracellularly bound to the cells.

The suspension from the first extraction was self-digested at 90°C for 24 h in the presence of SDS. This treatment resulted in the hydrolysis of most of the proteins in the extracts. Essentially complete proteolytic activity was retained after 2 h of incubation (> 90% of initial activity). Samples incubated for 10 h and 24 h retained 31.5% and 12.5% of the initial activity, respectively. Fig. 20 shows an SDS-polyacrylamide gel electrophoresis (PAGE) with the progressive degradation of the extracted proteins. Only a band at about 14 kDa could be detected after 24 h of incubation; this band increased during the first hour of incubation and remained stable throughout the rest of the incubation period. This product is smaller than other thermostable proteases (Cowan et al. 1987; Blumentals et al. 1990).

Optimum proteolytic activity of the self-digested extract was at 100°C (Fig. 21). Significant activity was detected from 60°C up to 125°C (maximum tested). Optimum pH was 7, and significant activity was detected from pH 5.5–8.5 (Fig. 21). The half-lives of the self-digested extract were 10 and 8 h at 90° and 100°C, respectively, although the half-life decreased dramatically to 19 min when autoclaved at 121°C. These results show much higher thermostability than archaelysin (Cowan et al. 1987) from *Desulfurococcus* (70-90 min at 95°C) and lower thermostability than the proteases studied by Blumentals et al. (1990) from *P. furiosus* (33 h at 98°C). However, Blumentals et al. (1990) reported that their protease was inactivated after 20 min at 120°C, whereas activity is still detectable after autoclaving the *T. peptonophilus* self-digested extract.

Thermophilic proteases usually are extremely resistant to denaturing agents (Cowan et al. 1987; Blumentals et al. 1990; Klingeberg et al. 1995). *T. peptonophilus* extracts also showed resistance to several detergents and denaturing agents, such as 1% SDS, 5% β-mercaptoethanol, 8 M urea, 0.2% sodium deoxycholate, 0.1% Triton X-100, and their combinations. Self-digested extracts were inhibited by the protease

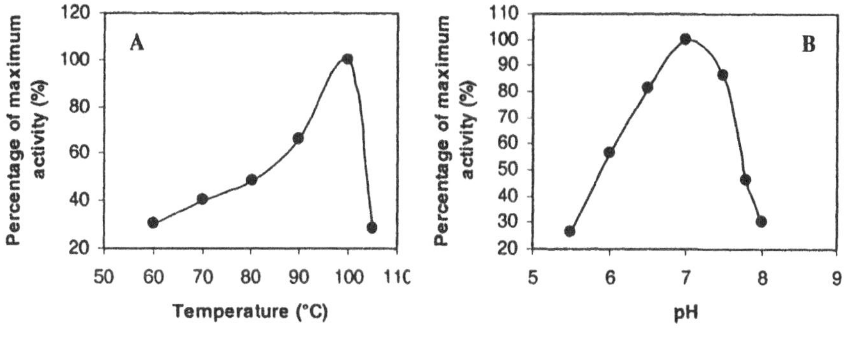

Fig. 21. Effect of temperature (**A**) and pH (**B**) on proteolytic activity by the self-digested extract from *Thermococcus peptonophilus*. The 100% activity was assayed at 100°C and pH 7. (From González et al. 1996b, with permission)

inhibitors phenylmethylsulfonyl fluoride (PMSF) and diisopropylfluorophosphate (DFP) (Table 7); iodoacetic acid showed partial inhibition, and disodium ethylenediaminetetraacetate (EDTA) and ethyleneglycol-bis (β-aminoethylether-*N, N, N', N'*-tetraacetic acid (EGTA)) had low effect on the proteolytic activity of *T. peptonophilus* extracts. These results suggest that the proteolytic activity of *T. peptonophilus* extracts could be mostly due to serine proteases. The partial inhibition by iodoacetic

Table 7. Effects of enzyme inhibitors on proteolytic activity of *T. peptonophilus* (OG1 strain) extracts

Inhibitor	Percent of inhibition
EDTA	11.7
EGTA	29.4
PMSF	75.0
DFP	88.5
Iodoacetic acid	48.6

From González et al. 1996b, with permission.
PMSF, phenylmethylsulfonyl fluoride; DFP, diisopropylfluorophosphate; EDTA, disodium ethylenediaminetetraacetate; EGTA, ethyleneglycol-bis (β-aminoethylether)-*N, N, N', N'*-tetraacetic acid.

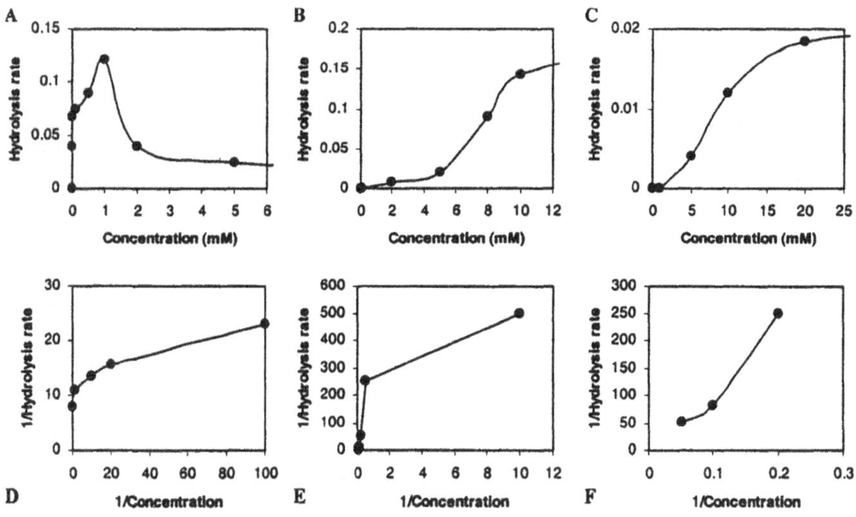

Fig. 22. Kinetics of proteolytic activity by *Thermococcus peptonophilus* self-digested extracts on several synthetic substrates. **A, D** *N*-Benzoyl-DL-arginine-*p*-nitroanilide. **B, E** L-leucine-*p*-nitroanilide. **C, F** L-lysine-*p*-nitroanilide. **D, E, F** These are reciprocal plots of **A, B, C**, respectively. **D, E** show positive cooperativity in **A, B**, respectively. Hydrolysis rates are given in optical density (OD) units per minute; concentrations are given in millimoles. (From González et al. 1996b, with permission)

acid suggests that HS group(s) might be close to, or partially involved in, the active site of these proteases. However, HS groups in *T. peptonophilus* extracts might not be involved in protein stability, as judged by the resistance to denaturing agents mentioned above.

High-affinity hydrolytic activity toward *N*-benzoyl-DL-arginine-4-nitroanilide was observed for *T. peptonophilus* extracts. Other hydrolyzed substrates were L-leucine-*p*-nitroanilide and L-lysine-*p*-nitroanilide, although the extracts showed low affinity towards these substrates (Fig. 22). These results demonstrate the existence of low-affinity aminopeptidase activity in *T. peptonophilus* extracts. In addition, kinetic studies showed a positive co-operativity toward the synthetic substrates *N*-benzoyl-DL-arginine-4-nitroanilide and L-leucine-*p*-nitroanilide (Fig. 22). Positive co-operativity was not observed when casein was the offered substrate (Fig. 23), possibly because of the presence of several hydrolyzable sites in the casein molecules versus only one hydrolyzable target in the synthetic substrates. *N*-Benzoyl-DL-arginine-4-nitroanilide is an optimum substrate for *T. peptonophilus* extracts; however, it is not known why these extracts did not hydrolyze the ester analog *N*-benzoyl-L-arginine ethyl ester. Similarly, the extracellular protease archaelysin (Cowan et al. 1987) did not hydrolyze any ester substrate tested. In contrast, the protease reported by Blumentals et al. (1990) from *P. furiosus* was active only toward *N*-benzoyl-*N*-arginine

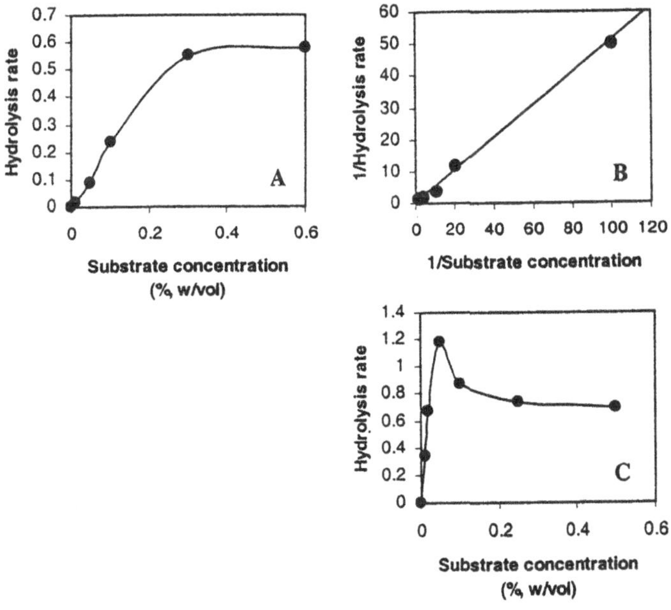

Fig. 23. Kinetics of proteolytic activity by *Thermo-coccus peptonophilus* self-digested extracts using casein as substrate. **A** Kinetic study on self-digested extracts; **B** Double-reciprocal plot of data shown in **A**, **C** Kinetic study carried out with cell suspensions of *T. peptonophilus*. (From González et al. 1996b, with permission)

ethyl ester and L-lysine-*p*-nitroanilide. An extracellular protease from *T. stetteri* (Klingeberg et al. 1995) also showed high hydrolysis on benzoyl-arginine-nitroanilide.

Thermococcus peptonophilus extracts showed inhibition at high synthetic substrate concentrations. High-substrate inhibition was also observed by Cowan et al. (1987) for archaelysin. When casein solutions were used for testing proteolytic activity in cell suspension, high-substrate inhibition was also observed, but we did not observe this phenomenon when the casein assays were carried out with the self-digested extracts (Fig. 23). This suggests the presence of regulatory mechanisms of the extracellular proteolytic activity in *T. peptonophilus*.

The physiological significance of the extracellular proteolytic activity in *T. peptonophilus* is clear. Let us remember that *T. peptonophilus* is able to grow only on peptides. It grows on proteinaceous substrates (e.g., tryptone, casein, yeast extract) and is unable to utilize any other carbon source. *T. peptonophilus* is also unable to grow on a mixture of 21 amino acids. Additionally, *T. peptonophilus* does not grow on completely hydrolyzed casein (acid hydrolysis, 1 M HCl, overnight at 80°C, followed by neutralization), but it grows optimally on partially digested casein (overnight protease treatment). These results show that *T. peptonophilus* requires peptides. It is likely that extracellular or membrane-bound proteases are associated with the transport of amino acids or small peptides into the cell. At this respect, efforts have been made to sequence the genes encoding for an oligopeptide membrane-binding protein found in the *T. peptonophilus* genome.

6.2.3 Survival and Viability of Thermophiles

Microorganisms live in changing environments. This is even more critical for hyperthermophiles owing to the variability of conditions at hydrothermal vents and the possibility of being dispersed away from their hot environments. Therefore microorganisms must adapt to survive under adverse circumstances. Among prokaryotes, several survival strategies have been reported (Roszak and Colwell 1987). Some microorganisms are able to form spores; non-spore-forming microorganisms are usually able to enter dormant physiological states that allow them to resist adverse conditions (Roszak and Colwell 1987). In nature, the largest proportion of cells are in dormant states (Roszak and Colwell 1987) waiting for better conditions to come. It has not been reported whether extreme thermophiles are able to develop different physiological states of survival under adverse environmental conditions. This has been extensively reported for mesophilic bacteria (Roszak and Colwell 1987; Postgate 1977). The objective of this study (González et al. 1996a) was to characterize a hyperthermophilic archaeon, *Thermococcus peptonophilus*, in various physiological states of survival.

Various physiological states can be defined as the result of using various counting procedures. Table 8 shows the correspondence between the counting protocols used in this study and the physiological states of the studied microorganism. Total cell counts lead to the highest estimates of cell numbers, as every visible cell in the

Table 8. Physiological states of survival identified by several counting procedures

Counting procedures	Physiological state
Most probable number	Culturable cells
Rhodamine staining—MPN	Active, nonculturable cells
Rhodamine staining	Active cells
DAPI–rhodamine staining	Inactive, nonculturable cells
DAPI staining	Total cells (viviform population)
DAPI staining–MPN	Nonculturable cells
DAPI staining, decrease over time [a]	Dead cells

From González et al. 1996a, with permission.

DAPI: 4',6'-diamidino-2-phenylindole; MPN, most probable number.

Minus sign indicates that the latter count was subtracted from the former. The first procedure listed generally gives higher counts.

[a] Maximum total cell number minus total cell number at a determined time.

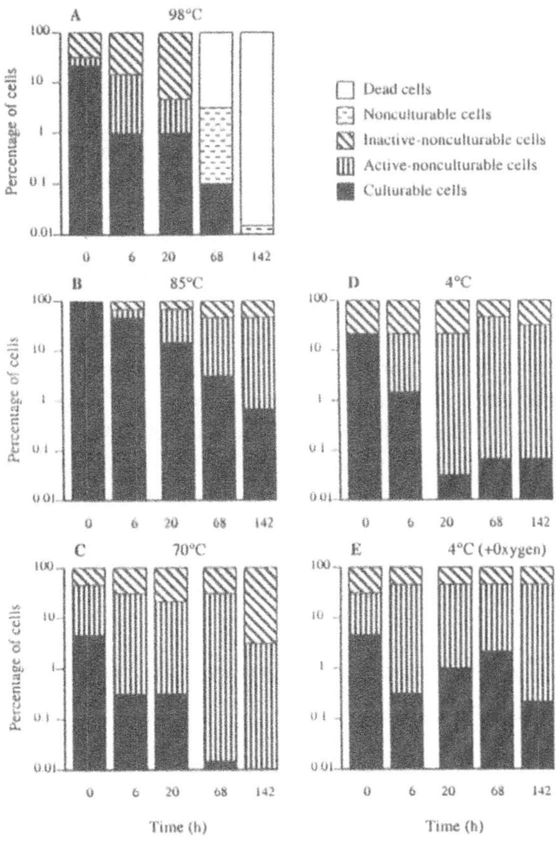

Fig. 24. Characterization of *Thermococcus peptonophilus* cells in several physiological states of survival during exposure to different conditions. Culturable, active nonculturable, inactive nonculturable, and dead cells are shown as percentages of the initial total cell number. Cells were exposed to **A** 98°C, **B** 85°C, **C** 70°C, **D** 4°C, and **E** oxygenated at 4°C. (From González et al. 1996b, with permission)

sample is accounted for. Rhodamine staining is dependent on its active incorporation by the cells. Therefore rhodamine-stained cells should be considered those showing a detectable active metabolism (Yu and McFeters 1994; González et al. 1996a). Counts of culturable cells are those accounting for the cells able to grow in the culture conditions used in the study; these are usually the lowest counts. Culturable cell numbers can be obtained by plate counts or by the most probable number (MPN). The MPN estimates the actual number of culturable microorganisms based on the number of tubes showing growth after inoculating and incubating a series of tubes and dilutions under optimum conditions for growth (Koch 1994; González 1996). In this study, *T. peptonophilus* cells could be classified in several states (Table 8): culturable cells, active nonculturable cells, active cells, inactive nonculturable cells, total cell number, nonculturable cells, and the dead cells that can no longer be detected.

At 85°, 70°, and 4°C, increases in the fraction of active, nonculturable cells were observed together with decreases in the fraction of culturable cells (Fig. 24). Throughout these survival experiments, the fraction of inactive, nonculturable cells represented only a small portion of the total cells present at these temperatures (Fig. 24); total cell numbers did not decrease under these conditions. At 98°C, in addition to a progressive decrease in the fraction of culturable cells, a decrease of total cell counts was observed. Thus, the number of dead cells increased during the survival experiments (Fig. 24). Some of these dead cells of *T. peptonophilus* could be visualized as "ghosts" or lysed cells in these samples (as rests of cell envelopes). This is the first study showing several physiological states during the survival of a strictly anaerobic, hyperthermophilic archaeon.

A temperature of 4°C, under anoxic or oxic conditions, appeared to be an adequate temperature for short-term storage of *T. peptonophilus* cells. It has also reported for other hyperthermophiles (Huber et al. 1990). The results in Fig. 24 showed that *T. peptonophilus* can survive at 4°C, maintaining a high proportion of active cells and an important fraction of culturable cells. Additionally, the presence of oxygen does not have a significant effect on the proportion of survivors of *T. peptonophilus* cells at 4°C. Because the temperature in the deep-sea is about 2°-4°C, *T. peptonophilus* may survive for long periods in deep-sea waters. This finding represents a possible mechanism for the dispersion of extreme thermophiles in marine waters and suggests that the deep-sea floor is a potential sink for hyperthermophiles. Thus, a possible place for retrieving hyperthermophiles could be the deep-sea floor, as suggested by recent studies (Canganella et al. 1997; Takai et al. 1999).

6.2.4 Pressure and Temperature Effects on *T. peptonophilus*

Current knowledge of the effects of high pressures on biological processes has been reviewed (Jaenicke 1981; Ludlow and Clark 1991; Bartlett 1992; Somero 1992; Kato, this volume); however, only a few studies have focused on the effect of pressure at high temperatures (Jaenicke et al. 1988; Bernhardt et al. 1988; Reysenbach and Deming 1991; Erauso et al. 1993; Pledger et al. 1994; Holden and Baross 1995) on specific thermophilic strains. In this study we analyzed the effects of high hydrostatic pressures and temperatures on the growth and viability of the hyperthermophile *Thermococcus peptonophilus* (Canganella et al. 1997).

To achieve these objectives, a unique, large-scale cultivation system called DEEP-BATH(deep-sea baro/thermophiles cultivation system) was used throughout the study. The DEEP-BATH cultivation system was developed at the Japan Marine Science and Technology Center (JAMSTEC) (Kyo et al. 1991; Moriya et al. 1995). This high-pressure/high-temperature bioreactor has been designed to work within a 0-300°C range and up to 68 MPa, and it is suitable for continued sampling without perturbing the culture.

Thermococcus peptonophilus has an optimal growth temperature of 85°C at atmospheric pressure (González et al. 1995). During cultivation at 30 MPa, the growth rate was enhanced (Figure 25). The optimal temperature for growth shifted to 90°-95°C at 45 MPa; the growth rate was significantly faster at pressures of 45 MPa or more than at atmospheric pressure for temperatures 90°C or higher. These results are in agreement with those presented for *Pyrococcus abyssi* (Erauso et al. 1993) and *Pyrococcus endeavori* (Pledger et al. 1994).

Cell numbers, ATP levels, and proteolytic activity in *T. peptonophilus* cells were studied during incubation at high temperatures and pressures. Cellular ATP levels were directly related to growth rate; that is, the higher the ATP level, the higher the growth rate (Figures 26). The cellular ATP content usually increased during the exponential phase of growth, reaching maximum values during the late-exponential phase before entering a stationary phase (Fig. 26). Once the cell number starts leveling off, the ATP level in *T. peptonophilus* cells sharply decrease. Proteolytic activity was estimated by hydrolysis of casein (González et al. 1996b). The highest proteolytic activity was observed during the early exponential phase, suggesting that an increased demand for nutrients was needed to reach the highest possible growth

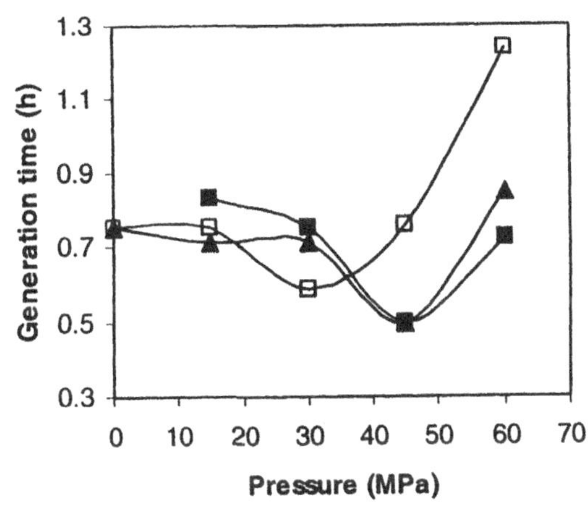

Fig. 25. Generation time of *Thermococcus peptonophilus* at various temperatures and hydrostatic pressures: 85°C (*open squares*), 90°C (*closed triangles*), and 95°C (*closed squares*). (From Canganella et al. 1997, with permission)

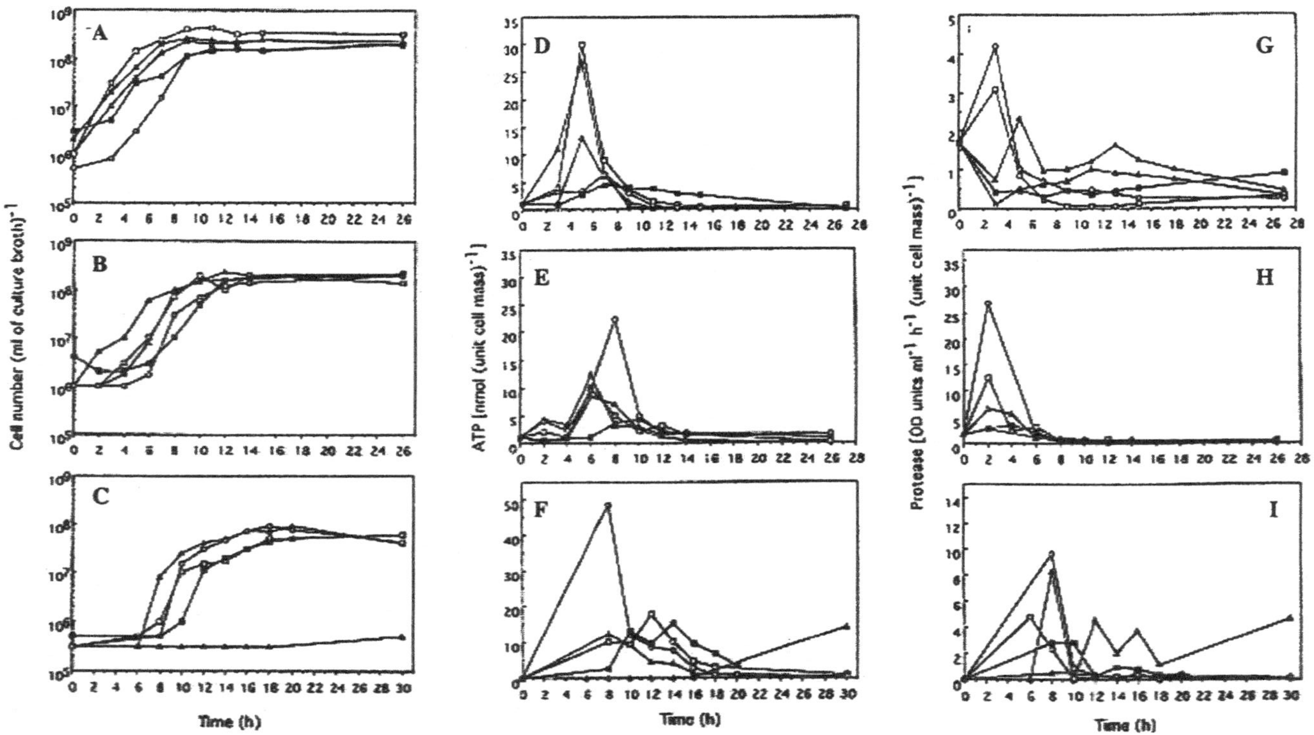

Fig. 26. Growth, ATP content, and proteolytic activity in *Thermococcus peptonophilus* cells at 85°C (**A**, **D**, **G**), 90°C (**B**, **E**, **H**), and 95°C (**C**, **F**, **I**) and under different hydrostatic pressures: 0.1 MPa (*open triangles*), 15 MPa (*closed triangles*), 30 MPa (*open squares*), 45 MPa (*open circles*), 60 MPa (*closed squares*). Unit cell mass: 10^8 cells/ml. (From Canganella et al. 1997, with permission)

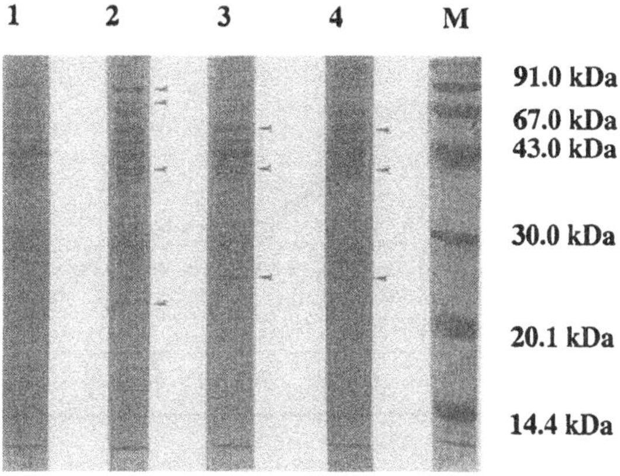

Fig. 27. SDS-PAGE of *Thermococcus peptonophilus* total cell proteins (10 μg) after cultivation at 90°C and 0.1 MPa (lane *1*), 60 MPa (lane 2), 95°C and 0.1 MPa (lane *3*), starvation for 132 h at 90°C at 0.1 MPa (lane *4*). Molecular mass marker (lane *M*) in kilodaltons. *Arrows* indicate the overproduced proteins as compared with lane *1*. (From Canganella et al. 1997, with permission)

rates (Fig. 26). The highest proteolytic activities were also observed at the optimal pressure for growth (45 MPa). Activity rapidly decreased during *T. peptonophilus* growth, reaching minimum levels as the stationary phase progressed. Proteolytic activity and ATP content patterns were similar (Fig. 26).

SDS-PAGE (Fig. 27) of cell extracts was performed to look for expression of distinct proteins under different growth conditions. Some proteins in the 22-, 40-, 70-, and 90-KDa range were induced after cultivation at 60 MPa (compare lanes 1 and 2 in Fig. 27). Temperatures close to the upper limit *for T. peptonophilus* cells also appeared to induce the production of certain proteins with molecular masses of 25, 40, and 55 KDa (compare lanes 1 and 3 in Fig. 27). Additionally, prolonged starvation (132 h) of *T. peptonophilus* cells led to the induction of specific proteins (compare lanes 1 and 4 in Fig. 27). These results suggest that elevated pressures and elevated temperatures can induce the production of specific proteins in response to culturing conditions. Responses equivalent to those of Fig. 28 have been reported during heat-shock experiments performed with *Pyrococcus endeavori* (Holden and Baross 1995).

Starvation also induced the production of specific proteins in *T. peptonophilus* (Fig. 27). Additional experiments exposing *T. peptonophilus* to prolonged incubation in nutrient-limited media showed that total cell numbers and ATP levels in *T. peptonophilus* cells decreased exponentially during starvation (Fig. 28). A more dramatic effect took place at atmospheric pressure than at 30 MPa. ATP levels in starved cells at atmospheric pressure and at 30 MPa were similar, although cells at 30 MPa showed a more progressive increase of ATP content after an initial decrease; ATP levels at the end of the experiment at 30 MPa were similar to the initial levels while the ATP levels at atmospheric pressure were about half the initial values. These results were confirmed by a higher number of culturable cells (MPN counts) at 30 MPa (2.5×10^5 cells/ml after 88 h) than at atmospheric pressure (5.7×10^4 cells/ml after 88 h at 0.1 MPa). Proteolytic activity at 30 MPa showed an slight increase during the initial 15 h of starvation, whereas cells at atmospheric pressure showed a

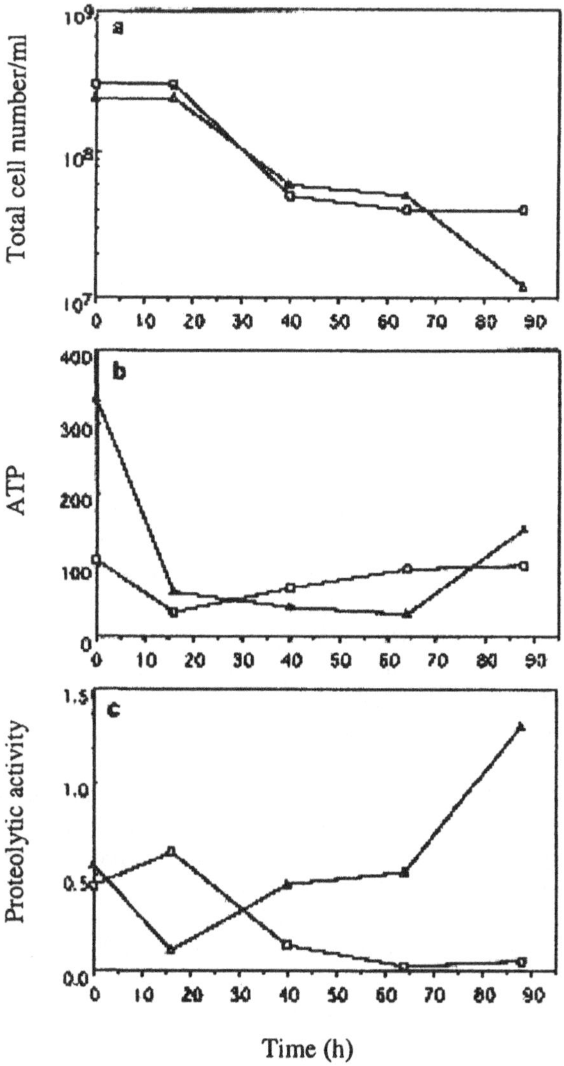

Fig. 28. Effect of starvation on *Thermococcus peptonophilus* at 90°C and various hydrostatic pressures: 0.1 MPa (*triangles*) and 30 MPa (*squares*). **A** total cell number per milliliter, **B** ATP content (nanomoles per unit cell mass), **C** Proteolytic activity (OD units per milliliter per hour per unit cell mass). Unit cell mass: 10^8 cells/ml. (From Canganella et al. 1997, with permission)

sharp decrease of proteolytic activity. An increase of proteolytic activity at atmospheric pressure from 15 h might be a result of cellular lysis over the starvation process.

The starvation process of *T. peptonophilus* at low temperature (10°C) was also studied (Fig. 29). After growth at 85°C, cells were incubated at 10°C for about 5 days. Total cell numbers increased during the incubation period, probably as a result of disaggregation of cell clumps. Protease activity showed high variability, but an overall decrease over time was observed; higher proteolytic activity values were always obtained under high pressure. ATP concentration per cell mass showed a dramatic difference between cells under pressure (45 MPa) and those at atmospheric

Fig. 29. Effect of starvation on *Thermococcus peptonophilus* at 10°C and various hydrostatic pressures: 0.1 MPa (*triangles*) and 45 MPa (*squares*). **A** total cell number/ml, **B** ATP content (nanomoles per unit cell mass), **C** proteolytic activity (OD units per milliliter per hour per unit cell mass). Unit cell mass: 10^8 cells/ml. (From Canganella et al. 1997, with permission)

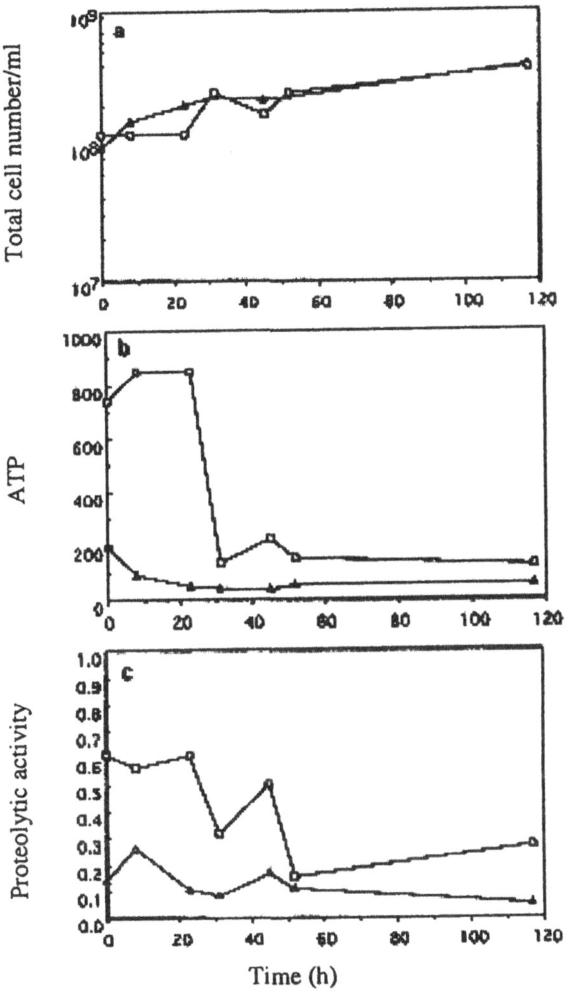

pressure (0.1 MPa). ATP levels under pressure were about eight-fold higher than those under atmospheric pressure. These differences between pressurized and nonpressurized cells were also reflected in the number of culturable cells (MPN counts). At the end of the experiment the number of culturable cells under pressure (3×10^8 cells/ml at 45 MPa) was about five-fold the number at atmospheric pressure (6.2×10^7 at 0.1 MPa). These results suggest that elevated pressure enhances the survival of *T. peptonophilus* under conditions of prolonged starvation at both high and low temperatures, but the effects are more obvious at lower temperatures. No significant decreases in viability were observed in starved cultures at low temperature and high hydrostatic pressures (45 MPa). These results indicate that exposure of the hyperthermophiles (i.e., *T. peptonophilus*) to a colder environment is better tolerated under high pressure; this is probably the case in deep-sea environments, where low temperatures and high pressures coexist.

Several studies have indicated that high pressure has a positive effect on the survival of *P. abyssi* and *P. endeavori* at supraoptimal temperatures (Pledger et al. 1994; Holden and Baross 1995). These results provide increasing evidence of a critical role of high pressure on the survival of deep-sea hyperthermophiles under natural conditions. Nevertheless, not every extreme thermophile isolated from the deep-sea has shown positive growth responses under high hydrostatic pressures; examples are *Desulfurococcus* strains (Jannasch et al. 1988) and *Thermaerobacter marianensis* (Takai et al. 1999).

In summary, the growth rate of *T. peptonophilus* is enhanced by high pressure up to 45 MPa. Moreover, high pressure (45 MPa) produced a shift from 85 to 90-95°C of the optimal temperature for *T. peptonophilus* growth. These properties defined *T. peptonophilus* as a barophilic microorganism (Kato, this volume).

6.3 Remarks and Closing Comments

A great portion of the effort carried out during this project has been focused on retrieval of novel thermophiles from the deep-sea (Kobayashi et al. 1994; González et al. 1994, 1995, 1998; Kwak et al. 1995; Grote et al. 1999; Takai et al. 1999). Results suggests the presence of an enormous microbial diversity (Fig. 16) of thermophilic microorganisms at the seafloor both at hydrothermal vents and far away from these hot environments (González et al. 1996a; Canganella et al. 1997; Takai et al. 1999). In fact, variations in medium composition often leads to the discovery of new species even when working on the same sampling sites. Scientists are starting to understand how extreme thermophiles thrive in hot environments. It is easily understood that research on this subject must continue to retrieve a significant portion of the high-temperature microbial diversity.

In addition to the retrieval, isolation, and characterization of deep-sea thermophiles, research has been conducted on some of the properties of those extreme thermophilic microorganisms. Enzymes and metabolism (Kobayashi et al. 1995; González et al. 1996b), viability and survival (González et al. 1996b), and pressure and temperature effects (Canganella et al. 1997) have been studied and the results are presented above. Further analyses of the intriguing characteristics of novel, already described extreme thermophiles will certainly improve our understanding of these exceptional microorganisms and thus promote their potential biotechnological applications.

References

Baross JA, Hoffman SE (1985) Submarine hydrothermal vents and associated gradient environments as sites for the origin and evolution of life. Origins Life 15:327-3450

Bartlett DH (1992) Microbial life at high pressures. Sci Prog 76:479-496

Bernhardt G, Jaenicke R, Ludemann H-D, Konig H, Stetter KO (1988) Does high pressure extend the temperature range of viability of thermophilic archaebacteria? Pressure-dependent growth enhancement and growth inhibition of *Methanococcus thermolithotrophicus*. Appl Environ Microbiol 54:1258-1261

Blumentals II, Robinson AS, Kelly RM (1990) Characterization of sodium dodecyl-sulfate-resistant proteolytic activity in the hyperthermophilic archaebacterium *Pyrococcus furiosus*. Appl Environ Microbiol 56:1992-1998

Böck A, Kandler O (1985) Antibiotic sensitivity of archaebacteria. In: Woese CR, Wolfe RS (eds) The bacteria, vol 8. Academic, Orlando, pp525-544

Brock TD (1978) Thermophilic microorganisms and life at high temperatures. Springer, New York

Brock TD, Madigan MT (1991) Biology of microorganisms. Prentice Hall, Englewood Cliffs, NJ

Canganella F, González JM, Yanagibayashi M, Kato C, Horikoshi K (1997) Pressure and temperature effects on growth and viability of the hyperthermophilic archaeon *Thermococcus peptonophilus*. Arch Microbiol 168:1-7

Consalvi V, Chiaraluce R, Politi L, Vaccaro R, De Rosa M, Scandurra R (1991) Extremely thermostable glutamate dehydrogenase from the hyperthermophilic archaebacterium *Pyrococcus furiosus*. Eur J Biochem 202:1189-1196

Corliss JB, Dymond J, Gordon LI, Edmond JM, von Herzen RP, Ballard RD, Green K, Williams D, Bainbridge A, Crane K, van Andel TH (1979) Submarine thermal springs on the Galapagos Rift. Science 203:1073-1083

Cowan DA, Smolenski KA, Daniel RM, Morgan HM (1987) An extremely thermostable extrecellular proteinase from a strain of the archaebacterium *Desulfurococcus* growing at 88°C. Biochem J 247:121-133

Dirmeier R, Keller M, Hafenbradl D, Braun F-J, Rachel R, Burggraf S, Stetter KO (1998) *Thermococcus acidaminovorans* sp. nov., a new hyperthermophilic alkalophilic archaeon growing on amino acids. Extremophiles 2:109-114

DiRuggiero J, Robb FT, Jagus R, Klump HH, Borges KM, Kessel M, Mai X, Adams MWW (1993) Characterization, cloning, and in vitro expression of the extremely thermostable glutamate dehydrogenase from the hyperthermophilic archaeon, ES4. J Biol Chem 268:17767-17774

Erauso G, Reysenbach A-L, Godfroy A, Meunier J-R, Crump B, Partensky F, Baross JA, Marteinsson V, Barbier G, Pace NR, Prieur D (1993) *Pyrococcus abyssi* sp. nov., a new hyperthermophilic archaeon isolated from a deep-sea hydrothermal vent. Arch Microbiol 160:338-349

Fiala, G, Stetter KO (1986) *Pyrococcus furiosus* sp. nov. represents a novel genus of marine heterotrophic archaebacteria growing optimally at 100C. Arch Microbiol 145:56-61

González JM (1996) A general purpose program for obtaining most probable number tables. J Microbiol Methods 26:215-218

González JM, Kato C, Horikoshi K (1995) *Thermococcus peptonophilus* sp. nov., a fast-growing, extreme thermophilic archaebacterium isolated from deep-sea hydrothermal vents. Arch Microbiol 164:159-164

González JM, Kato C, Horikoshi K (1996b) Membrane-bound, sodium dodecyl sulfate-resistant proteolytic activity from the extreme thermophilic archaeon *Thermococcus peptonophilus*. J Mar Biotechnol 4:159-164

González JM, Kato C, Horikoshi K (1996a) Culturability and survival of an extreme thermophile isolated from deep-sea hydrothermal vents. Arch Microbiol 166:64-67

González JM, Robb FT, Ammerman JW, Maeder DL, Yanagibayashi M, Tamaoka J, Kato C (1998) *Pyrococcus horikoshii* sp. nov., a hyperthermophilic archaeon isolated from a hydrothermal vent at the Okinawa Trough. Extremophiles 2:123-130

González JM, Sato T, Kato C, Horikoshi K (1994) Isolation and characterization of hyperthermophilic archaebacteria from Southwestern Pacific hydrothermal vents. JAMSTEC J Deep Sea Res 10:471-480

Grote R, Li L, Tamaoka J, Kato C, Horikoshi K, Antranikian G (1999) *Thermococcus siculi* sp. nov., a novel hyperthermophilic archaeon isolated from a deep-sea hydrothermal vent at the Mid-Okinawa Trough. Extremophiles 3:55-62

Higuchi S, Kobayashi T, Kimura K, Horikoshi K, Kudo T (1997) Molecular cloning, nucleotide sequence and expression in *Escherichia coli* of hyperthermophilic glutamate dehydrogenase gene from *Thermococcus profundus*. J Ferment Bioeng 83:405-411

Holden JF, Baross JA (1995) Enhanced thermotolerance by hydrostatic pressure in the deep-sea hyperthermophile *Pyrococcus* strain ES4. FEMS Microbiol Ecol 18:27-34

Horikoshi K (1991) General view of alkaliphiles and thermophiles. In: Horikoshi K, Grant WD (eds) Superbugs: microorganisms in extreme environments. Springer, New York, pp3-14

Huber R, Stoffers P, Cheminee JL, Richnow HH, Stetter KO (1990) Hyperthermophilic archaebacteria within the crater and open-sea plume of erupting MacDonald seamount. Nature 345:179-181

Hudson RC, Ruttersmith LD, Daniel RM (1993) Glutamate dehydrogenase from the extremely thermophilic archaebacterial isolate AN1. Biochim Biophys Acta 1202:244-250

Jaenicke R (1981) Enzymes under extremes of physical conditions. Annu Rev Biophys Bioeng 10:1-67

Jannasch HW, Mottl M (1985) Geomicrobiology of deep-sea hydrothermal vents. Science 229:717-725

Jannasch HW, Wirsen CO, Molyneaux SJ, Langworthy TA (1988) Extremely thermophilic fermentative archaebacteria of the genus *Desulfurococcus* from deep-sea hydrothermal vents. Appl Environ Microbiol 54:1203-1209

Jones WJ, Leigh JA, Mayer F, Woese CR, Wolfe RS (1983) *Methanococcus jannaschii* sp. nov., an extremely thermophilic methanogen from a submarine hydrothermal vent. Arch Microbiol 136:254-261

Kawarabayasi Y, Sawada M, Horikawa H, Haikawa Y, Hino Y, Yamamoto S, Sekine M, Baba S, Kosugi H, Hosoyama A, Nagai Y, Sakai M, Ogura K, Otsuka R, Nakazawa H, Takamiya M, Ohfuku Y, Funahashi T, Tanaka T, Kudoh Y, Yamazaki J, Kushida N, Oguchi A, Aoki K, Kikuchi H (1998) Complete sequence and gene organization of the genome of a hyper-thermophilic archaebacterium, *Pyrococcus horikoshii* OT3. DNA Res 5:55-76.

Kengen SWM, Stams AJM (1994) Formation of L-alanine as a reduced end product in carbohydrate fermentation by the hyperthermophilic archaeon *Pyrococcus furiosus*. Arch Microbiol 161:168-175

Klingeberg M, Galsinsky B, Sjoholm C, Kasche V, Antranikian G (1995) Purification and properties of a highly thermostable, sodium dodecyl sulfate-resistant and stereospecific proteinase from the extremely thermophilic archaeon *Thermococcus stetteri*. Appl Environ Microbiol 61:3098-3104

Klump H, DiRuggiero J, Kessel M, Park J-B, Adams MWW, Robb FT (1992) Glutamate dehydrogenase from the hyperthermophile *Pyrococcus furiosus*. J Biol Chem 267:22681-22685

Kobayashi T, Higuchi S, Kimura K, Kudo T, Horikoshi K (1995) Properties of glutamate dehydrogenase and its involvement in alanine production in a hyperthermophilic archaeon, *Thermococcus profundus*. J Biochem 118:587-592.

Kobayashi T, Kwak YS, Akiba T, Kudo T, Horikoshi K (1994) *Thermococcus profundus* sp. nov., a new hyperthermophilic archaeum isolated from a deep-sea hydrothermal vent. Syst Appl Microbiol 17:232-236.

Koch AL (1994) Growth measurement. In: Gerhardt P, Murray RGE, Wood WA, Krieg NR (eds) Methods for general and molecular bacteriology. American Society for Microbiology, Washington, DC, pp248-277

Kwak YS, Kobayashi T, Akiba T, Horikoshi K, Kim YB (1995) A hyperthermophilic sulfur-reducing archaebacterium, *Thermococcus* sp. DT1331, isolated from a deep-sea hydrothermal vent. Biosci Biotechnol Biochem 59:1666-1669

Kyo M, Tuji T, Usui H, Itoh T (1991) Collection, isolation and cultivation system for deep-sea microbes study: concept and design. Oceans 1:419-423

Ludlow JM, Clark DS (1991) Engineering considerations for the application of extremophiles in biotechnology. Crit Rev Biotechnol 10:321-345

Miroshnichenko ML, Bonch-Osmolovskaya EA, Neuer A, Kostrikina NA, Chernych NA, Alekseev VA (1989) *Thermococcus stetteri* sp. nov., a new extremely thermophilic marine sulfur-metabolizing archaebacterium. Syst Appl Microbiol 12:257-262

Moore S, Stein WH (1963) Chromatographic determination of amino acids by the use of automatic recording equipment. Methods Enzymol 6:819-831

Moriya K, Inada T, Kyo M, Horikoshi K (1995) Large-scale fermentation under high hydrostatic pressure using a newly developed deep-sea baro/thermophilic collection and cultivation system. J Mar Biotechnol 2:175-177

Neu HC, Heppel LA (1965) The release of enzymes from *Escherichia coli* by osmotic shock and during the formation of spheroplasts. J Biol Chem 240:3685-3692

Neuer A, Jannasch HW, Belkin S, Stetter KO (1990) *Thermococcus litoralis* sp. nov.: a new species of extremely thermophilic marine archaebacteria. Arch Microbiol 153:205-207

Ohshima T, Nishida N (1993) Purification and properties of extremely thermostable glutamate dehydrogenases from two hyperthermophilic archaebacteria, *Pyrococcus woesei* and *Pyrococcus furiosus*. Biosci Biotechnol Biochem 57:945-951

Pace NR (1991) Origin of life—facing up to the physical setting. Cell 65:531-533

Pledger RJ, Crump BC, Baross JA (1994) A barophilic response by two hyperthermophilic, hydrothermal vent Archaea: an upward shift in the optimal temperature and acceleration of growth rate at supra-optimal temperatures by elevated pressure. FEMS Microbiol Ecol 14:233-242

Postgate JR (1977) Death in macrobes and microbes. In: Gray TRG, Postgate JR (eds) The survival of vegetative microbes. Cambridge University Press, Cambridge, pp1-19

Reysenbach A-L, Deming JM (1991) Effects of hydrostatic pressure on growth of hyperthermophilic archaebacteria from the Juan de Fuca Ridge. Appl Environ Microbiol 57:1271-1274

Robb FT, Park J-B, Adams MWW (1992) Characterization of an extremely thermostable glutamate dehydrogenase: a key enzyme in the primary metabolism of the hyperthermophilic archaebacterium, *Pyrococcus furiosus*. Biochim Biophys Acta 1120:267-272

Roszak DB, Colwell RR (1987) Survival strategies of bacteria in natural environments. Microbiol Rev 51:365-379

Schönheit P, Schäfer T (1995) Metabolism of hyperthermophiles. World J Microbiol Biotechnol 11:26-57

Somero GN (1992) Adaptations to high hydrostatic pressure. Annu Rev Physiol 54:557-577

Stetter KO (1986) Diversity of extremely thermophilic archaebacteria. In: Brock TD (ed) Thermophiles: general, molecular, and applied microbiology. Wiley, New York, pp39-74

Stetter KO (1995) Microbial life in hyperthermal environments: microorganisms from exotic environments continue to provide surprises about life's extremities. ASM News 61:285-290

Takai K, Inoue A, Horikoshi K (1999) *Themoaerobacter marianensis* gen. nov., sp. nov., an aerobic extremely thermophilic marine bacterium from the 11,000 m deep Mariana Trench. Int J Syst Bacteriol 49:619-628

Takami H, Inoue A, Fuji F, Horikoshi K (1997) Microbial flora in the deepest sea mud of the Mariana Trench. FEMS Microbiol Lett 152:279-285

Tunnicliffe V (1991) The biology of hydrothermal vents: ecology and evolution. Oceanogr Mar Biol Annu Rev 29:319-407

Wayne LG, Brenner DJ, Colwell RR, Grimont PAD, Kandler O, Krichevaky MI, Moore LH, Moore WEC, Murray RGE, Stackebrandt E, Starr MD, Truper HG (1987) Report of the ad hoc committee on reconciliation of approaches of bacterial systematics. Int J Syst Bacteriol 37:463-464

Woese CR (1987) Bacterial evolution. Microbiol Rev 51:221-271

Yu FT, McFeters GA (1994) Rapid in situ assessment of physiological activities in bacterial biofilms using fluorescent probes. J Microbiol Methods 20:1-10

Zeikus JG, Ben-Bassat, Hegge PW (1980) Microbiology of methanogenesis in thermal, volcanic environments. J Bacteriol 143:432-440

Zillig W, Holz I, Janekovic D, Schafer W, Reiter WD (1983) The archaebacterium *Thermococcus celer* represents a novel genus within the thermophilic branch of the archaebacteria. Syst Appl Microbiol 4:88-94

Zillig W (1992) The order Thermococcales. In: Balous A, Truper HG, Dworkin M, Harder W, Schelerfer K-H (eds) The prokaryotes, 2nd edn, vol 1. Springer-Verlag, New York, pp702-706

7 Deep-Sea Psychrophiles

Tetsuo Hamamoto

Psychrophilic and psychrotrophic bacteria in the deep-sea environment have been isolated and characterized. More than 100 psychrotrophic and 9 psychrophilic bacteria were isolated from deep-sea sediment samples. To seek possibilities to apply the isolates in biotechnology and to analyze their physiological adaptation mechanisms to low temperature, the enzymes they produce and their membrane fatty acid compositions were studied. Among a range of these bacteria, a series of strains that produce enzymes including amylases and lipases were found. Most of the examined enzymes showed significant retention of activity at low temperature and low thermostability. Analysis of total cell membrane fatty acid composition demonstrated a larger presence of unsaturated fatty acids than in their mesophilic counterparts. The growth temperature of the psychrophiles also affects the ratio of unsaturated fatty acids. The results illustrate their physiological adaptation to the permanently cold deep-sea environment.

7.1 Introduction

Bacteria capable of growing at low temperatures (0°C or lower) are defined as psychrophiles and psychrotrophs. Psychrophiles are bacteria that have an optimal temperature for growth lower than 15°C, a maximal temperature for growth lower than 20°C, and a minimal temperature for growth at 0°C or lower. Psychrotrophs are those that can grow at low temperatures but have higher optimal growth temperatures (Morita 1975). Technological advances in the field of marine technology dur-

Key words. Psychrophile, Psychrotroph, Amylase, Protease, Cold adaptation, Deep sea, Membrane fatty acid, Polyunsaturated fatty acid ing the 1990s have made it possible to study in detail characteristics of the deep sea and its floor (Myers and Anderson 1992). Because the temperature of deep-sea water is constant at about 3°C (Morita 1975; Jannasch and Taylor 1984; Austin 1988) isolation of psychrophilic and psychrotrophic bacteria from deep-sea mud samples was carried out. Recent progress in the research of psychrophiles has allowed those microorganisms to be exploited biotechnologically and has permitted a basic understanding of life at low temperatures (Russell 1998; Russell and Hamamoto 1998). This chapter describes characteristics of psychrophilic and psychrotrophic bacteria isolated from the deep-sea during the DEEPSTAR research program of the Japan Marine Science and Technology Center (JAMSTEC). The properties of their enzyme producion and the cellular fatty acid composition of the isolates were analyzed to observe their biological adaptation to cold temperatures.

7.2 Isolation and Characterization of Deep-Sea Psychrophiles and Psychrotrophs

A number of psychrotrophic and psychrophilic bacteria from collected deep-sea sediment samples were isolated in the DEEPSTAR program (Hamamoto and Horikoshi

Table 1. Properties of isolated representative psychrophiles and psychrotrophs

Strain	Origin	Depth (m)	G+C (%)	Growth characteristics at various temperatures (°C): Temperature/generation time (h)		
				Minimum temp. [a]	Optimum temp.	Maximum temp.
2311	Japan Trench	6500	44	–2.5 / 24	13 / 3	19 / 7
2312	Japan Trench	6500	45	–2.5 / 19	14 / 2	18 / 3
2313	Japan Trench	6500	43	–2.5 / 12	8 / 4.5	16 / 14
29-1	Sea of Japan	447	44	–2.5 / 30	8 / 6	20 / 32
29-6	Sea of Japan	447	42	≤ 5 / 16	7 / 4	9 / 18
81413	Sea of Japan	289	44	≤ 5 / 16	11 / 3.5	19 / 26
814-4	Sea of Japan	289	41	≤ 5 / 26	13-16 / 2.5	23 / 8
3021	Sagami Bay	1160	ND	≤ 5 / 52	13-15 / 6	17 / 8.5
4-3	Sagami Bay	1160	45	0 / 24	22.5 / 1	22.5 / 1
5405	Suruga Bay	2625	ND	0 / 12	10 / 2	18 / 6
5501	Suruga Bay	2485	ND	0 / 10	18 / 6	20 / 6
5502	Suruga Bay	2485	ND	0 / 10	13 / 4	18 / 12
5710	Suruga Bay	2220	ND	≤ 5 / 34	4-8 / 6	11 / 10

ND, not determined
All the strains are irregular rods, motile, glucose-fermentative, catalase-positive, oxidase-positive, O-129 sensitive; they require 0.2 M NaCl but not seawater.
[a] Generation times at –5°C are shown when grown at –5°C.

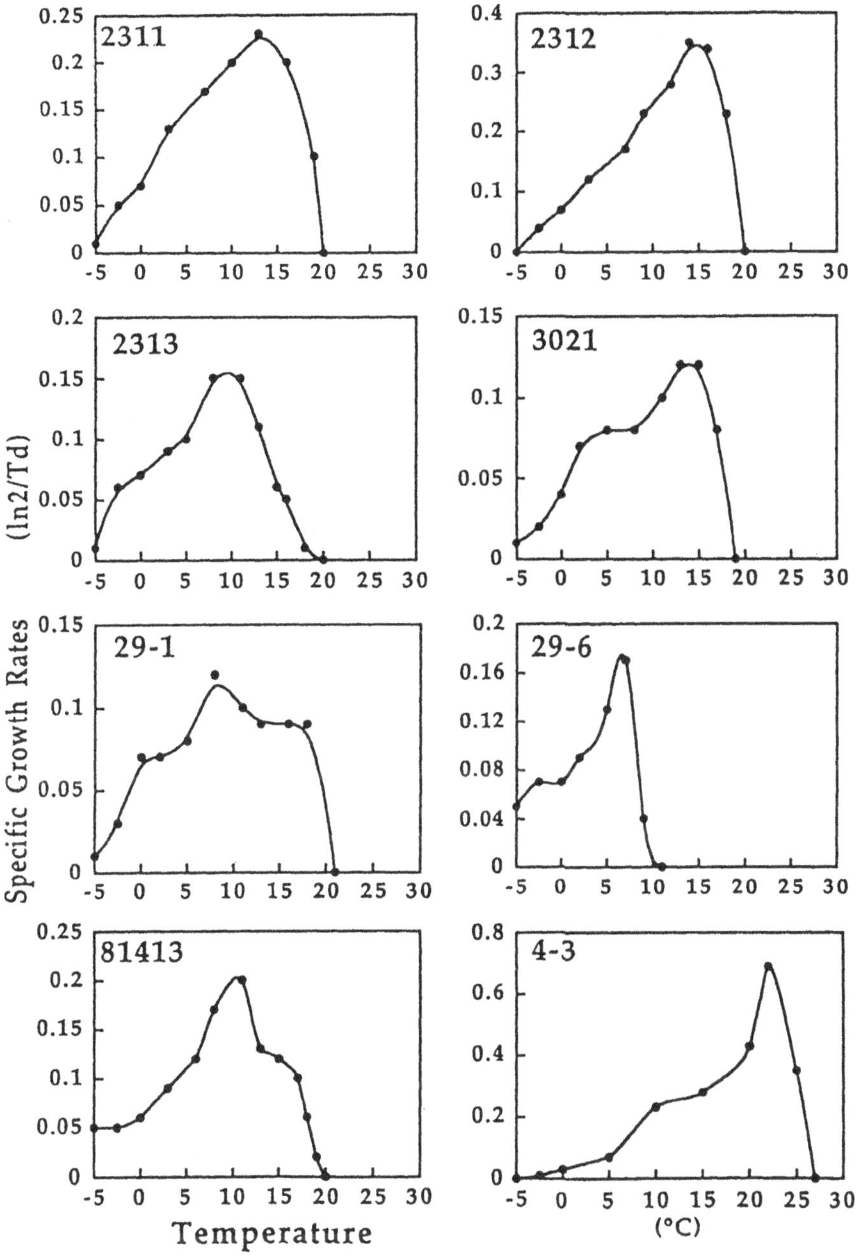

Fig. 1. Effect of temperature on the growth of the psychrophilic and psychrotrophic bacteria isolated from sea sediment samples. The growth at various temperatures was expressed as the specific growth rate, *ln2*/generation time (*Tcl*) (From Hamamoto and Horikoshi 1993, with permission)

1993). Deep-sea sediment samples were collected when it was possible to collect bottom mud samples during such dives. The isolates shown in Table 1 are those isolated from the samples collected during a series of dives by the Shinkai 2000 and the Shinkai 6500 submersibles (Table 1). Sea-sediment samples were collected during dives at depths of 6500 m in the Japan Trench (40°07'N, 144°12'E; water temperature 1.6°C), 1160 m in Sagami Bay (35°00'N, 139°14'E, water temperature 2.8°C), and 289 and 447 m in the Sea of Japan (35°48'N, 132°22'E, water temperature not measured; and 43°34'N, 139°33'E, water temperature 0.5°C, respectively) during 1990 and 1991.

Growth rates were measured by counting cell numbers of cultures at various temperatures in liquid LB medium containing 3% NaCl (1% Difco tryptone, 0.5% Difco yeast extract, 3% NaCl). The results are shown in Fig. 1 as specific growth rates at various temperatures. Strains 2311, 2312, 2313, 3021, 81413, and 29-6 were psychrophiles; strains 4-3 and 29-1 were considered psychrotrophs owing their maximum growth temperatures of 25° and 21°C, respectively, which are higher than that for psychrophiles (20°C). Because all the isolates exhibited growth capability at -2.5°C, they were considered adapted to their marine habitat, which is a typical permanently low-temperature environment with temperatures at 3°C or lower. The deep-sea bacteria were able to grow efficiently at low temperatures, illustrating their physiological adaptation to the permanently cold deep-sea environment.

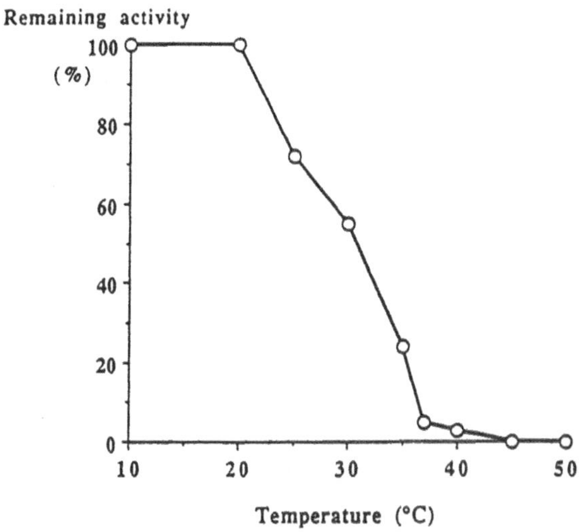

Fig. 2. Effect of temperature on the stability of the amylase from *Vibrio* sp. strain 4-3. The enzyme reaction mixture without substrate was incubated for 1 h at various temperatures. The reaction mixtures were then added to substrate (soluble starch) solution and incubated at 30°C for 10 min. Activity was measured by the amount of reducing sugar released. (From Hamamoto and Horikoshi 1991, with permission)

7.3 Enzyme Production and Activity

7.3.1 Amylases

Two amylases from psychrotrophic bacteria isolated from sediment samples have been studiedin the DEEPSTAR research project. Amylases were purified from culture supernatants by combinations of column chromatography. Amylase activity was measured by the amount of reducing sugar formed from soluble starch by the Somogyi-Nelson method (Hamamoto and Horikoshi 1991). The amylases were purified from the culture supernatants of two psychrotrophic isolates, *Vibrio* spp. strains 4-3

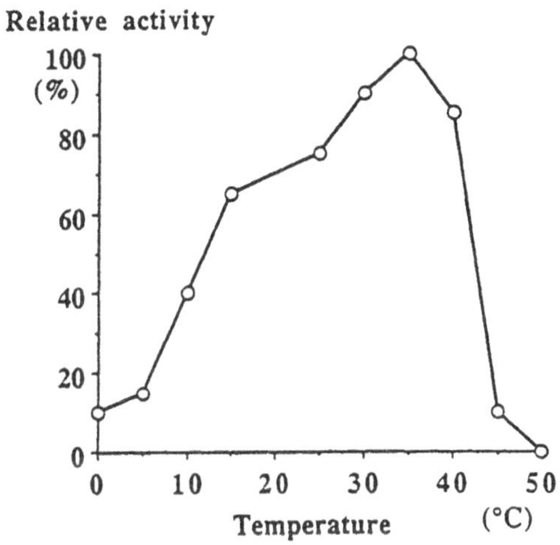

Fig. 3. Effect of temperature on the activity of the amylase from *Vibrio* sp. strain 4-3.
The enzyme reaction mixture without substrate was preincubated for 30 min at various temperatures. Substrate (soluble starch) solution was then added to the reaction mixtures and incubated for 5 min. Activity was measured by the amount of reducing sugar released. (From Hamamoto and Horikoshi 1991, with permission)

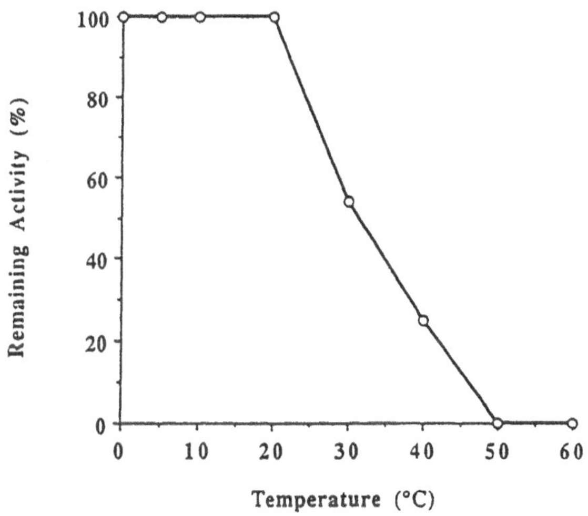

Fig. 4. Effect of temperature on the stability of the amylase from *Vibrio* sp. strain 814-4
Amylase activity was measured by the same method as that described in Fig. 3. (From Hamamoto and Horikoshi 1993, with permission)

(Hamamoto and Horikoshi 1991) and 814-4 (Hamamoto and Horikoshi 1993). Strain 4-3, collected at Sagami Bay, was found to produce an amylase with a molecular mass estimated to be 70 kDa. The amylase activity of purified enzyme was destroyed after 1 h of incubation at 45°C, and inactivation was observed at temperatures as low as 25°C after 1 h of incubation (Fig. 2). The effect of temperature on amylase activity was examined after a 10-min incubation with the purified amylase and is shown in Fig. 3. The amylase showed maximum activity at 35°C, retaining about 40% and 10% of maximal activity at 10° and 0°C, respectively.

Another amylase with a molecular weight of 45 kDa was purified from the culture supernatant of a psychrotroph, identified as *Vibrio* sp. strain 814-4, isolated from a marine sediment sample collected in the Sea of Japan. The amylase hydrolyzed amylopectin and soluble starch but did not hydrolyze cyclodextrins or pullulan. The activity of purified enzyme was destroyed after a 30 -min incubation at 50°C, and inactivation was observed at a temperature as low as 30°C after a 30-min incubation (Fig. 4). The effect of temperature on the purified amylase activity is

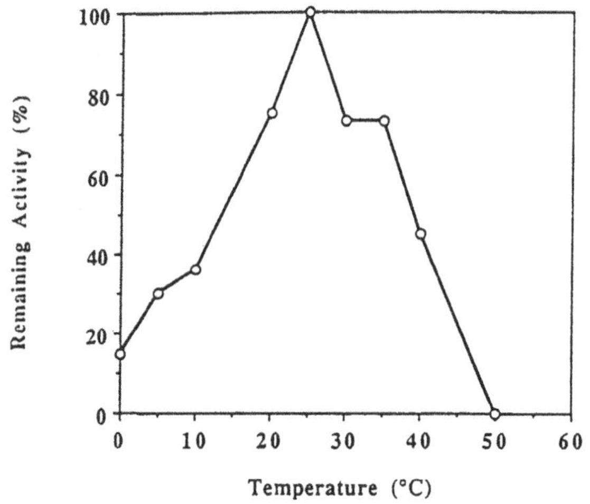

Fig. 5. Effect of temperature on the activity of the amylase from *Vibrio* sp. strain 814-4
Amylase activity was measured by the same method as that described in Fig. 4. (From Hamamoto and Horikoshi 1993, with permission)

Table 2. Characteristics of amylases from deep-sea psychrotrophic bacteria

Characteristics	No. 4-3	No. 814-4
Molecular mass	66 kDa	45 kDa
Optimum pH	6.0-8.5	6.0-7.8
Activity was lost completely	45°C, 1 h	50°C, 30 min
Optimum temperature	35°C	25°C
Activity retained (% of maximum)		
At 10°C	40%	35%
At 0°C	10%	15%

All the characteristics were observed with purified enzymes.

shown in Fig. 5. The amylase had a maximum activity at 25°C. About 35% of the activity was observed at 10°C, and 15% of the optimal activity remained at 0°C.

The retention of significant activity at low temperatures and the low thermostability of these amylases are similar to the properties of other enzymes isolated from psychrotrophic microorganisms (Nakajima et al. 1974, Mitchell et al. 1985), suggesting that these enzymes may provide useful tools for biochemical study of thermolability, and it may be possible to apply these enzymes when processing starch at low temperatures. The characteristics of these amylases are summarized in Table 2.

7.3.2 Lipases

Lipase activities produced in culture supernatants were measured according to the method of Heymann and Mentlein (1981) with *p*-nitrophenyl (*p*-NP)laurate as a substrate, and the increase of hydrolyzed *p*-NP was measured by the increase in absorbance at 405 nm. For reactions containing *n*-hexane (50%, v/v), vigorously shaken reaction mixtures were centrifuged, and the absorbance of the lower aqueous phase was measured at 405 nm. Four strains produced *p*-NP laurate-hydrolyzing enzymes in culture media. The four activities were completely lost after a 10-min incubation at 40°C.

As these activities could be observed in the presence of *n*-hexane, the effects of temperature on activities in aqueous and *n*-hexane-containing reaction mixtures were compared (Table 3). At -10°C, enzyme reactions occurred at rates of 16%-37% of those at the optimal temperatures in the *n*-hexane-containing reaction mixtures. The

Table 3. Effect of temperature on enzymatic activities of culture supernatants of lipase-producing strains

Strain	Activity (%)							
	-10°C	0°C	10°C	20°C	30°C	40°C	50°C	60°C
5405								
a	ND	ND	ND	66	100	53	21	9
h	37	49	98	100	85	34	15	ND
5501								
a	ND	ND	ND	60	100	80	53	7
h	18	39	100	68	64	18	11	ND
5502								
a	ND	ND	ND	36	100	48	40	0
h	16	68	95	100	68	21	0	ND
5710								
a	ND	ND	ND	27	100	81	65	0
h	28	44	72	100	83	61	44	ND

Residual activities were calculated taking the activity at the optimum temperature as 100%.
a, activity in aqueous reaction mixture; h, activity in *n*-hexane/water (50%, v/v) reaction mixtures; ND, not done.

optimal temperatures for enzyme activity for all four strains decreased by 10°–20°C when *n*-hexane was present in the reaction mixtures. However, only 1.5%, 3.7%, 2.3%, and 6.9% of the activities obtained in the aqueous reaction mixtures were observed in the presence of *n*-hexane for strains 5401, 5501, 5502, and 5710, respectively. Enzymes excreted by these bacteria displayed low optimal temperature and low thermal stability, which are the main characteristics of cold-adapted enzymes. Particularly, lipase in reactions containing *n*-hexane showed activity at temperatures as low as -10°C. Although the low-temperature optima and the low thermostability reported here were observed only with six enzymes, characterization of more enzymes from psychrophilic and psychrotrophic bacteria would provide useful tools for the biochemical study of psychrophily. It may also be possible to apply these enzymes in various chemical reactions at low temperature.

7.3.3 Protease

A psychrotrophic bacterium identified as *Vibrio* sp. strain 5709, which showed a maximum growth temperature of 20°C and were able to grow at 0°C, produced a protease of molecular weight 35 000 daltons, was stable in the pH range 5.0-9.0, and worked optimally at pH 8.5-9.0 (Hamamoto et al. 1995a). Protease activity was observed in culture supernatants of *Vibrio* sp. strain 5709 in liquid LB medium after 5 days at 5°C. The protease was precipitated with ammonium sulfate (50%-80% (w/v) saturation), and the precipitate was dialyzed against 20 mM Tris-HCl buffer pH 8.5. The dialyzate was subjected to diethylaminoethyl (DEAE)-Toyopearl 650M (Toyo Soda Co.) column chromatography equilibrated with 20 mM Tris-HCl buffer pH 8.5. Protease was eluted with an NaCl gradient (0-500 mM) at an NaCl concentration of 100 mM. A single active peak was concentrated and then purified further using Mono Q chromatography with an NaCl gradient (0-500 mM) at an NaCl concentration of 250 mM. The purity of the enzyme was confirmed with Superdex 200 HR 10/30 (Pharmacia) gel filtration chromatography. The homogeneity of the activity was confirmed also with sodium dodecyl sulfate polyacrylamide gel electrophoresis (SDS-PAGE). The results of the purification are summarized in Table 4. The overall

Table 4. Purification of the protease

Substrate	Volume (ml)	Total protein (mg)	Total activity (kU)	Specific activity (U/mg protein)	Recovery (%)
Culture supernatant	1880	2124.0	663	313	100
Ammonium sulfate ppt.	14	43.0	449	10 491	68
DEAE-Toyopearl	156	11.0	233	20 821	35
MonoQ	18	1.4	72	51 214	11
Superdex 200	28	1.0	56	56 000	8

DEAE, diethylaminoethyl.

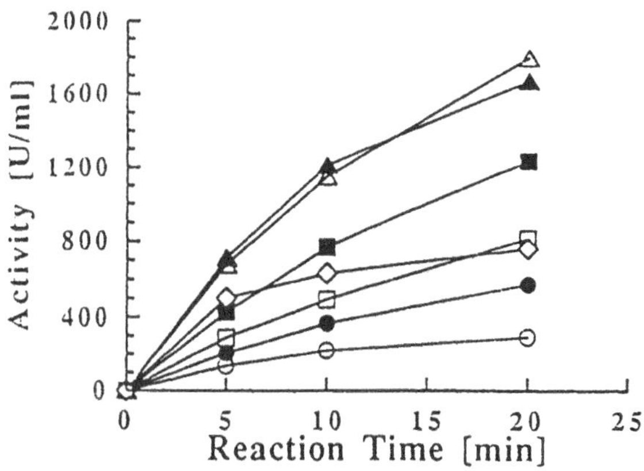

Fig. 6. Effect of temperature on protease activity. Protease activities are shown for 20 min at 0°C (*open circles*), 10°C (*closed circles*), 20°C (*open squares*), 30°C (*closed squares*), 40°C (*open triangles*), 45°C (*closed triangles*) and 50°C (*open diamonds*). (From Hamamoto et al. 1995a, with permission)

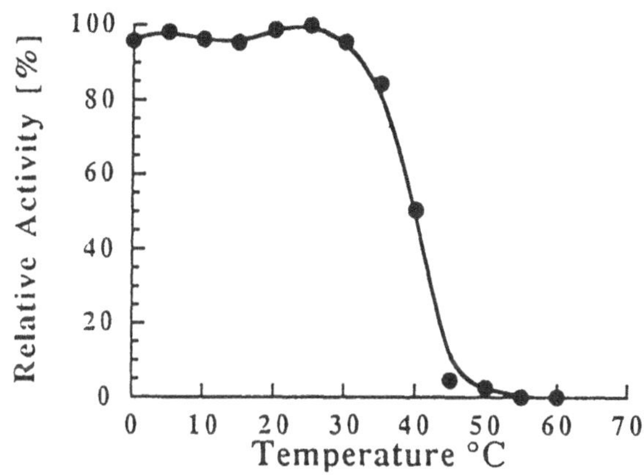

Fig. 7. Effect of temperature on the stability of protease. The protease activity was measured after incubation for 20 min at the indicated temperature. Activity is expressed relative to the activity after incubation at 25°C (100%). (From Hamamoto et al. 1995a, with permission)

recovery was 8.5% with a purification factor of 178. The molecular weight of the enzyme deduced by the SDS-PAGE and Superdex 200 gel filtration chromatography was 3.5×10^4 daltons. Activity was optimal at 40°C and was completely lost after 20 min incubation at 45°C. Considerable activity was retained at low temperatures: 16%, 32% and 45% of the maximum activity at 0°, 10°, and 20°C, respectively. Protease activity was measured by the amount of tyrosine residue hydrolyzed and released from casein measured by a modified Folin method (Takami et al. 1990). The concentration of protein was measured with the BCA protein assay (Pierce). One unit of activity was defined as the activity that released 1 mg of tyrosine from casein per minute.

The effect of temperature on the activity of the protease is shown in Fig. 6. The enzyme had an optimal temperature for activity at around 40°–45°C. At 0°, 10°, and 20°C the enzyme maintained 16%, 32%, and 45%, respectively, of its activity at optimum temperature. Little activity was observed after 20 min at 45°C due to thermal inactivation (Fig. 7).

The optimal pH for the activity was 8–9 (Fig. 8), and the enzyme was stable between pH 5–9 after 20 min incubation (Fig. 9). The activity was affected by the presence of 10 mM phenylmethylsulfonylfluoride (PMSF) but not by 10 mM ethylendiaminetetraacetic acid (EDTA) (Table 5). The results suggest that this enzyme is an alkaline serine protease.

The activity was inhibited by 10 mM Zn^{2+} (13.1%) and Cd^{2+} (5.4%), and it was affected by the presence of 1 mM Hg^{2+} (37.1%). The protease showed the characteristics of a typical psychrotrophic enzyme.

The enzyme's optimal temperature for activity was 40°C, and 45% of the optimal activity was maintained at 20°C. Similar optimum activity at low temperatures

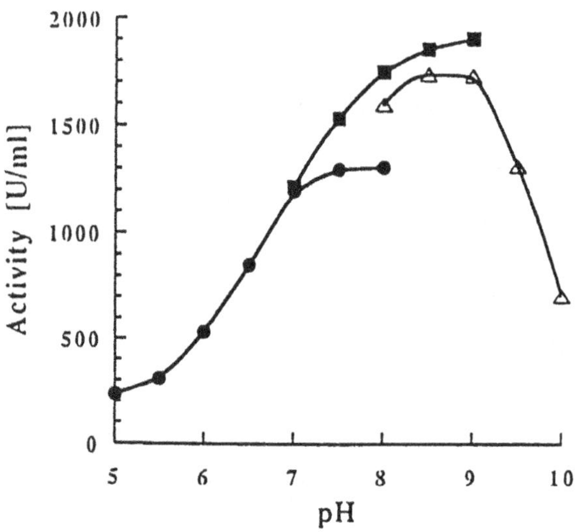

Fig. 8. Effect of pH on the protease activity.
The effect of pH on the activity of protease was measured with Tris-HCl (*circles*), glycine-NaCl-NaOH (*triangles*), and phosphate buffer (*circles*). (From Hamamoto et al. 1995a, with permission)

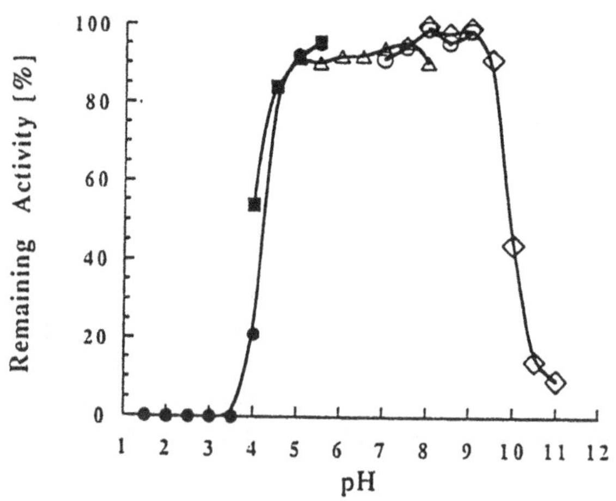

Fig. 9. Effect of pH on the stability of protease. Protease activity was measured at pH 8.5 after incubation for 20 min at the indicated pH levels with sodium acetate-HCl (*filled circles*); acetate-sodium acetate (*closed squares*); phosphate (*triangles*); Tris-HCl (*open circles*); and glycine-NaCl-NaOH buffer (*diamond*). (From Hamamoto et al. 1995b, with permission)

Table 5. Effects of metal ions and substances on protease activity

Ions/Substance	Concentration (mM)	Remaining activity (%)
Cd^{2+}	10	5.4
Cu^{2+}	10	51.2
Sr^{2+}	10	68.5
Fe^{3+}	10	38.0
Mg^{2+}	10	74.6
Ba^{2+}	10	58.2
Zn^{2+}	10	13.1
Hg^{2+}	1	37.1
Urea	800	124.9
IAA	10	98.8
PMSF	10	0.5
EDTA	10	101.3

IAA, iodoacetic acid; PMSF, phenylmethylsulfonylfluoride.

were observed with other proteases from psychrophiles and psychrotrophs (Nakajima et al. 1974; Margesin and Schinner 1991; Ray et al. 1992; Hamamoto et al. 1994a). The temperature profiles of psychrophiles and psychrotrophs are of scientific interest. In addition, their enzymes, like the enzyme presented here, are new tools for biotechnological applications in various fields (Gounot 1991).

7.4 Low Temperature Adaptation

7.4.1 Membrane Fatty Acid Composition

Physiological adaptations of these psychrophilic and psychrotrophic microorganisms are thought to be regulated by diverse biochemical characteristics (Gounot 1991). In particular, adaptations of metabolic catalysts, membrane transport systems, and membrane fluidity are known to play key roles in psychrophily (Morita 1975; Russell 1984). To analyze the membrane fluidity maintenance mechanism, the cellular fatty acid composition of the psychrophilic *Vibrio* sp. strain 5710 isolated from a deep-sea sediment sample was analyzed (Hamamoto et al. 1994a).

The change in the fatty acid composition by the growth was examined using a psychrophile isolated from a deep-sea sediment sample that could not grow above 11°C, designated strain 5710. The bacterium was grown in 10 ml of modified LB medium containing 3% NaCl in 24 mm diameter test tubes; the tubes were shaken constantly at 100 rpm in waterbaths filled with 20%-50% anticoolant in water (v/v) when necessary. Cell growth was monitored by the increase of optical density at 660 nm. In temperature shift experiments, cultures grown to the mid-log phase (OD at 660 was 0.4-0.5) were transferred directly to a waterbath set at the designated shift temperature. The growth of the psychrophile *Vibrio* sp. strain 5710 was monitored at

Table 6. Total cellular fatty acid compositions of strain 5710 grown at various temperatures

Fatty acid type	Fatty acid composition (mean % ± SD) [a]							
	–5°C		0°C		5°C		10°C	
	Log (0.3)	Sta (0.9)	Log (0.3)	Sta (0.8)	Log (0.4)	Sta (1.0)	Log (0.3)	Sta (0.9)
14:0	12.1 ± 0.4	10.6 ± 1.8	13.3 ± 0.4	12.6 ± 0.4	14.1 ± 0.5	19.5 ± 0.1	16.1 ± 0.8	11.3 ± 1.2
14:1t	11.2 ± 0.5	8.0 ± 2.4	9.3 ± 0.5	8.1 ± 0.6	9.0 ± 0.5	5.1 ± 0.1	11.0 ± 0.5	6.8 ± 0.9
16:0	4.7 ± 0.2	9.8 ± 3.3	5.7 ± 0.3	8.9 ± 1.3	8.0 ± 0.3	11.9 ± 0.2	7.1 ± 0.2	12.0 ± 0.7
16:1t	7.2 ± 1.9	6.4 ± 0.4	9.1 ± 0.1	7.9 ± 0.4	7.3 ± 0.1	6.7 ± 1.4	8.3 ± 0.2	8.2 ± 0.3
16:1c	36.5 ± 1.1	38.6 ± 1.7	32.5 ± 1.0	38.9 ± 0.5	33.3 ± 0.7	40.9 ± 1.7	33.7 ± 0.2	39.4 ± 1.3
17:0cyclo	4.0 ± 0.9	2.4 ± 0.2	4.9 ± 0.1	3.5 ± 0.1	4.0 ± 0.7	4.0 ± 0.5	3.5 ± 0.1	3.8 ± 0.0
18:1	0.8 ± 0.0	3.0 ± 1.6	0.5 ± 0.5	1.4 ± 0.3	2.9 ± 0.2	3.0 ± 0.0	0.4 ± 0.7	2.6 ± 0.6
18:2	0.8 ± 0.1	0.5 ± 0.1	1.3 ± 0.2	0.6 ± 0.0	0.8 ± 0.1	0.6 ± 0.1	0.8 ± 0.7	0.7 ± 0.0
18:3	2.1 ± 0.3	1.4 ± 0.1	2.7 ± 0.1	2.0 ± 0.0	2.5 ± 0.2	1.8 ± 0.2	1.6 ± 0.1	1.4 ± 0.1
22:6	16.4 ± 1.2	16.2 ± 1.1	15.1 ± 1.8	12.8 ± 0.5	12.8 ± 0.1	13.6 ± 0.2	11.7 ± 0.8	10.5 ± 0.1

From Hamamoto et al. 1994b, with permission.

[a] Log, cellular fatty acid composition at mid-log phase; Sta, of early-stationary phase. The numbers in parentheses indicate the optical density of the culture at the time of collection.

various temperatures. Strain 5710 showed no growth above 13°C and had its optimum growth temperature at 4° or 8°C (Table 1). Growth was observed even at -5°C, the lowest temperature the growth could be examined. Generation times were 35, 9, 6, and 10 h at -5°, 0°, 5°, and 10°C. The maximum growth density between 0° and 8°C was 1.0–1.2 in turbidity measured at 660 nm. At -5°C the maximum cell density was 0.35 after 87 h of cultivation.

For extraction of total cellular fatty acids, cells were harvested by centrifugation at 5000 *g* for 10 min, freeze-dried, and stored at -80°C. Total lipids from 20–50 mg of the freeze-dried whole cells were extracted and acid-esterified. The resulting methyl esters were extracted and concentrated. Samples were taken from three cultures. The extracted fatty acid methyl esters (FAMEs) were analyzed using gas-liquid chromatography (GLC). The FAMEs separated by GLC were identified by comparing their retention times with those of authentic standards, except 17:0cyclo, which was confirmed with a FAME sample from *Pseudomonas fluorescens* (obtained in our laboratory), which was known to contain 17:0cyclo. Cultures grown to their mid-log (optical density at 660 nm: 0.35-0.55) and early stationary phases at -5°, 0°, 5°, and 10°C were used to determine the bacterial total cellular fatty acid composition (Table 6). The overall cellular fatty acid composition matched those of psychrophilic *Vibrio* species previously investigated. The cells at both growth stages at all temperatures contained 9-*cis*-hexadecenoic acid (16:1c, palmitoleic acid) as a major fatty acid along with tetradecanoic acid (14:0, myristic acid), 9-*trans*-tetradecenoic acid (14:1t, myristelaidic acid), hexadecanoic acid (16:0, palmitic acid), 9-*trans*-hexadecenoic acid (16:1t, palmitelaidic acid), cyclopropane hexadecanoic acid (17:0cyclo), octadecanoic acid (18:0, stearic acid), 9-*cis*-octadecenoic acid (18:1, oleic acid), 9,12-*cis*,*cis*-octadienoic acid (18:2, linoleic acid), 9, 12,15-all *cis*-octatrienoic acid (18:3, linolenic acid), and 4,7,10,13,16,19-all *cis*-docosahexaenoic acid (22:6, DHA).

Culture temperature and growth phase affected the fatty acid composition. Among the detected FAMEs, the most significant difference caused by different culture temperatures was observed in the amount of 22:6 in the cells during both log and stationary phases. In the cells at the mid-log phase, those grown at -5°C contained 16.4% of 22:6 as a proportion of the total cellular fatty acids, whereas the amount decreased as the growth temperature increased and the cells grown at 10°C contained only 11.7. The same tendency was observed with the FAME compositions obtained from cells during the early-stationary phase. In contrast, higher proportions of 14:0 and 16:0 were observed at the higher growth temperatures during both the mid-log phase and the early stationary cells, except that the cells grown at 10°C contained a smaller proportion of 14:0 than those in the 0°C and 5°C cultured cells. Other FAME contents were not affected significantly by the growth temperature. There was also a difference in the FAME caused by the growth phases. The proportions of 16:0, 16:1c, and 18:1 increased from that in the mid-log phase cells to that in cells at the early stationary phase. The increase was not affected by the growth temperature. In contrast, 14:1t, 18:2, and 18:3 decreased in proportion as the cultivation proceeded. The psychrophilic *Vibrio* sp. strain 5710 was shown to have a fatty acid composition typical of marine *Vibrio* species, except that strain 5710 contained 22:6 as a polyun-

Table 7. Effect of temperature shifts on cellular fatty acid composition of 5710

Fatty acid type	Fatty acid composition with initial culture temperature					
	0°C			10°C		
	Log [a]	Sta [b]	Shift from 0° to 10°C [c]	Log	Sta	Shift from 10° to 0°C
14:0	13.3	12.6	13.5	16.1	11.3	11.9
16:0	5.7	8.9	10.5	7.1	12.0	10.4
22:6	15.1	12.8	12.7	11.7	10.5	14.0

From Hamamoto et al. 1994b, with permission.
Cellular fatty acid compositions at [a] mid-log phase, [b] early-stationary phase, and [c] early stationary phase in the culture with a temperature shift.

saturated fatty acid (PUFA). It has been demonstrated that 22:6 increases as a function of decreasing temperature in 5710, which suggests that regulation of PUFA ratios play an important role in the adaptation of the membrane to growth temperature in psychrophilic bacteria.

Moreover, an effect of temperature shift on the proportion of FAMEs was observed (Table 7). With a temperature shift from 10°C to 0°C, the contents of 22:6 exhibited a significant increase, whereas little difference was observed following a shift from 0°C to 10°C. In contrast, a decrease in 16:0 was observed after a shift from 10°C to 0°C, and an increase in the 16:0 content occurred after a temperature shift from 0°C to 10°C. The increase of 16:0 after a temperature shift from 0°C to 10°C was 1.6% compared with that of nonshifted cells, whereas the shift from 10°C to 0°C produced a decrease of 1.6%. The proportions of other FAMEs were slightly increased or decreased by a temperature shift, although the changes were in the same direction.

In summary, a temperature shift from 10°C to 0°C caused an increase in 22:6 content with no significant increase in other FAMEs, supporting the major adaptation involving the PUFA. In addition, the amounts of 14:0 and 16:0 were higher in cells grown at higher temperatures, suggesting that the cells grown at higher temperature adapted to the growth temperature by increasing their amounts of saturated fatty acids. In the case of cells grown at 10°C, which was close to the maximum growth temperature for strain 5710, the 14:0 content was not as high as expected. This finding was confirmed by two separate experiments. It may be another adaptation mechanism to the high temperature that led the cells to decrease short chain fatty acids to maintain the fluidity of the membrane. However, there were no significant temperature-dependent changes in relative proportion with 16:1*trans* and 16:1*cis* in strain 5710, which is therefore considered to be less important in the temperature adaptation of this organism. The cellular fatty acid composition of strain 5710 showed, on one hand, the existence of PUFAs as a characteristic of marine psychrophiles; on the other hand, the changes in the FAME composition by growth temperatures were smaller than in previously investigated psychrophiles (DeLong and Yayanos 1986; Fukunaga and Russell 1990). The smaller difference in strain 5710 may be due to the narrower range of the growth-permissible temperature. Alternatively, the narrow growth temperature range might be due to a reduced ability to control membrane fluidity.

A change in FAME composition was observed between the mid-log phase and the early stationary phase. Starvation of a psychrophilic marine *Vibrio* induced qualitative and quantitative changes in fatty acids (Oliver and Stringer 1984). Likewise, the major fatty acid, 16:1c, in strain 5710 grown at 5°C increased from 33.3% to 40.9%. The change suggested that not only severe starvation stress but also weak starvation caused by consumption of some specific nutrient(s) through the batch growth cycle could affect the cellular fatty acid composition. The significant increase of the major component might be a preparation mechanism for the bacterium until the next encounter with a nutrient-rich condition. In the natural low-nutrient marine environment, it would be in the starved or near-starved condition most of the time. It might therefore be important for marine psychrophiles to have the capacity to prepare rapidly for the opportunity to flourish whenever possible in the natural environment.

7.4.2 Analysis of PUFA Presence Among Psychrophiles

The most extreme and interesting characteristics of low-temperature adaptation in psychrophiles are the presence of PUFAs, including eicosapentaenoic acid (EPA, 20:5) and docosahexaenoic acid (DHA, 22:6). Therefore the cellular fatty acid compositions of marine psychrophilic and psychrotrophic vibrios isolated from deep-sea sediment samples and the correlation between bacterial fatty acid compositions and cardinal growth temperatures were analyzed (Hamamoto et al. 1995b).

Cell growth was monitored by the increase of optical density at 660 nm. Other characteristics of the isolates, DNA G+C contents, sensitivity to the *Vibrio*-stat reagent (O-129), and biochemical tests were performed as previously described (Hamamoto and Horikoshi 1991). Production of proteases and amylases was observed by the halo-forming ability of the colonies on the plates containing 0.1% casein and 0.1% starch azure in the modified LB agar medium at 5°C, respectively. Growth was also observed on a minimal medium containing sodium citrate 0.3%; sodium glutamate 0.2%; NaCl 2.5%; K_2HPO_4 1.4%; KH_2PO_4 0.4%; $MgSO_47H_2O$ 0.02%; $FeCl_3$ 0.0005%; agar 2.0%, and a trace element mixture (1 mg/l each of $ZnSO_4$, $CuSO_4$, $MnCl_2$, $CoSO_4$, and $NiSO_4$ at final concentration).

The cardinal temperatures for the growth of the marine vibrios were monitored by the increase of turbidity in the modified LB liquid medium (Table 8). All the strains were sensitive to *Vibrio*-stat reagent O-129; they were catalase-positive, oxidase-positive, gram-negative motile *Vibrio*-like rods. The G+C contents of the total DNA of seven strains were 41-45 mol%, which coincided with those of other known vibrios. Ten strains produced amylase, and one produced a protease activity in agar medium. Four isolates grew on the minimum medium. These characteristics were used to discriminate each isolate. Strains 5710 and 2311 showed no growth at 20°C. Sixteen strains showed their maximum growth temperature at 20°C, and three strains could grow up to 25°C. Four other strains could grow at higher temperatures (up to 30°C for one strain and up to 37°C for three strains.

Cultures grown to their mid-log phase were used to determine of bacterial total cellular fatty acid compositions (Table 9). All the isolates contained 9-*cis*-hexadecenoic acid (16:1c, palmitoleic acid) as a major fatty acid together with dodecanoic acid (12:0, lauric acid), tetradecanoic acid (14:0, myristic acid), 9-*trans*-tetradecenoic acid (14:1t, myristelaidic acid), hexadecanoic acid (16:0, palmitic acid), cyclopropane heptadecanoic acid (17:0cyclo), octadecanoic acid (18:0, stearic acid), and 9-

Table 8. Characteristics of the 22 isolates

Strain	Origin	G+C content	Maximum growth temperature (°C)	Growth on MM	Exoenzyme Production	Presence of PUFA [a] EPA	DHA
Group I							
5710	Suruga Bay (2220 m)	NT	15	+		–	22.7 [c]
2311	Japan Trench (6500 m)	44	15	–		–	17.4
2312	Japan Trench (6500 m)	45	20	–		–	15.8
2313	Japan Trench (6500m)	43	20	–	Amylase	–	11.8
3021	Sagami Bay (1160 m)	NT	20	–	Amylase	–	10.3
5401	Suruga Bay (2625 m)	NT	20	+		–	8.1
5403	Suruga Bay (2625 m)	NT	20	–		–	9.0
5501	Suruga Bay (2485 m)	NT	20	–	Amylase	–	17.0
5601	Suruga Bay (2355 m)	41	20	–	Amylase	–	16.6
5605	Suruga Bay (2355 m)	NT	20	–		–	7.4
5703	Suruga Bay (2220 m)	NT	20	–	Amylase	–	18.6
5705	Suruga Bay (2220 m)	NT	20	–		–	21.5
Group II							
29-1	Sea of Japan (447 m)	44	20	–	Amylase	19.7	–
814-4	Sea of Japan (289 m)	41	20	–	Amylase	17.9	–
5709	Suruga Bay (2220 m)	NT	20	–	Protease	14.2	–
4-3	Sagami Bay (1160 m)	45 [b]	25	–	Amylase	17.9	–
5402	Suruga Bay (2625 m)	NT	25	+	Amylase	17.7	–
Group III							
5602	Suruga Bay (2355 m)	NT	25	–	Amylase	–	–
5502	Suruga Bay (2485 m)	NT	30	–		–	–
6001	Ryukyu Trench (5425m)	NT	37	–		–	–
6102	Ryukyu Trench (5110m)	NT	37	–		–	–
6301	Ryukyu Trench (5110m)	NT	37	+		–	–

From Hamamoto et al. 1995a, with permission.

NT, not tested; MM, minimal mediam; EPA, eicosapentenoic acid; DHA, docosahexaenoic acid.

[a] Numbers indicated molecular weight percent of extracted fatty acid methyl esters (FAMEs) of each poly unsaturatel fatty acid (PUFA) in total FAMEs.

[b] Data are from Hamamoto and Horikoshi 1993

[c] Data are from Hamamoto and Horikoshi 1991

cis-octadecenoic acid (18:1, oleic acid). Except for the two strains with the highest growth maximum temperatures (6102 and 6301), all of the isolates contained octadecatrienoic acid (18:3, positions of double bonds were not determined). Twelve isolates that showed the lowest maximum growth temperatures (group I) also contained 9-*cis*-tetradecenoic acid (14:1, myristoleic acid), 9-*trans*-hexadecenoic acid (16:1t, palmitelaidic acid), and 4,7,10,13,16,19-all *cis*-docosahexaenoic acid (22:6, DHA). Five isolates with slightly higher maximum growth temperatures (20°-25°C) consisted of a group of bacteria (group II) containing 5,8,11,14,17-all *cis*-eicosapentaenoic acid (20:5, EPA). Five isolates with the highest maximum growth temperatures (25°-37°C) contained no PUFAs as their cellular fatty acids (group III).

The above results show that psychrophilic and psychrotolerant vibrios possess fatty acid compositions typical of marine *Vibrio* species, except that those with lower cardinal temperatures for growth contain PUFAs. The presence of PUFAs in

Table 9. Fatty acid compositions of the isolates

Strain	12:0	14:0	14:1t	14:1c	16:0	16:1t	16:1c	17:0 cyc.	18:1	18:2	18:3	20:5	22:6
Group I													
5710 [a,b]		16.1	11.0		7.1	8.3	33.7	3.5	0.4	0.8	1.6		11.7
2311 [a]	0.6	11.6	6.2	0.6	10.0	9.1	32.9	2.9	1.3		2.4		17.4
2312	0.6	12.8	6.0	0.8	7.2	7.4	36.5	3.2	0.3		1.8		15.8
2313	1.5	15.1	4.2	tr	16.4	6.9	40.1	2.3	1.8				11.8
3021	1.5	15.9	6.0	4.7	13.4	11.9	35.3	2.8	1.6		0.7		10.3
5401	4.1	15.0	13.1	1.0	5.0	7.1	33.9	3.0	0.6		1.5		8.1
5403	3.3	13.8	8.6	1.4	6.6	9.9	38.6	3.6	1.5		1.5		9.8
5501	0.4	7.5	3.4	0.5	10.3	6.1	36.6	3.8	4.4		1.6		17.0
5601	0.3	8.8	3.7	0.6	11.3	6.8	39.3	2.6	2.1		2.0		16.6
5605	1.9	16.3	13.5	2.1	5.5	5.7	37.2	3.1	1.0		2.1		7.4
5703	0.6	21.8	16.6	1.1	4.9	4.1	26.9	2.3	tr		1.5		18.6
5705	1.1	6.0	3.4	0.6	12.5	4.8	35.5	2.9	5.0		1.2		21.5
Group II													
29-1	4.6	7.7	4.5		12.5		39.5	3.8	3.3		1.3	19.7	
814-4	4.2	5.2	2.5		12.3		37.9	3.9	5.6		1.1	17.9	
5709	4.4	11.7	9.5		12.4		32.8	4.9	2.1			14.2	
4-3	3.5	3.7	1.6		16.8		34.5	3.7	11.8		1.6	17.9	
5402	2.7	4.5	1.9		18.1		39.0	2.3	8.7		1.1	17.7	
Group III													
5602	10.2	11.2	5.4		19.5		38.1	3.0	4.1	3.2	2.6		
5502	5.7	9.9	1.2		24.8		40.2	4.2	10.2		1.9		
6001	4.0	1.6	3.1		9.5		53.4	0.8	23.1		0.9		
6102		2.9			11.2	3.6	24.1	1.2	49.0				
6301	4.7	1.8	3.5		9.3		53.0	0.9	23.1				

From Hamamoto et al. 1995b, with permission.

Fatty acid compositions are expressed as molecular weight percent of FAMEs.

[a] Cultured at 10°C.

[b] Data are from Hamamoto et al. 1994b.

prokaryotes was first reported by DeLong and Yayanos (1986) in deep-sea barophiles and *Vibrio marinus* MP-1. They also showed that 22:6 increased as a function of temperature decrease in *V. marinus*. Similar PUFAs (20:5) were found in Antarctic bacteria, although the amounts were low (Nichols et al. 1993).

As demonstrated here, further analyses of FAMEs of psychrophilic and psychrotolerant deep-sea vibrios divided them into three groups: I, II, and III. PUFAs were observed in two groups of bacteria with lower cardinal growth temperatures (groups I and II), suggesting that it is one of the characteristics of psychrophiles.

Isolates in group I contained 14:1c, 16:1t, and 22:6 in addition to the FAMEs which were found in common in all the examined strains. The amounts of 22:6 ranged between 21.5% and 7.4% of total fatty acids., which are comparable to those obtained by DeLong and Yayanos (1986) from barophilic bacteria, although it is higher than those of Antarctic bacteria (up to 3.3%) (Nichols et al. 1993). These bacteria contained two unusual 9-*trans*-unsaturated fatty acids (14:1t and 16:1t) which were known to be present in the psychrophilic *Vibrio* ABE-1 (Okuyama et al. 1991). In *Vibrio* sp. strain 5710, 16:1t does not appear to be important for temperature adaptation (Hamamoto et al. 1994b). However, considering that the presence of 16:1t was limited to the group of bacteria with the lowest growth temperature limits, it is possible that 16:1t plays a part in the low-temperature adaptation of these isolates.

Six isolates in group II whose maximum growth temperatures were at 20°-25°C contained 20:5 but not 14:1c and 16:1t, which were characteristic of the strains in group I. Members of group III with the highest maximum temperature for growth (25°-37°C) possessed FAME compositions similar to those of group II, except for the absence of 20:5.

Cold adaptation of psychrophiles and psychrotolerants can be reviewed in terms of physiological and biochemical mechanisms (Russell 1990). Inactivation of an enzyme, or a system such as protein synthesis machinery or energy regeneration, can be a cause of the upper growth temperature limit. In contrast, the lower growth temperature limit can be determined by the freezing properties of aqueous solution inside the cell. However, one of the major determinants of the growth temperature range of psychrophiles is considered to be adaptive changes in cellular lipids, which affect regulation of membrane fluidity and permeability (Russell and Fukunaga 1990). I have shown that those marine vibrios with low maximum growth temperatures (i.e., 20°-25°C) contain PUFAs (20:5 or 22:6), in contrast to those with higher temperature limits . PUFAs may play a role in facilitating biochemical functions in membranes by maintaining their fluidity, which at the same time limit bacterial tolerance to elevated temperatures. Indeed, our research found that the presence of PUFAs correlated with a low value for the upper growth temperature limit. The highest maximum growth temperature of the isolates containing PUFAs was 20°-25°C, which coincided with the maximum temperature limit Morita (1975) used to define psychrophiles. These results support the definition of psychrophiles from a biochemical point of view, at least among deep-sea vibrios. It is suggested that the presence of PUFAs may be an important chemotaxonomic tool for discriminating psychrophiles from psychrotolerants.

7.5 Final Comments

I have provided an outline of the research in the DEEPSTAR program on psychrophilic and psychrotrophic microorganisms from the deep-sea. Technological advances and molecular biological knowledge accumulation allowed the research on deep sea microorganisms including psychrophiles. However, the investigation of psychrophilic microorganisms has lagged behind that of thermophiles. Since the DEEPSTAR research program focused on psychrophiles as one of its research targets in 1991, the decade of the 1990s has seen a great advance in understanding the physiological adaptation mechanism to low temperature (Ohkuma et al. 1996; Russell 1997) and the biotechnological potential of psychrophiles (Russell 1998; Russell and Hamamoto 1998). It is hoped that the recent increase of awareness of the importance of psychrophilic organisms and their biochemical components in biotechnological applications, such as the COLDZYME project in the European Commission (Russell 1998), will bring us many benefits.

References

Austin B (1988) Marine microbiology. Cambridge University Press, Cambridge

DeLong EF, Yayanos AA (1986) Biochemical and ecological significance of novel bacterial lipids in deep-sea prokaryotes. Appl Environ Microbiol 51:730-737

Fukunaga N, Russell NJ (1990) Membrane lipid composition and glucose uptake in two psychrotolerant bacteria from Antarctica. J Gen Microbiol 136:1669-1673

Gounot A-M (1991) Bacterial life at low temperature: physiological aspects and biotechnological implications. J Appl Bacteriol 71:386-397

Hamamoto T, Horikoshi K (1991) Characterisation of an amylase from a psychrotrophic *Vibrio* isolated from a deep-sea mud sample. FEMS Microbiol Let 84:79-84

Hamamoto T, Horikoshi K (1993) Deep-sea microbiology research within the Deepstar program. J Mar Biotechnol 1:119-124

Hamamoto T, Kaneda M, Horikoshi K, Kudo T (1994a) Characterization of a protease from a psychrotroph, *Pseudomonas fluorescens* 114. Appl Environ Microbiol 60:3878-3880

Hamamoto T, Kaneda M, Kudo T, Horikoshi K (1995a) Characterization of a protease from psychrophilic *Vibrio* sp. strain 5709. J Mar Biotechnol 2:219-222

Hamamoto T, Takata N, Kudo T, Horikoshi K (1994b) Effect of temperature and growth phase on fatty acid composition of the psychrophilic *Vibrio* sp. strain no. 5710. FEMS Microbiol Lett 119:77-82

Hamamoto T, Takata N, Kudo T, Horikoshi K (1995b) Characteristic presence of polyunsaturated fatty acids in marine psychrophilic vibrios. FEMS Microbiol Lett 129:51-56

Heymann E, Mentlein R (1981) Carboxylesterases-amidases. Methods Enzymology. 77:333-344

Jannasch HW, Taylor CD (1984) Deep sea microbiology. Annu Rev Microbiol 38:487-514

Margesin R, Schinner F (1991) Characterization of a metalloprotease from psychrophilic *Xanthomonas maltophila*. FEMS Microbiol Lett 79:257-262

Mitchell P, Yen HC, Mathemeier PF (1985) Properties of lactate dehydrogenase in a psychrophilic marine bacterium. Appl Environ Microbiol 49:1332-1334

Morita RY (1975) Psychrophilic bacteria. Bacteriol Rev 39:144-167

Myers SF, Anderson A (1992) Microbes from 20,000 feet under the sea. Science 255:28-29

Nakajima M, Mizusawa K, Yoshida F (1974) Purification and properties of an extracellular proteinase of psychrophilic *Escherichia freundii*. Eur J Biochem 44:87-96

Nichols DS, Nichols PD, McMeekin TA (1993) Polyunsaturated fatty acids in Antarctic bacteria. Antarc Sci 5:149-160

Ohkuma M, Ohtoko K, Takada N, Hamamoto T, Usami R, Kudo T, Horikoshi K (1996) Characterization of malate dehydrogenase from deep-sea psychrophilic *Vibrio* sp. strain no. 5710 and cloning of its gene. FEMS Microbiol Lett 137:247-252

Okuyama H, Okajima N, Sasaki S, Higashi S, Murata N (1991) The *cis/trans* isomerization of the double bond of a fatty acid as a strategy for adaptation to changes in ambient temperature in the psychrophilic bacterium. Biochim Biophys Acta 1084:13-20

Oliver JD, Stringer WF (1984) Appl Environ Microbiol 47:461-466

Ray MK, Uma Devi K, Seshu Kumar G, Shivaji S (1992) Extracellular protease from the Antarctic yeast *Candida humicola*. Appl Environ Microbiol 58:1918-1923

Russell NJ (1984) Mechanisms of thermal adaptation in bacteria: blueprints for survival. Trends Biochem Sci 9:108-112

Russell NJ (1990) Cold adaptation of microorganisms. Philos Trans R Soc Lond B Biol Sci 326:595-611

Russell NJ (1997) Psychrophilic bacteria—molecular adaptations of membrane lipids. Comp Biochem Physiol 118A:489-493

Russell NJ (1998) Molecular adaptations in psychrophilic bacteria: potential for biotechnological applications. Adv Biochem Eng Biotechnol 61:1-21

Russell NJ, Fukunaga N (1990) A comparison of thermal adaptation of membrane lipids in psychrophilic and thermophilic bacteria. FEMS Microbiol Rev 75:171-182

Russell NJ, Hamamoto T (1998) Psychrophiles. In: Horikoshi K, Grant WD (eds) Extremophiles: microbial life in extreme environments. John Wiley, New York, pp 25-45

Takami H, Akiba T, Horikoshi K (1990) Characterization of an alkaline protease from *Bacillus* sp. No. AH-101. Appl Microbiol Biotechnol 33:519-523

8 Organic Solvent-Tolerant Microorganisms

Akira Inoue

The deep-sea bed is a unique environment that experiences extremely high pressures and low temperatures. Microorganisms living there have developed particular characteristics that allow them to thrive at such extremes. In studies aimed at improving our understanding of microbial adaptation to the deep-sea environment, we have isolated and characterized a number of microorganisms from samples of deep-sea mud obtained by the manned submersible Shinkai 6500. This vehicle, which is operated by the Japanese Marine Science and Technology Center (JAMSTEC), has the ability to submerge to a depth of 6500 m. It was believed that unknown microbes we have not been able to isolate yet must be thriving in the deep-sea floor, and we attempted to isolate these microbes, especially those that are organic solvent tolerant, from the deep-sea mud samples. Such microorganisms could prove useful for new biotechnology application, such as two-phase (water/organic solvent) bioreacter systems.

Organic solvent tolerant microorganisms can grow in the presence of a large amount of organic solvent. Most organic solvents are generally biotoxic and kill most microorganisms at low concentrations (about 0.1%). Although some microorganisms, including the genus *Pseudomonas* (Kitagawa 1956), *Achromobacter* (Worsy and Williams 1975) and *Nocardia* (Raymond et al. 1967) can assimilate organic solvents, their tolerance for the solvent is less than 0.3%. We determined that organic solvent-tolerant microorganisms could grow with more than 10% organic solvent, such as toluene or benzene, and looked for such microorganisms in all the Japanese islands. We have reported the discovery of a variant strain of *Pseudomonas putida*, strain IH-2000, which is a first isolate of organic solvent tolerant bacteria, isolated

Key words. Organic solvent, Benzene, Toluene, Oil degrading, *Pseudomonas*, Log P, Deep-sea, Hydrocarbon degrading from the mud samples of Kumamoto Prefecture, Kyushu Island (Inoue and Horikoshi 1989). Strain IH-2000 was capable of growing in culture media containing more than 50% (v/v) toluene.

Surprisingly, we observed that many organic solvent-tolerant microbes exist even in the deep-sea. Therefore we are focusing on organic solvent-tolerant microbes and their industrial application.

8.1 Distribution of Solvent-Tolerant Microorganisms

To obtain information about the presence of solvent-tolerant microorganisms, we had tried to compare the isolation frequency of organic solvent-tolerant microorganisms from land soil and abyssal mud, respectively. As shown in Table 1, the isolation frequency in land soil was $1.0 \times 10^{-8}$to 3.6×10^{-8}. In contract, the isolation frequency in abyssal mud samples was 350×10^{-8} to 700×10^{-8}. From these experiments we found that 100 times as many organic solvent-tolerant microorganisms could be isolated from deep-sea mud samples as from soil samples taken from land.

Table 1. Comparison of the isolation frequency of organic solvent-tolerant microbes from different sources

Source	Total microbes ($\times 10^5$ cells/g)	Solvent-tolerant microbes (cells/g)	Isolation frequency ($\times 10^{-8}$)
Land soil	2800 – 11 500	7 – 16	1.0 – 3.6
Abyssal mud	54 – 109	19 – 52	350 – 700

Table 2. Organic solvent-tolerant microorganisms in the author's laboratory

Strain	Property	Source	Reference
Organic solvent tolerant bacteria			
DS-711	Degrades crude oil	Suruga Bay, 1945 m	Moriya and Horikoshi 1993a
DS-944	Utilizes sulfur	Sagami Bay, 1168 m	Moriya and Horikoshi 1993b
DS-1906	Degrades polyaromatic hydrocarbons	Sagami Bay, 1168 m	Abe et al. 1995
ST-1	Degrades cholesterol	Okushiri Ridge, 1963 m	Moriya et al. 1995
Organic solvent-tolerant yeast			
Y-40	Hydrocarbon-degrading yeast	Sagami Ridge, 1200 m	Fukumaki et al. 1994

As a result, we have discovered many useful organic solvent-tolerant strains from the deep-sea environment. Some of these isolates from our laboratory are shown in Table 2.

8.2 A Benzene-Tolerant Hydrocarbon-Degrading Bacteria

Because various organic solvents (OSs), such as benzene, toluene, and *p*-xylene, are known to be components of petroleum, and the deep-sea is the ultimate receptor of polluted petroleum hydrocarbons, we focused on deep-sea bacteria that showed the characteristics of halotolerant growth in addition to OS tolerance (Moriya and Horikoshi, 1993a, 1993b). We also studied whether isolation of deep-sea bacteria having such characteristics could improve the degradation of polluting petroleum hydrocarbons in marine environments.

The main purpose of our study was to elucidate the mechanisms of OS tolerance and hydrocarbon degradation by OS-tolerant microorganisms; hence, various kinds of microorganism were needed. We used the log p of the OSs as a parameter for their degree of toxicity. Log p is defined as the common logarithm of a partition coefficient (p) of a given OS between equimolar *n*-octanol and water. Inoue and Horikoshi (1989) showed that certain microorganisms were able to grow in the presence of OSs whose log p is higher than a particular value. Toluene (log p 2.7) and *p*-xylene (log p 3.1) are so toxic that only a few bacteria tolerant of these OSs have been isolated. In the case of benzene (log p 2.1), no tolerant bacteria have yet been reported.

8.2.1 Isolation and Identification of OS-Tolerant Hydrocarbon-Degrading Bacteria

We developed a novel, simple, efficient method for isolating of a benzene-tolerant bacterium from deep-sea samples. To isolate benzene-tolerant bacteria that also exhibited the characteristic of halo-tolerant growth, 1-ml samples of sediment was transferred to test tubes containing 3 ml artificial seawater. Benzene 1 ml (20% v/v) was then added, and the test tubes were plugged with silicone-rubber stoppers. The cultures were incubated at 4°C for a week in a test tube shaker. Then 5 ml (50% v/v) of kerosene was added, and the cultures were incubated for another 5 days at 20°C. After incubation, the cultures were aseptically transferred to separating funnels and allowed to stand for 30 min. The upper layers were carefully separated from the seawater layers, and a portion of each upper layer was directly spread onto M-I agar salt medium comprising 1 g of proteose peptone no. 3 (Difco), 0.5 g of phytone peptone (BBL), 1 g of $CaCl_2 \cdot 2H_2O$, 0.1 g of $MgCl_2 \cdot 6H_2O$, and 0.05 g of Na_2SO_3 per liter of distilled water at pH 7.0 and incubated for 2 days at 30°C. For microbial growth M-II medium, which contained 5 g of proteose peptone no. 3, 2.5 g of phytone peptone, 1 g of $CaCl_2 \cdot 2H_2O$, 0.1 g of $MgCl_2 \cdot 6H_2O$, and 0.05 g of Na_2SO_3 per liter of distilled water and a modified LB-medium (Inoue and Horikoshi 1991b) (tentatively named LB-1 medium in this study) were used. To prepare the salt medium, the above media were supplemented with NaCl at 11.8% w/v unless otherwise stated. For preparation of agar medium, agar was added at 1.5%. The pH was adjusted to 6.5-7.0 by 1 N HCl or 1 N NaOH. Artificial seawater consisting of 35 g of NaCl, 10 g of $MgCl_2 \cdot 6H_2O$, 1.0 g of $CaCl_2 \cdot 2H_2O$ and 0.7 g of KCl per liter of dis-

Table 3. Content of *n*-alkanes in 10 ml of kerosene used in this study

Hepatane	(n-C_7)	43 mg
Octane	(n-C_8)	48 mg
Nonane	(n-C_9)	232 mg
Decane	(n-C_{10})	292 mg
Undecane	(n-C_{11})	345 mg
Dodecane	(n-C_{12})	759 mg
Tridecane	(n-C_{13})	583 mg
Tetradecane	(n-C_{14})	382 mg
Pentadecane	(n-C_{15})	216 mg
Hexadecane	(n-C_{16})	45 mg

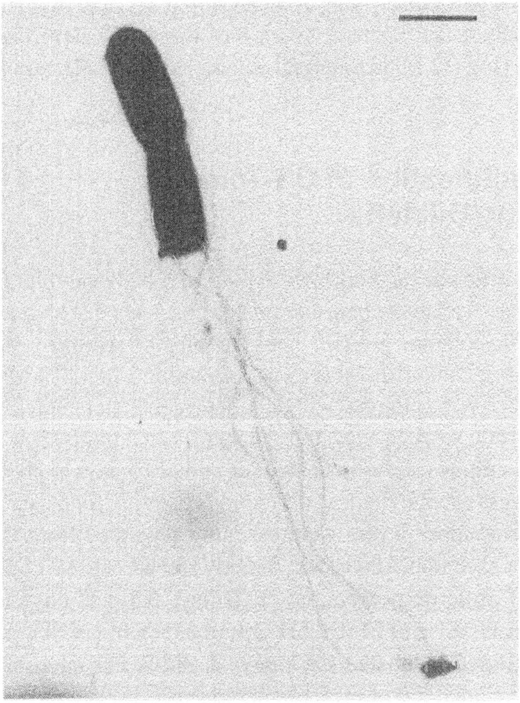

Fig. 1. Electron micrograph of a benzene-tolerant *Flavobacterium* sp. DS-711 cell. Bar 1 μm

tilled water was used for the preparation of concentrated cell suspensions. As the *n*-alkane substrate for the microbial degradation experiment, kerosene was used at the indicated concentration. The content of *n*-alkanes having 7-16 carbon atoms in 10 ml of kerosene is listed in Table 3.

After treatment of the deep-sea sediment samples collected from a depth of 1945 m in Suruga Bay with 20% v/v benzene followed by 50% v/v kerosene, 56 candidate colonies grown on selective medium that combined the characteristics of halotolerant growth and benzene tolerance were isolated as potential hydrocarbon degraders. These candidates belonged to various genera, including *Arthrobacter* sp., *Bacillus* sp., *Acinetobacter* sp., *Pseudomonas* sp., *Flavobacterium* sp., and *Vibrio* sp. (data not shown). The candidates could also grow in the presence of 1% benzene (data not shown). Growth of the 56 isolates on M-I salt medium with 5% v/v of benzene was

Table 4. Characteristics of strain DS-711

Morphological characteristics	
Form	Rods
Size (μm)	0.5 × 1.0 to 0.5 × 2.0
Gram stain	Negative
Motility	+
Flagellum	Peritrichous
G+C DNA (mol%)	67.4
Cultural characteristics	
Aerobiosis	+
Temperature (°C)	7 – 42 (optimal 37)
pH	5 – 9.0
NaCl (mol)	0 – 3.0 (optimal 1.0)
Biochemical characteristics	
Production of pigment	+ (orange color)
Oxidase	+
Catalase	+
Phosphatase	+
O-F test	Oxidative
Poly-β-hydroxylate granules	–
Denitrification	–
Hydrolysis of gelatin	–
Tween 80	+
Resistance to antibiotics (10μg)	
Streptomycin	+
Penicillin	+
Josamycin	+
Erythromycin	+
Lincomycin	+
Nalidixic acid	+
Kanamycin	+
Ampicillin	–
Tetracycline	–
Chloramphenicol	–
Novobiocin	–

further examined. Finally, the bacterial strain that showed the highest growth in the presence of 5% v/v benzene among above 56 isolates was selected as a hydrocarbon degrader that also exhibited the characteristics of halotolerant growth and benzene tolerance. This strain was identified based on its morphological and biochemical characteristics, as outlined in volume I of *Bergey's Manual of Systematic Bacteriology* (Krieg et al. 1984). Analysis of the G+C content of DNA was done by the method of Tamaoka and Komagata (1984). The growth temperature of selected isolates was determined using a temperature gradient bio-photorecorder (TN-112D, Advantec). Cells of the isolate had peritrichous flagella and showed motility. The isolate formed orange colonies on M-I and M-II salt media. It was gram-negative, non-spore-forming, strictly aerobic, oxidase-positive, and catalase-positive. The isolate showed resistance to many antimicrobials including streptomycin, penicillin, lincomycin, nalidixic acid, josamycin, and kanamycin and a susceptibility to novobiocin. Based on its morphological and biochemical characteristics, as shown in Fig. 1 and Table 4, in addition to its characteristic of benzene tolerance the selected strain seemed to be a variant strain of *Flavobacterium* sp. and was tentatively named strain DS-711

8.2.2 OS Tolerance of Hydrocarbon-Degrading Bacteria

The OS tolerance of the *Flavobacterium* sp. strain DS-711 was compared with control strains, *Escherichia coli* strain IFO 3806, *Bacillus subtilis* strain IFO 3009, and the toluene-tolerant *Pseudomonas putida* strain IH-2000. Tested strains were cultured on M-II salt medium and LB-I medium in the presence of OSs such as benzene, toluene, and *p*-xylene. Both *E. coli* IFO 3806 and *B. subtilis* IFO 3009 were unable to grow not only in the presence of 1% benzene but also 1% toluene and 1% *p*-xylene, as shown in Table 5.

The toluene-tolerant strain IH-2000 showed considerable growth in the presence

Table 5. Comparison of organic solvent tolerance between strain DS-711 and soil strains

Solvent (log *p*)	Growth (OD_{660}) [a]			
	Strain DS-711	*P. putida* IH-2000	*E. coli* IFO 3806	*B. subtilis* IFO 3009
No additions	1.36	2.98	3.05	2.95
Benzene 1% (2.1)	1.33	NG	NG	NG
Benzene 5%	1.08	NG	NG	NG
Toluene 10% (2.7)	1.18	2.88	NG	NG
p-Xylene 10% (3.1)	1.06	2.64	NG	NG

Bacteria were grown on M-II salt medium and LB-I medium with each solvent tested (benzene, toluene, and *p*-xylene) at the indicated concentration (v/v%). Growth was determined by measuring the optical density at 660nm (OD_{660}).
NG, no growth.
[a] *Pseudomonas putida, Escherichia coli, Bacillus subtilis.*

of 10% toluene and 10% *p*-xylene, but it could not grow in the presence of more than 1% benzene. In contrast to IH-2000, DS-711 could grow in the presence of not only 10% toluene and 10% *p*-xylene but also 5% benzene. The tolerance of strain DS-711 to various toxic OSs in addition to benzene, toluene, and *p*-xylene, including saturated and unsaturated aliphatic hydrocarbons, alicyclic hydrocarbons, aromatic hydrocarbons, alcohol, and ethers, was further examined. As shown in Table 6, DS-711 tolerated various toxic OSs, though others such as methoxytoluene, propanol, butanol, heptanol, octylalcohol, benzylether, cyclohexanone, and ethylacetate were not tolerated. Subsequently, we investigated the correlation between the log p value of the OSs and growth on the OSs using the OS-tolerant strains DS-711 and IH-2000, on the basis of the data in Table 6. The growth limiting log p of strain IH-2000 was 2.4; that is, this strain could grow on OSs whose log p is higher than 2.4. However,

Table 6. Solvent tolerance of strain DS-711

Inhibitation of growth and solvents in group	Growth [a] at solvent conc. (v/v)	
	10%	5%
No inhibition	+	+
Hydrocarbons (pentane, hexane heptane, octane, isooctane, nonane, decane, dodecane, 2-pentane, 2-hexane, 1-octene, 1,3-pentadiene, 1,7-octadiene, cyclohexane, methylcyclopentane, cyclohexane, methylcyclohexane, benzene, toluene, p-xylene, nitrobenzene)		
Ketones (acetone)		
Ethers (butylether, diethylether)		
Miscellaneous (dimethylformamide, dimethylsulfoxide, acetonitrile)		
Slight inhibition	±	+
Hydrocarbons (1-dodecene, styrene)		
Alcohols (methanol)		
Ethers (hexylether, propyleneoxide)		
Mpderate inhibition	–	±
Hydrocarbons (butylcyclohexane, cyclooctane, fluorobenzene, propylbenzene, chlorobenzene, o-dichlorobenzene, bromobenzene)		
Alcohols (ethanol, decylalcohol)		
Ketones (2-heptanone)		
Inhibition	–	–
Hydrocarbons (methoxytoluene)		
Alcohols (propanol, butanol, heptanol, octylalcohol)		
Ethers (benzylether, butylvinylether)		
Ketones (methylethylketone, 2-pentanone, 2-hexanone, cyclohexanone)		
Miscellaneous (chloroform, ethylacetane, acetal)		

The initial cell concentration was 1×10^6 cells/ml. Incubation was done for 5 days at 35°C.
[a] +, growth (> 5×10^8 cells/ml); ±, slight growth (> initial cells concentration); –, no growth.

strain DS-711, which showed growth in the presence of benzene (log p 2.1), did not grow in the presence of heptanol (log p 2.4). Some OSs, such as hexylether (log p 5.1), cyclooctane (log p 4.5), propylbenzene (log p 3.8), *o*-dichlorobenzene (log p 3.6), and styrene (log p 2.9) had an inhibitive effect on the growth of DS-711.

8.2.3 Hydrocarbon Degradation Experiment

Degradation of *n*-alkanes in kerosene and aromatic hydrocarbons by strain DS-711 was compared with that of the control strains, *Alteromonas* sp. DS-201 and *Pseudomonas putida* IH-2000. To measure their ability to degrade *n*-alkanes in kerosene and aromatic hydrocarbons, one loop of strains was inoculated in 5000-ml flasks containing 1000 ml M-II salt medium and LB-1 medium. Incubation was carried out for 2 days at 30°C. After incubation the cultures were centrifuged at 12 000 rpm and washed two times with artificial seawater and sterile distilled water to prepare a 15-

Table 7. Hydrocarbon degradation by solvent-tolerant strains

Hydrocarbon	Degradation (%)		
	P. putida IH-2000	*Alteromonas* sp. no. DS-201	DS-711
n-Alkanes			
n-C_7	28.0	30.5	97.0
n-C_8	30.9	30.3	82.1
n-C_9	35.7	32.4	82.1
n-C_{10}	10.6	9.7	71.4
n-C_{11}	11.8	10.8	68.0
n-C_{12}	40.2	41.4	86.7
n-C_{13}	46.2	45.3	85.8
n-C_{14}	39.9	44.6	88.0
n-C_{15}	35.7	39.0	92.9
n-C_{16}	40.2	38.1	80.5
Aromatic hydrocarbons			
Benzene	NG	NG	0
Toluene	0	NG	0
p-Xylene	0	NG	0

Degradation (%) was calculated by the following equation.

$$\text{Degradation (\%)} = \frac{\begin{array}{c}\text{Residual hydrocarbon conc.}\\ \text{of an additional control}\end{array} - \begin{array}{c}\text{Residual hydrocarbon conc.}\\ \text{of inoculated samples}\end{array}}{\text{Residual hydrocarbon conc. of an additional control}} \times 100$$

The residual *n*-alkane contents (mg) in 10-ml kerosene after treatment of an additional control are: *n*-C_7, 34; *n*-C_8, 39; *n*-C_9, 190; *n*-C_{10}, 245; *n*-C_{11}, 270; *n*-C_{12}, 653; *n*-C_{13}, 501; *n*-C_{14}, 332; *n*-C_{15}, 188; *n*-C_{16}, 39.
NG, no growth.

ml concentrated cell suspension. The cell suspensions were inoculated into 100 ml fresh M-II salt medium and LB-1 medium at an initial concentration of 1 × 10^7 or 1 × 10^9 cells/ml and 10 ml (10% v/v) of kerosene and 1 ml (1% v/v) of aromatic hydrocarbon were overlaid individually. When the stepwise addition of the concentrated cell suspension was complete, the suspensions were added after 2 days and 3 days. An additional noninoculated control was run for measurement through evaporation. Inoculated samples and the additional control were incubated with shaking at 100 rpm for a week at 30°C. At the end of the incubation, cultures were transferred to separating funnels for extraction and analysis of hydrocarbons. The percentage degradation of hydrocarbons is given as the difference between the additional control and the inoculated samples. The total *n*-alkane content was made up of the total content of *n*-alkanes having 7-16 carbon atoms. Cultures on kerosene and aromatic hydrocarbons were centrifuged. Supernatants were transferred to separating funnels for hydrocarbon extraction. Residual kerosene and aromatic hydrocarbon oil droplets were separated from the lower aqueous layer. Then 50 ml of benzene containing 0.1 g of *n*-eicosane as an internal standard was added to the kerosene oil droplets. Kerosene was extracted with benzene as described in detail elsewhere (Waker and Colwell 1974; Walker et al. 1975). Residual aromatic hydrocarbon was extracted with 4-ml of *n*-decane. Any remaining water was removed by drying over Na_2SO_4 Anhydrous extracts were transferred to a vial and stored at -20°C until gas chroma-

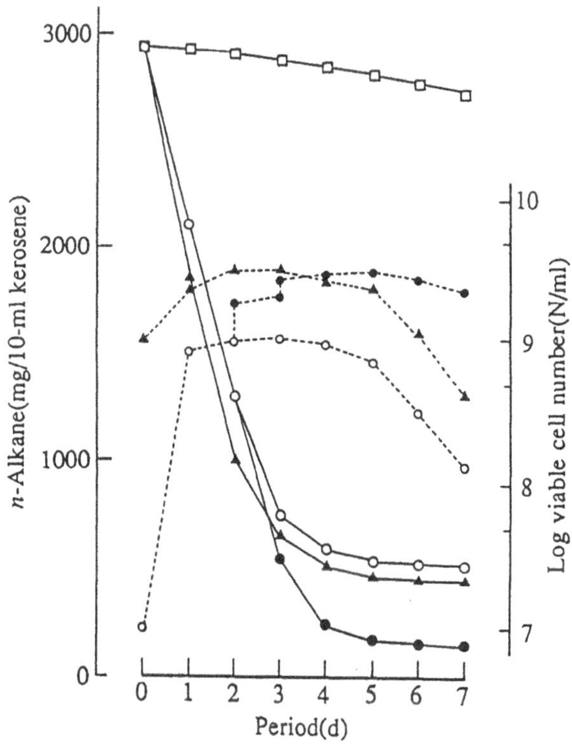

Fig. 2. Changes in *n*-alkane contained in kerosene using the benzene-tolerant strain DS-711 for 7 days at 30°C. *Squares*, uninoculated control; *open circles*, initial cell concentration at 1 × 10^7 cells/ml; *triangles*, initial cell concentration at 1 × 10^9 cells/ml; *filled circles*, stepwise addition of concentrated cell suspension; *solid lines*, residual *n*-alkane; *broken lines*, viable cell concentration

tography (GC) analysis was performed.

Gas chromatographs were obtained on a Shimadzu model GC-14A gas chromatograph equipped with a single flame ionization detector. A glass column (2.1 m × 3.2 mm i.d.) packed with 10% silicone OV-101 on 60/80 mesh chromosorb was used. The carrier gas, nitrogen, was run through the column at a rate of 40 ml/min; and the temperature was programmed from 35° to 350°C at 10°C/min. Individual peak areas were quantified. Column efficiency and detector response were measured with a mixture of *n*-alkanes with carbon atom numbers C_7–C_{16} as standards. The system yielded a relative error of less than 2.5-6.0 for all components.

As shown in Table 7, all the strains used in this experiment could degrade *n*-alkanes in kerosene but did not degrade aromatic hydrocarbons. However, *n*-alkane degradation by strain DS-711 was elevated remarkably compared to that of the control strains, DS-201 and IH-2000. In particular, percentage degradation of *n*-decane and *n*-undecane by both control strains was approximately 10%, whereas that by strain DS-711 was 71.4% and 68.0%, respectively. DS-711 showed the highest *n*-alkane degradation among the strains used, with the degradation of *n*-alkanes by the bacterium reaching approximately 85% after 7 days.

The influence of the initial cell concentration on *n*-alkane degradation and the effect of stepwise addition of concentrated cell suspension on *n*-alkane degradation were subsequently examined using strain DS-711. The total *n*-alkane concentration in 10 ml of kerosene and the viable cell concentration were measured daily until the concentration reached a constant value. When the initial cell concentrations were fixed at 1×10^7 and 1×10^9 cells/ml, strain DS-711 showed good growth in the presence of 10 v/v % kerosene, as shown in Fig. 2. After 1 day the initial cell concentrations of strain DS-711 reached 2×10^9 and 6×10^9 cells/ml, respectively. After 4 days the viable cell concentration of both treatments decreased gradually. The initial *n*-alkane content in 10 ml of kerosene (2945 mg) decreased gradually and reached a constant value of approximately 500 mg for both treatments. When stepwise addition of concentrated cell suspension was done, the initial *n*-alkane also decreased and reached a constant value of approximately 100 mg.

8.2.4 Discussion

Fifty-six bacterial strains that transferred to the kerosene layer after treatment of deep-sea sediments with benzene followed by kerosene were truly OS-tolerant hydrocarbon degraders. The bacteria obtained belonged to various genera, including *Arthrobacter* sp., *Bacillus* sp., *Acinetobacter* sp., *Pseudomonas* sp., *Flavobacterium* sp., and *Vibrio* sp., showing that our isolating method facilitated isolation of a variety of benzene-tolerant bacteria species.

When garden soil samples were treated with benzene followed by kerosene under the same conditions, the number of isolates obtained from the hydrocarbon oil layer was significantly less than in the case of the deep-sea samples (data not shown). Prior to this study, we compared the viable cell concentrations of microorganisms from deep-sea and soil samples and found that the microbial concentration of the soil samples was approximately 1-100 times higher than that of deep-sea sediment samples. Consequently, our results indicated that OS-tolerant microorganisms are obtainable from deep-sea sediment samples with high frequency compared with soil samples. Possibly, the number of OS-tolerant microorganisms that exist in the deep-sea is far higher than that in soil.

The deep-sea bacterial strain *Flavobacterium* sp. DS-711, which was isolated by our novel, simple, efficient method, exhibited excellent benzene tolerance. Inoue and Horikoshi (1991b) showed that the toluene-tolerant strain IH-2000 could grow on media containing 50% OSs but could not grow with a 5% concentration of solvents such as acetone and cyclohexanone.

Inoue and Horikoshi (1989; 1991a) also used the log p parameter to indicate the toxicity of OSs. Log p, which is defined as the logarithm of a solvent's partition coefficient in a standard octanol: water mixture, was used as a quantitative index of solvent polarity in a quantitative structure-activating relationship analysis (Hansch et al. 1963; Hansch and Fujita 1964). A previous study showed that certain microorganisms can grow in the presence of an OS whose log p is higher than a specific value (Inoue and Horikoshi 1989). We examined the correlation between log p and growth among OS-tolerant strains on the basis of the data in Table 6. Strain IH-2000 could grow on OSs whose log p was higher than that of heptanol (log p 2.4), whereas strain DS-711 showed no growth or only slight growth on some OSs, such as hexylether, cyclooctane, propylbenzene, *o*-dichlorobenzene, styrene, and heptanol whose log p was higher than that of benzene. Accordingly, we believe that using the parameter log p to assess the toxicity of OSs is not applicable to all OS-tolerant microorganisms. The OS tolerance mechanism of strain DS-711 seemed to differ considerably from that of the toluene-tolerant strain IH-2000.

Because microorganisms that are able to grow in the presence of 5% benzene have not yet been found, it will be of interest to compare the cell surface components of the benzene-tolerant strain DS-711 with those of strain IH-2000. We tried to isolate a benzene-tolerant plasmid from DS-711, but no plasmid was obtained. As strain DS-711 harbors no plasmid DNA, the OS tolerance of this bacterium is thought to be dominated by information in genomic genes. Generally, gram-negative bacteria show higher OS tolerance than gram-positive ones (Inoue and Horikoshi 1989, 1991a). At present, we are investigating the mechanism of OS tolerance from two aspects: cell surface components and genetics. The discovery of a benzene-tolerant bacterium from the deep-sea is also valuable from the view point of microbial ecology.

Subsequently, hydrocarbon degradation by the benzene-tolerant DS-711 was compared with that of the control strains, DS-201 and the toluene-tolerant IH-2000. As shown in Table 7, with a shaking culture the degradation of *n*-alkanes in kerosene by the benzene-tolerant DS-711 was greater than that of our selected control strains, DS-201 and IH-2000. In particular, degradation of *n*-decane and *n*-undecane, which were hardly degraded by the control strains, were markedly elevated. However, these strains, including DS-711, showed a percentage degradation of *n*-decane and *n*-undecane lower than that of other *n*-alkanes such as *n*-heptane, *n*-octane, *n*-nonane, *n*-dodecane, *n*-tridecane, *n*-tetradecane, *n*-pentadecane, and *n*-hexadecane. This was thought to be due to lower activity of the emulsifying and solubilizing substance in regard to *n*-decane and *n*-undecane in kerosene. A study of the relationship between hydrocarbon degradation and hydrocarbon emulsification and solubilization is now in progress.

We are examining the identification and characterization of the hydrocarbon-emulsifying and solubilizing substance in addition to elucidating OS tolerance and hydrocarbon degradation. Interestingly, neither DS-711 nor IH-2000 degraded aromatic hydrocarbons, such as benzene, toluene and *p*-xylene. Hydrocarbon degradation is associated with assimilation. Accordingly, these results indicated that tolerance to aromatic hydrocarbons, such as benzene, toluene and *p*-xylene, is distinguishable from tolerance to *n*-alkanes at least. The *n*-alkane degrading ability of this bacterium will be useful for the bioremediation of oil spills, especially marine oil spills. Because of the OS tolerance of the bacterium is also available for commercial application in industrial biotransformation processes that involve the use of water-insoluble organic substrates, such as biocatalysis in a water–OS two-liquid-phase system, discovery of the benzene-tolerant strain DS-711 is worthy of note.

When the influence of the initial cell concentration on *n*-alkane in kerosene was examined, no remarkable differences of *n*-alkane degradation between the two treatments were observed, as shown in Fig. 2. A gradual decrease of viable cell concentration occurred during both treatments. As *n*-alkane seemed to be converted to CO_2 and water (see below), this phenomenon is possibly caused by the toxicity of by-products other than *n*-alkane produced during hydrocarbon degradation. When stepwise addition of concentrated cell suspension was done, the total *n*-alkane degradation reached more than 90%. We found more efficient degradation of *n*-alkanes by this treatment than with the above treatments. Though data are not shown, breakdown products were not evident from the gas chromatography (GC) profiles. The conclusion drawn from these results is that *n*-alkane was assimilated in cell material and converted to CO_2 or products not detectable in the GC system employed. Essentially, all of the available *n*-alkanes were converted to CO_2. Accordingly, these results also indicated that the benzene-tolerant strain DS-711 had the characteristic of degrading *n*-alkanes, which are major components of petroleum, without the influence of OSs.

8.3 Benzene-Tolerant Sulfur-Degrading Bacteria

The availability of organic solvent (OS)-tolerant microorganisms for industrial biotransformation processes that involve the use of water-insoluble organic substrates as biocatalysts in water-OS two-liquid-phase systems emphasizes the importance of the characterization of these microorganisms. We have maintained our interest in the deep-sea environment to isolate OS-tolerant microorganisms with characteristics suitable for their application in industrial biotransformation. We attempted to isolate an OS-tolerant bacterium from the deep-sea to degrade sulfur compounds in the presence of OS (Morita and Horikoshi 1993b).

8.3.1 Isolation and Identification of OS-Tolerant Sulfur-Degrading Bacteria

Deep-sea sediment slurry samples were aseptically collected by sterile core samplers from a depth of 1168 m in Sagami Bay (35° N, 139° 13'30" E) during dive 586 of the manned submersible Shinkai 2000 in 1991. Screening the OS-tolerant sulfur-degrading bacteria was done by a modification of the method described previously (Moriya and Horikoshi 1993a). Benzene (5 ml) was added to 5 ml of deep-sea sediment slurry, which was then incubated at 25°C for 7 days. After this preincubation the benzene layer was separated from the seawater layer, and the former was spread onto M-II agar medium (Moriya and Horikoshi 1993a) with 5.8% NaCl, 1% $Na_2S_2O_3$, and 0.002% phenol red. The plates were incubated at 37°C for 7 days. The sulfur-degrading ability could be detected directly on the plates because of the clear zone that forms around colonies that produce sulfuric acid from $Na_2S_2O_3$, caused by a pH decrease. Thus, microorganisms in which a clear zone formed around colonies were selected and purified.

One was designated strain DS-994. This bacterium was gram-positive, spore-forming, aerobic, catalase-positive, oxidase-negative, gelatin-liquefying, and motile. It is rod-shaped (0.3-5.0 × 0.6-3.0 μm) and has flagella, as shown in Fig. 3. From its morphological and physiological properties and its characteristics of halophilic growth, sulfur utilization, and OS tolerance, DS-994 seems to be a variant strain of *Bacillus* sp. Strain DS-994 displayed optimal growth at an initial pH 7.0, NaCl concentration of 12%, and 37°C.

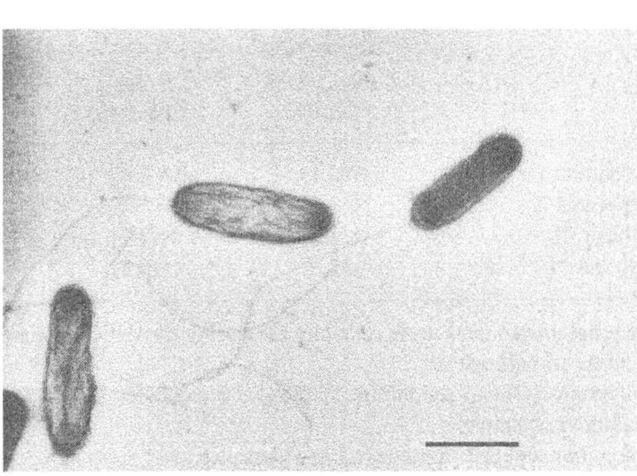

Fig. 3. Electron micrograph of the benzene-tolerant *Bacillus* DS-994 cell. *Bar* 1 μm

8.3.2 Solvent Tolerance of Sulfur-Degrading Bacteria

To investigate the influence of OSs on the growth of strain DS-994, it was inoculated into a test tube containing 5 ml of LB-II medium supplemented with 12% NaCl and OS, after which it was incubated at 37°C. LB-II medium contained 5 g of yeast extract (Difco, MI, USA), 10 g of tryptone (Difco), 1 g of $CaCl_2 \cdot 2H_2O$, and 0.1 g of $MgCl_2 \cdot 6H_2O$ per liter of distilled water. The initial pH of the media was adjusted to 7.0 by NaOH.The results are shown in Table 1. When incubated for 5 days, the strain showed good growth in the presence of 5% benzene, 10% toluene, and 10% *p*-xylene, respectively. In this OS tolerance experiment, *Escherichia coli* strain IFO 3806, *Bacillus subtilis* strain IFO 3009 and a representative OS-tolerant *Pseudomonas putida* strain IH-2000 (Inoue and Horikoshi 1989, 1991a) were employed as controls and incubated on LB-1 medium and LB-II medium supplemented with 0.5% NaCl in the presence of OS.

The *E. coli* and *B. subtilis* strains did not grow in the presence of any OS, whereas the *P. putida* strain did grow in the presence of toluene and *p*-xylene but not benzene (Table 8). Subsequently, the effect of benzene on the halophilic growth of this bacterium was further examined; it showed growth in the range of 0.01%–29.50% NaCl without influence from 5% v/v benzene (data not shown). Inoue and Horikoshi (1989, 1991a) used a partition coefficient) of organic solvents for a toxicity index. Log p is defined as the common logarithm of p of a given organic solvent between *n*-octanol and water (Hansch et al. 1963; Hansch and Fujita 1964). The toxicity of benzene (log p 2.1) is greater than that of the other organic solvents used in this study (toluene log p 2.7, *p*-xylene log p 3.1). Generally, gram-positive bacteria have lower OS tolerance than Gram negative ones (Inoue and Horikoshi 1989, 1991a). Therefore, the benzene tolerance of the gram-positive strain DS-994 is particularly significant. Inoue and Horikoshi (1989, 1991a) also showed that OS-tolerant micro-

Table 8. Bacterial growth on organic solvents

Organic solvent	Growth [a]			
	Bacillus sp. DS-994	*P. putida* IH-2000	*E. coli* IFO 3806	*B. subtilis* IFO 3009
No addition	+(3.1)	+(3.8)	+(3.5)	+(3.6)
Benzene 5%	+(2.5)	–	–	–
Toluene 10%	+(2.6)	+(3.7)	–	–
p-Xylene 10%	+(3.0)	+(3.8)	–	–

Bacterial growth was determined by measuring the optical density at 660 nm, represented by the numbers in parentheses.
Soil strains (OD_{660}) and strain DS-994 were incubated for 5 days at 37°C on LB-I and LB-II medium, respectively.
Log *p*: benzene 2.1; toluene 2.7; *p*-xylene 3.1.
[a] +, growth; –, no growth.

organisms, including the toluene-tolerant *P. putida* strain IH-2000, can grow in the presence of OSs whose log p is higher than the specific value. However, our previous study (Moriya and Horikoshi 1993a) indicated that the mechanism of OS tolerance of a benzene-tolerant bacterium was different from that of the toluene-tolerant *P. putida* strain IH-2000. Accordingly, investigations into the tolerance of strain DS-994 to various kinds of OSs, the mechanism of its OS tolerance, and the differences of OS tolerance between gram-positive and gram-negative bacteria are now in progress.

8.3.3 Degradation of Sulfur Compound

To confirm sulfuric acid production from $Na_2S_2O_3$ by strain DS-994, incubation was done on LB-II medium supplemented with 12% NaCl and 10% $Na_2S_2O_3$. When a 1-day fresh culture of DS-994 was inoculated into the medium and cultured aerobically for 10 days under the optimal growth conditions, the strain showed good growth, and the initial pH value of the culture broth gradually decreased below pH 5.0. Sulfuric acid from sulfur compounds was measured qualitatively by high-performance liquid chromatography (HPLC). HPLC was performed on a TSK gel-IC-Anion-PW column (4.6 mm × 5.0 cm) (Tosoh, Tokyo, Japan) with 3 mM sodium dibasic phosphate solution as an isocratic mobile phase at a flow rate of 0.8 ml min^{-1}. Sulfate ion was detected with a Waters 431 conductivity detector (Waters Associates, Milford, MA, USA). The total sulfur content of organic sulfur compounds was measured on a Perkin-Elmer organic elemental analyzer 2400 II CHNS/O Analyzer (Perkin-Elmer GmbH, Uberlingen, Germany). From the results of HPLC analysis of the culture broth at the end of the cultivation it was clear that strain DS-994 produced sulfuric acid from $Na_2S_2O_3$ (data not shown), and the sulfuric acid production seemed to cause the decrease of pH.

Because strain DS-994 showed OS tolerance and the ability to produce sulfuric acid from $Na_2S_2O_3$, its sulfur-degrading ability was further investigated in the presence of kerosene with or without 5% benzene. Kerosene and diesel oil were used as the model petroleum. When needed, the model petroleum was supplemented with OSs at the concentrations indicated. Strain DS-994 showed good growth in the presence of kerosene. Furthermore the results of HPLC analysis indicated that DS-994 changed the sulfur of $Na_2S_2O_3$ to sulfuric acid in the presence of kerosene with or without 5% benzene (data not shown). When kerosene was supplemented with 10% of other organic solvents, sulfuric acid production was also observed (data not shown). Thus, strain DS-994 has great ability to degrade $Na2S_2O_3$ as an inorganic sulfur compound in the presence of an organic solvent.

To investigate its ability to utilize organic sulfur compounds such as dibenzothiophene, thiophene, and ethylmethyl sulfide, which are present in petroleum, strain DS-994 was incubated in a water-model petroleum two-liquid-phase system. Dibenzothiophene (DBT), thiophene (T) and ethylmethyl sulfide (EMS) were used as exogenous organic sulfur compounds; 0.2% of each organic sulfur compound was added to the model petroleum as total sulfur content. Sterilized seawater (4.5

Table 9. Utilization of organic sulfer compounds by strain DS-994

Organic sulfer compound	Total sulfer concentration (%)		
	Control 1	Control 2	Sample
Dibenzothiophene	0.2 (0)	0.195 (2.5)	0.175 (12.5)
Thiophene	0.2 (0)	0.195 (2.5)	0.180 (10.0)
Ethylmethylsulfide	0.2 (0)	0.190 (5.0)	0.170 (15.0)

Numbers in parentheses represent the percent utilization.

ml) was employed as the water layer. One-day fresh culture of DS-994 (500 ml) was centrifuged and washed two times with sterilized seawater. The concentrated cell suspension was prepared with 10 ml of sterilized seawater. To the water layer, 0.5 ml of concentrated cell suspension was added at approximately 1×10^9 cells/ml. Finally, 5 ml diesel oil with or without 5% OS was overlaid above the water layer. Two additional controls were run: (1) without inoculation and (2) with organic sulfur compounds suspended in the water layer without model petroleum and OS.

When incubated for 10 days under optimal growth conditions, strain DS-994 showed good growth on LB-II medium with 12% NaCl. As shown in Table 9, the initial total sulfur concentrations of DBT, T, and EMS in the diesel oil decreased to 0.175%, 0.18%, and 0.17%, respectively. The percent sulfur degradation of the organic sulfur compounds solubilized in the diesel oil was higher than that of control 2. In this experiment, no difference in the percent utilization of organic sulfur compounds solubilized in the diesel oil was observed with or without OS.

Furthermore, when vigorous shaking was applied, marked emulsification of the model petroleum was observed. Probably these results indicated that strain DS-994 was in contact with diesel oil droplets, which dispersed in the growth medium and oxidized the sulfur component to sulfuric acid without significant influence of the OS in the model petroleum. Thus, sulfur degradation by this OS-tolerant bacterium was observed in a water-model petroleum two-liquid-phase system.

One drawback of the bacterium is its sulfuric acid production, which is an undesirable characteristic due to the corrosive action of sulfuric acid on pipes and tanks used in the process of desulfurization. Consequently, further studies on intermediates derived from the sulfur metabolism of the bacterium seem necessary to enable critical desulfurization in a water-OS two-liquid-phase system.

8.4 Organic Solvent-Tolerant Marine Cholesterol-Degrading Bacterium

Microbial cholesterol conversion occurs in aqueous media even when the reactants, such as steroids, are barely soluble in water. The reactants are often in a solid state in such media, and so the reaction rate is usually hindered by the limited availability of the substrate. For industrial microbial cholesterol degradation processes, some detergents, such as Tween and Span, have been mainly used to (1) form stable suspensions of the water-insoluble substrate during fermentation and (2) prevent simultaneous formation of solid particles. Although this detergent addition helps to increase the degradation rate, it does not prevent the simultaneous process of solid particle formation. Accordingly, as a reactant reservoir to keep the product and the substrate in a soluble form until the degradation is complete, the use of OSs is thought to be a logical way to prevent simultaneous formation of the solid particles.

We attempted cholesterol degradation in a water-OS two-liquid-phase system by an OS-tolerant marine bacterium isolated from the deep-sea (Moriya et al. 1995).

8.4.1 Isolation and Identification of OS-Tolerant Cholesterol-Degrading Bacteria

Screening OS-tolerant bacteria that possessed the ability to degrade cholesterol was done by modifying our previously described method (Moriya and Horikoshi 1993a). Deep-sea sediment slurry samples were aseptically collected by sterile core samplers from 1963 m depth of the Okushiri Ridge at 43°35' North latitude and 139°45 ' East longitude (639th dive of the manned submersible Shinkai 2000 in 1992). The samples (100 ml) were treated with the addition of 100 ml of benzene and kept for 7 days at ambient temperature. After treatment, the upper benzene layer was spread onto M-II agar medium (Moriya and Horikoshi 1993a) supplemented with 5% NaCl and 0.1% cholesterol. Incubation was for 5 days at 30°C. Fifty colonies that formed clear zones due to cholesterol degradation on M-II agar medium were picked up and purified as candidates for benzene-tolerant cholesterol degraders.

The candidate that showed the best growth and cholesterol degradation in the presence of 5% (v/v) benzene containing cholesterol 1 mg/ml was selected and tentatively designated strain ST-1. This isolate was gram-positive, non-spore-forming, strictly aerobic, catalase-positive, oxidase-positive, and nonmotile. It had a typical rodoccus growth cycle. Its cell wall peptidoglycan contains lysine. Optimal growth was at pH 6.0-8.0 and 35°C, respectively. The mole percent of G+C is 62. Based on the above morphological and biochemical characteristics (Jones and Collins 1984), ST-1 seemed to be a strain of *Arthrobacter* sp.

8.4.2 Solvent Tolerance of Cholesterol-Degrading Bacteria

To compare the OS tolerance and cholesterol conversion ability of the selected isolate, *Arthrobacter symprex* strain ATCC 6946, *Flavobacterium dehydrogenans* strain ATCC 13930, *Mycobacterium* sp. ATCC 29472, and *Mycobacterium smegmatis* strain ATCC 12549 were employed as control cholesterol degraders. Strains ATCC 13930 (Bocren and Laane 1987) and ATCC 29472 (Flygare and Larsson 1987) were used as representative androstene producers. These five strains were individually inoculated into each 10 ml of modified LB-II medium (Moriya and Horikoshi 1993a) at

Table 10. Growth in the presence of various organic solvents [a]

Strain	Organic solvents					
	Benzene (log *p* 2.1)	Toluene (log *p* 2.9)	*p*-Xylene (log *p* 3.3)	*n*-Hexane (log *p* 3.9)	*n*-Decane (log *p* 6.0)	*n*-Dodecane (log *p* 7.0)
ST-1	+	+	+	+	+	+
Arthrobacter symplex ATCC 6946	–	–	–	–	+	+
Flavobacterium dehydrogenans ATCC 13930	–	–	–	–	+	+
Mycobacterium sp. ATCC 29472	–	–	–	–	+	+
M. smegmatis ATCC 12549	–	–	–	–	+	+

[a] +, growth; –, no growth. Cell growth was determined by measuring viable cell concentration with and without organic solvent.

approximately 1×10^6 cells/ml and then incubated in the presence of 0.5 ml of various OSs with vigorous shaking for 3 days at 30°C. Six solvents including aromatic hydrocarbons benzene, toluene, and *p*-xylene and the aliphatic hydrocarbons *n*-hexane, *n*-decane, and *n*-dodecane, were used to measure the tolerance to and the effect of these OSs in a 1:20 (v/v) OS-water biphasic system. After incubation, cell growth was determined by measuring viable cell concentration by M-II and LB-II agar plate counts.

As shown in Table 10, the control strains could grow in either *n*-decane (log p 6.0) or *n*-dodecane (log p 7.0) but could not grow in the presence of *n*-hexane (log p 3.9) or aromatic hydrocarbons. In contrast to these controls, strain ST-1 could grow in benzene, toluene, *p*-xylene (log p 3.1), or *n*-hexane in addition to *n*-decane and *n*-dodecane. The control strains and strain ST-1 grew in the presence of 50% (v/v) *n*-decane and *n*-dodecane. Strain ST-1 also grew even in the presence of 10% (v/v) toluene, 10% *p*-xylene, and 10% *n*-hexane in addition to 5% benzene. The final viable cell concentration of strain ST- I reached approximately 3×10^9 cells/ml with and without OSs. Strain ST-1 was halotolerant and showed good growth in the presence of OS without any effects of NaCl in the range 0-12% (data not shown).

8.4.3 Cholesterol Degradation of OS-Tolerant Bacteria

The cholesterol degradation of strain ST-1 and control strains in a buffer-OS two-liquid-phase system was examined. Ten liter volumes of 1-day cultures of individual strains were centrifuged and washed twice with the same volume of sterilized sodium phosphate buffer (100 mM, pH 7.0). The washed and concentrated cells were suspended in 200 ml of the buffer. In batch experiments, 50 ml of the cell suspension was added to a 5-liter jar fermenter containing 950 ml of sodium phosphate buffer at approximately 1×10^9 cells/ml, and 1 liter of OS containing cholesterol at 1 mg/ml was overlaid with 1 liter of the above buffer containing cell suspension. Incubation was done at 30°C for 2 days with agitation at 200 rpm. The agitation was stopped and the culture broth was kept standing for 2 hours after the batch experiment described above. The upper solvent layer was removed and analyzed for cholesterol degradation. An organic solvent containing cholesterol was added to the buffered lower layer, and this lower layer was treated in the same way as above. The above procedure was repeated in five successive bath experiments.

Both cholesterol and androsta-1,4-diene-3,17-dione (ADD), which is one of the androstenes derived from cholesterol, were measured by HPLC of the two layers at 2-day intervals. The OS phase was evaporated, and the residue was dissolved in ethylacetate. The steroids in the ethylacetate solution were then determined by HPLC using a C_{18} column and a mobile phase composed of acetonitrile and water (50:50. v/v). When necessary to ensure identification, HPLC runs were repeated with samples spiked with standards (cholesterol, cholestenone, and ADD). Three additional controls were run: (1) without inoculation; (2) with cholesterol suspended in buffer layer without OS; and (3) with cholesterol suspended in buffer containing 0.1% Tween 80.

The results of a batch experiment are shown in Table 11. When strain ST-1 was incubated in the presence of benzene, toluene, *p*-xylene, or *n*-hexane, the degradation the initial cholesterol (1 mg/ml) was 22%, 22%, 24%, and 56%, respectively. In the experiments with control 2 and control 3, the cholesterol degradation was 68% and 78%, respectively. In these experiments we also detected ADD produced from cholesterol qualitatively. Therefore, this bacterium had the ability to cleave the side

Table 11. Cholesterol degradation in two-liquid-phase system

Solvent	Cholesterol degredation (%)				
	ST-1	*Arthrobacter symplex* ATCC 6946	*Flavobacterium dehydrogenans* ATCC 13930	*Mycobacterium* ATCC 29472	*Mycobacterium smegmatis sp.* ATCC 12549
Water (control 2)	68 (22)	44	52 (12)	49 (10)	53
Tween 80 (control 3)	78 (40)	55	60 (18)	59 (19)	62
Two-liquid system					
Benzene	22 (9)	5	7 (ND)	5 (ND)	4
Toluene	22 (9)	7	9	7	4
p-Xylene	24 (10)	8	9	8	6
n-Hexane	56 (54)	20	38 (44)	32 (40)	38
n-Decane	88 (70)	59	66 (60)	66 (62)	69
n-Dodecane	92 (78)	67	74 (65)	73 (66)	77

Values in parentheses indicate the percent production of anbrosta-1,4-dione-3,17-dione (ADD) derived from cholesterol.
ND, not detected.

chain of cholesterol presumably due to the activity of a cholesterol dehydrogenase. In contrast, the control strains could scarcely degrade cholesterol in the presence of aromatic hydrocarbons but could degrade cholesterol in an aqueous medium and in a medium supplemented with either *n*-hexane, *n*-decane, or *n*-dodecane. However, the percentages of both cholesterol degradation and ADD production by the control strains were lower than those obtained by strain ST-1 in aqueous media and a two-liquid-phase system. When the batch experiment was successively repeated five times using *n*-dodecane, cholesterol degradation was 92% and 90% in the first and second batches, respectively (data not shown). The degradation efficiency of cholesterol decreased gradually in the third to fifth batches. Changes in the viable cell concentration of the five strains used for batch experiments were measured by conventional agar-plate counts at daily intervals. When incubated in the presence of benzene, toluene, or *p*-xylene, 1×10^9 cells/ml of the initial viable cell concentration of strain ST-1 decreased gradually to approximately 3×10^8 cells/ml after a day and then remained constant. However, the viable cell concentrations of control strains decreased abruptly within a day, and no colony formation was observed on the agar plates after 2 days. When incubated in *n*-hexane, the cell viability of strain ST-1 was more than 50%, but that of controls was less than 20% at the end of the incubation. After incubation in aqueous media and either *n*-decane or *n*-undecane, all strains tested remained at the initial viable cell concentration.

The marked effect of organic solvent on cholesterol degradation and ADD production might stem from the differences in cell viability. In general, a positive correlation is found between the hydrophobicity of organic solvents and the nontoxicity for the biocatalyst. Organic solvents with a log p value above 4 are highly hydrophobic and generally show no toxic effects on biocatalysts. The log p value is a measure for hydrophobicity suitable for characterization of organic solvents. Accordingly, biocatalysts derived from aromatic hydrocarbon-tolerant microorganisms are of interest for further study. Enzymes associated with cholesterol degradation of strain ST-1, which shows tolerance to aromatic hydrocarbons such as benzene, toluene, and *p*-xylene, are also an attractive object to study in detail.

8.5 OS-Tolerant Polyaromatic Hydrocarbon-Degrading Bacteria

We have focused on the deep-sea environment to obtain an OS-tolerant bacterium that possesses the useful characteristics for application in industry (Abe et al. 1995). We describe the polyaromatic hydrocarbon-degrading ability of an OS-tolerant bacterium isolated from the deep-sea.

Table 12. Utilization of GP MicroPlate substrates by organic solvent-tolerant marine bacterium DS-1906 [a]

Substrate	Result	Substrate	Result
α-Cyclodextrin	+	β-Cyclodextrin	+
Dextrin	+	Glycogen	+
Inulin	+	Mannan	–
Tween 40	–	Tween 80	–
N-Acetyl-D-grucosamine	+	*N*-Acetyl-D-mannosamine	–
Amygdalin	–	L-Arabinose	+
D-Arabitol	–	Cellobiose	+
D-Fructose	+	L-Fucose	–
D-Galactose	–	D-Galactorinic acid	–
Gentiobiose	+	D-Gluconic acid	–
α-D-Glucose	+	*m*-Inositol	–
α-D-Lactose	–	Lactulose	–
Maltose	+	Maltotriose	+
D-Mannitol	–	D-Mannose	+
D-Melibiose	–	α-Methyl-D-galactoside	–
β-D-Galactoside	–	3-Methylglucose	+
α-Methyl-D-glucoside	–	β-Methyl-D-glucoside	–
α-Methyl-D-mannoside	–	Palatinose	+
D-Psicose	+	D-Raffinose	–
L-Rhamnose	–	Salicin	–
Sedoheptulosan	–	D-Sorbitol	–
Stachyose	–	Sucrose	+
D-Tagatose	–	D-Trehalose	+
Turanose	+	Xylitol	–
D-Xylose	+	Acetic acid	–
α-Hydroxybutyric acid	–	β-Hydroxybutyric acid	–
γ-Hydroxybutyric acid	–	*p*-Hydroxyphenyl acetic acid	–
α-Ketovaleric acid	+	L-Malic acid	
Methylpyruvate	+	Monomethyl succinate	–
Propionic acid	–	Pyruvic acid	+
Succinic acid	+	*N*-Acetyl-L-glutamic acid	+
Alaninamide	–	D-Alanine	+
L-Alanine	+	L-Alanyl-glycine	+
L-Asparagine	+	L-Glutamic acid	+
Glycyl-L-glutamic acid	+	L-Pyroglutamic acid	–
L-Serine	+	2,3-Butanediol	–
Glycerol	+	Adenosine	+
Inosine	+	Thymidine	+
Uridine	+	Adenosine-5'-monophosphate	+
Thymidine-5'-monophosphate	+	Uridine-5'-monophosphate	+
Fructose-6-phosphate	+	Glucose-1-phosphate	–
Glucose-6-phosphate	+	D,L-α-Glycerol phosphate	+

[a] Only the substrates with positive results are shown.
+, positive; –, negative

8.5.1 Isolation and Identification of OS-Tolerant Polyaromatic Hydrocarbon-Degrading Bacteria

Screening of OS-tolerant polyaromatic hydrocarbon-degrading bacteria was carried out by a slight modification of the method described previously (Moriya and Horikoshi 1993a). Deep-sea sediment slurry samples were aseptically collected from a depth of 1168 m in Sagami Bay, which is situated at 34°99' 8" N and 139°13'5" E, during dive 2K-654 of the manned submersible Shinkai 2000 in 1992. Deep-sea sediment slurry (5 ml) was transferred to flasks containing 10 ml of artificial seawater. Benzene was overlaid on the aqueous layer (1:1), and these flasks were then closed with silicone-rubber stoppers. The cultures were incubated with shaking at 200 rpm for a week at 30°C. After incubation the benzene layer separated from the artificial seawater, and a portion of the benzene layer was spread directly onto plates of M-II and LB-II media (Moriya and Horikoshi 1993a, 1993b) containing 1% naphthalene. The plates were incubated for a week at 30°C. Microorganisms that formed clear zones around their colonies were selected and purified.

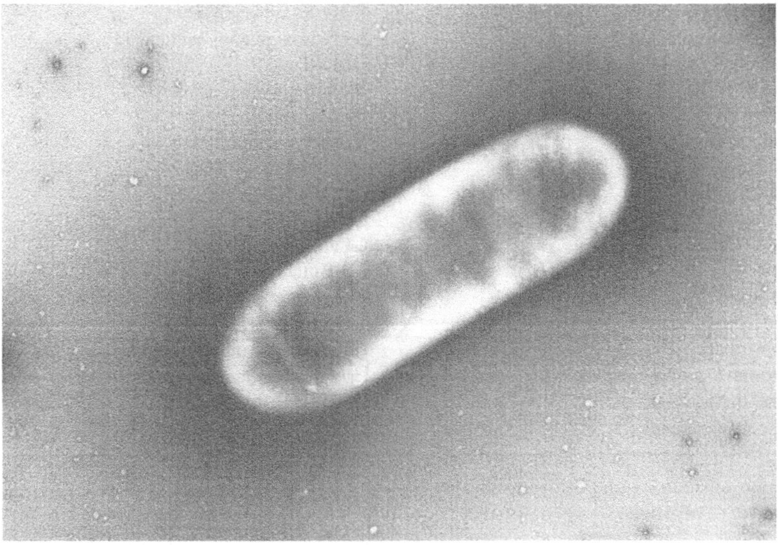

Fig. 4. Electron micrograph of the organic solvent-tolerant bacterium DS-1906. Strain DS-1906 was gram-positive, spore-forming, aerobic nonmotile, rod-shaped, and had flagella.

The isolate was selected as a polyaromatic hydrocarbon degrader that also exhibited organic solvent tolerance; it was named strain DS-1906. Strain DS-1906 was gram-positive, spore forming, aerobic, oxidase- and catalase-positive, gelatin-liquefying, and nonmotile. The ability of this strain to utilize various organic substances is shown in Table 12, with only the substrates that gave positive result listed. It is rod-shaped (1.0-2.0 × 2.0-4.0 μm) and has flagella (Fig 4). The effects of temperature (5°-50°C), NaCl concentration (0-3.5 M), and pH (2-11) on the growth of strain DS-1906 during a 24-h incubation using LB-II medium are shown in Fig 5. DS-1906

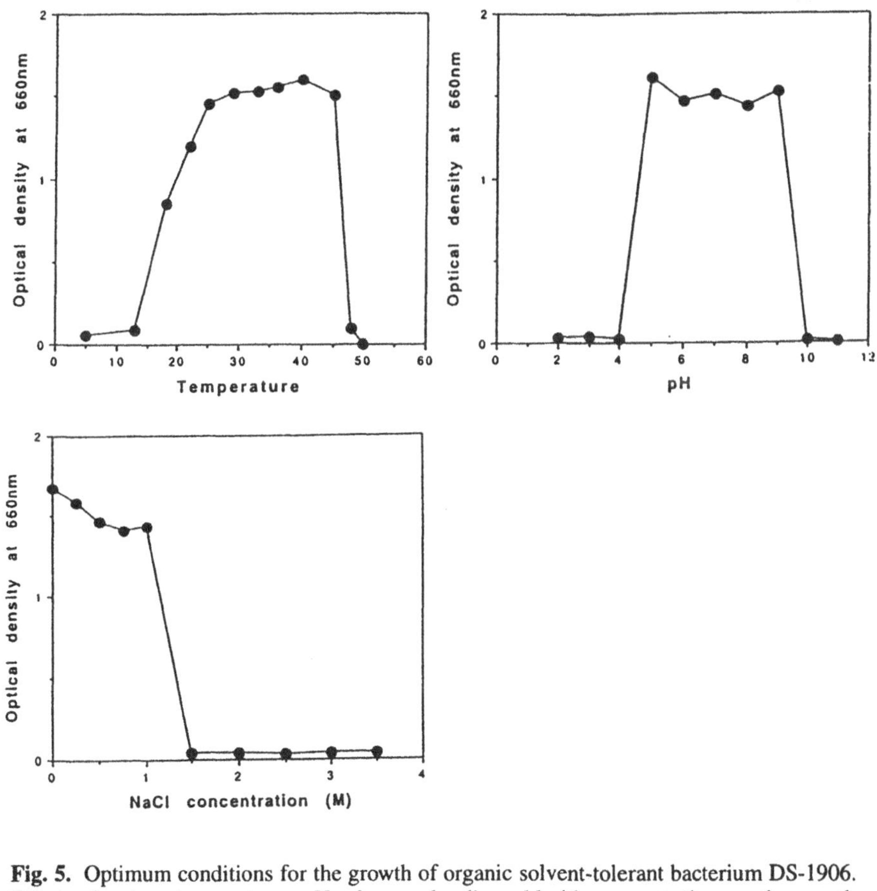

Fig. 5. Optimum conditions for the growth of organic solvent-tolerant bacterium DS-1906. Effects of various temperatures, pH values, and sodium chloride concentrations on the growth of the bacterium were studied using LB-II medium. Strain DS-1906 grew well at a wide range of temperatures (20°–45°C) with an optimum at 40°C. This strain also grow well at a wide range of pH values (pH 5–9), with a slightly acidic value (pH 5–6) enabling best growth. NaCl concentrations of 0–0.5 M enabled good yields by this strain

grew well at a wide temperature range (20°-45°C) with the optimum at 40°C (Fig. 5). This strain also grew well at a wide range of pH (5-9), and optimum pH for growth was pH 5-6 (Fig. 5). It grew well at NaCl concentrations over the range of 0-0.5 M (Fig. 5). Based on the biochemical and morphological characteristics, in addition to its polyaromatic hydrocarbon-degrading ability and organic solvent tolerance, this strain appeared to be a strain of *Bacillus* sp.

8.5.2 Organic Solvent Tolerance of *Bacillus* sp. DS-1906

The OS tolerance of the strain DS-1906 was compared with that of control strains *Escherichia coli* IFO 3806, *Bacillus subtilis* IFO 3009, and the toluene-tolerant *Pseudomonas putida* IH-2000. These strains were individually incubated in culture tubes containing 5 ml of LB-II medium with shaking at 120 rpm for 2 days at 30°C. Cultures were seeded into 5 ml of fresh medium to an optical density of 0.03 at 660 nm (OD_{660}), and various OSs were added to each tube at 10% concentration. Incubation was then carried out for 3 days at 30°C. After incubation, the OS tolerance of each strain was estimated by determining cell growth (OD_{660}). Various hydrocar-

Table 13. Comparison of organic solvent tolerance between strain DS-1906 and soil strains [a]

Solvent		Tolerance			
	Log *p*	*E. coli* IFO 3806	*B. subtilis* IFO 3009	*P. putida* IH 2000	DS-1906
n-Dodecane	7.0	–	–	+ (1.90)	+ (2.42)
n-Decane	6.0	–	–	+ (1.90)	+ (2.40)
n-Nonane	5.0	–	–	+ (1.88)	+ (2.35)
n-Hexylether	5.1	–	–	+ (1.90)	+ (2.30)
n-Octane	4.9	–	–	+ (1.83)	+ (2.30)
Isooctane	4.8	–	–	+ (1.80)	+ (2.27)
Cyclooctane	4.5	–	–	+ (1.80)	–
Diphenylether	4.2	–	–	+ (1.80)	+ (2.25)
n-Hexane	3.9	–	–	+ (1.78)	+ (2.25)
n-Propylbenzene	3.8	–	–	+ (1.75)	–
o-Dichlorobenzene	3.6	–	–	+ (1.75)	–
Cyclohexane	3.4	–	–	+ (1.70)	+ (2.10)
Ethylbenzene	3.3	–	–	+ (1.70)	–
p-Xylene	3.1	–	–	+ (1.65)	–
Styrene	2.9	–	–	+ (1.65)	–
Toluene	2.8	–	–	+ (1.58)	–
1-Heptanol	2.4	–	–	+ (1.53)	–
Benzene	2.0	–	–	–	+ (1.80)

The initial cell concentration was 1×10^5 cells/ml. Incubation was for 3 days at 30°C. Growth was determined by measuring the optical density at 660 nm (OD_{660}).
+, growth (> 5×10^8 cells/ml); –, no growth.

bons, including benzene, 1-heptanol, toluene, styrene, *p*-xylene, ethylbenzene, cyclohexane, *o*-dichlorobenzene, *n*-propylbenzene, *n*-hexane, diphenylether, cyclooctane, isooctane, *n*-octane, *n*-hexylether, *n*-nonane, *n*-decane, and *n*-dodecane, were employed as organic solvents. Naphthalene, fluorene, phenanthrene, anthracene, pyrene, chrysene, and 1,2-benzopyrene were used as polyaromatic compounds.

The results are shown in Table 13. When incubated for 3 days, the *E. coli* and *B. subtilis* strains did not grow in the presence of any of the organic solvents tested. *P. putida* strain IH-2000 grew in the presence of all the organic solvents except for benzene (Table 13). In contrast to these strains, strain DS-1906 grew even in the presence of 10% benzene and some other solvents, but it could not grow in the presence of many other OSs, such as 1-heptanol, toluene, styrene, *p*-xylene, and cyclooctane.

8.5.3 Polyaromatic Hydrocarbons Degradation by *Bacillus* sp. DS-1906

Degradation of naphthalene by DS-1906 was compared with that by the control strains in a medium-OS two-liquid-phase system. Bacteria were incubated in 9 ml of LB-II medium with 1 ml of OSs containing 1% naphthalene. Two additional controls were prepared: (1) without inoculation; and (2) without organic solvent. Incubation was for 1 week at 30°C with shaking at 200 rpm. At the end of the incubation, the remaining polyaromatic hydrocarbons were extracted and analyzed. The percent degradation of polyaromatic hydrocarbons was given as a difference between the control and the inoculated samples. Naphthalene was extracted twice with benzene containing 0.01% *n*-tetracosane as an internal standard. Residual water in the benzene extracts was removed with anhydrous sodium sulfate. The anhydrous extracts were transferred to a sample vial and stored at -20°C before gas chromatographic (GC) analysis on a Shimadzu model GC-14B gas chromatograph equipped with a flame ionization detector (FID) and a 25-m fused silica capillary column (CBJ1; Shimadzu, Tokyo, Japan). The separation conditions were as follows: injector temperature 250°C; detector temperature 320°C; column temperature program 50°-280°C at 23°C/min., holding at 280°C for 5 min. The carrier gas was helium at a flow rate of 6 ml/min.

We observed a 1-day latency period before growth in the presence of *n*-hexane, and then DS-1906 grew rapidly. After the 1-day of latency period, residual naphthalene decreased gradually. After 7 days the value of residual naphthalene reached about 50% (Fig. 6). As shown in Table 14, the control strains did not degrade polyaromatic hydrocarbons, whereas that by strain DS-1906 was 48.3% in the presence of *n*-hexane. Furthermore, strain DS-1906 could degrade various polyaromatic compounds, such as fluorene, phenanthrene, anthracene, pyrene, chrysene, and 1,2-benzopyrene in a medium-OS (benzene) two-liquid-phase system (Table 15). When polyaromatic hydrocarbons were solubilized in an organic solvent, the percent degradation was kept at a constant value, in contrast to the highly variable degradation in the absence an organic solvent. The percent degradation in the presence of an organic solvent was higher than that obtained from cultures in which polyaromatic hydrocarbons were suspended in the medium without an organic solvent.

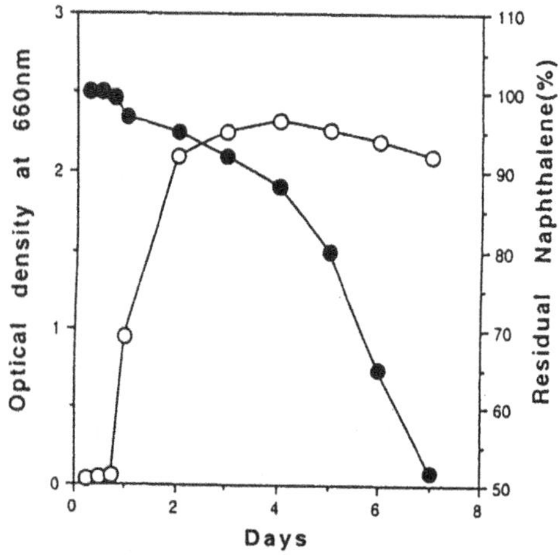

Fig. 6. Degradation of naphthalene by DS-1906 in a medium–*n*-hexane two-liquid-phase system for a week at 30°C. *Open circles*, optical density at 660 nm; *filled circles*, residual naphthalene

Table 14. Polyaromatic hydrocarbons degradation by solvent-tolerant strains

Solvent	Degradation (%)			
	E. coli IFO 3806	*B. subtilis* IFO 3009	*P. putida* IH 2000	DS-1906
n-Decane	NG	NG	ND	ND
n-Octane	NG	NG	ND	33.2
Isooctane	NG	NG	ND	28.4
Hexane	NG	NG	ND	48.3
Cyclohexane	NG	NG	ND	26.3
Toluene	NG	NG	ND	NG
Benzene	NG	NG	ND	17.2
No OS	ND	ND	ND	11.7-35.5

Degradation of polyaromatic hydrocarbons and naphthalene was examined in the medium–organic solvent (OS) two-liquid-phase system. Bacteria were incubated in 9 ml of LB-II medium containing 1 ml of organic solvents (10%) containing 1% naphthalene for 7 days at 30°C.

$$\text{Degradation (\%)} = \frac{\text{residual polyaromatic hydrocarbons conc. of an additional control} - \text{residual polyaromatic hydrocarbons conc. of inoculated samples}}{\text{residual polyaromatic hydrocarbons conc. of an additional control}} \times 100$$

NG, no growth; ND, no degradation.

Table 15. Polyaromatic hydrocarbon degradation by solvent-tolerant strain DS-1906

Substrate	Degradation (%)	
	Without OS	With OS
Naphthalene	11.7-35.5	17.2
Fluorene	10.5-32.0	17.4
Phenanthrene	10.2-28.7	16.8
Anthracene	2.6-28.5	15.2
Pyrene	2.3-26.7	13.2
Chrysene	3.0-25.8	14.5
1,2-Benzopyrene	2.7-24.8	12.6

Degradation was calculated by the following equation:

$$\text{Degradation (\%)} = \frac{\begin{array}{c}\text{residual polyaromatic}\\ \text{hydrocarbons conc. of an}\\ \text{additional control}\end{array} - \begin{array}{c}\text{residual polyaromatic}\\ \text{hydrocarbons conc. of}\\ \text{inoculated samples}\end{array}}{\text{residual polyaromatic hydrocarbons conc. of an additional control}} \times 100$$

OS, organic solvent (benzene).

8.5.4 Discussion

We isolated an OS-tolerant bacterium from deep-sea sediment samples after treatment with 50% benzene. This bacterial strain DS-1906, was an OS-tolerant polyaromatic compounds degrader. It displayed tolerance to a variety of organic solvents at a high concentration (> 10% v/v). Moreover, strain DS-1906 showed polyaromatic compound-degrading ability in the presence of an organic solvent. It was able to degrade 48.3% naphthalene solubilized in *n*-hexane. The amounts of naphthalene thus degraded by this bacterium was higher than when the naphthalene was suspended in the medium without organic solvent. When cultures were shaken, marked emulsification of the organic solvent was observed. This result possibly indicated that the OS-tolerant bacterium was in contact with organic solvent dispersed in the medium, which could allow the bacterium to degrade the polyaromatic hydrocarbon substrates efficiently.

Although we tried to isolate OS-tolerant plasmids from this OS-tolerant bacterium (unpublished data), no plasmids were obtained. Accordingly the OS tolerance is thought to be encoded by information in genomic genes.

Inoue and Horikoshi (1989; 1991a) used a partition coefficient (p) or organic solvents as a toxicity index. Log p is defined as the common logarithm of p of a given organic solvent between *n*-octanol and water (Hansch et al. 1963; Hansch and Fujita 1964). Strain IH-2000 could grow on organic solvents with log p values higher than that of 1-heptanol (log p 2.4). Strains DS-711 (Moriya and Horikoshi 1993a), DS-944 (Moriya and Horikoshi 1993b), and DS-1906 showed excellent OS tolerance. However, these deep-sea bacteria had different OS tolerance than that seen with strain IH-2000, as shown Table 2. Generally, gram-positive bacteria have lower OS tolerance than gram-negative bacteria (Inoue and Horikoshi 1989, 1991a). Therefore, the benzene tolerance of gram-positive strain DS-1906 is particularly significant.

We suspect that the mechanism of OS tolerance of the benzene-tolerant bacterium DS-1906 is different from that of the toluene-tolerant *P. putida* strain IH-2000. The mechanism of OS tolerance is thought to be due not only to the differences between cell surface components but also to the existence of an OS-tolerant enzyme or OS-deactivating enzymes. It is likely that OS-emulsifying and -solubilizing substances also play a part in diminishing the toxicity of the organic solvent.

8.6 A Organic Solvent Hydrocarbon-Degrading Marine Yeast

We have reported the isolation, characteristics, and applications such as the degradation of polluting petroleum by microorganisms that tolerate organic solvents (Aono et al. 1991a, 1992; Nakajima et al. 1992; Moriya and Horikoshi 1993a, 1993b; Fukumaki et al. 1994). Some such microorganisms can degrade hydrocarbons and organic sulfur compounds in the presence of organic solvents (Moriya and Horikoshi 1993a, 1993b). The organisms might be used for the microbial conversion of water-insoluble substances. We attempted to isolate and investigate a marine yeast that tolerates organic solvents and explore its hydrocarbon-degrading ability.

8.6.1 Isolation and Solvent Tolerance of Marine Yeast from Deep-Sea

Deep-sea sediment samples were collected aseptically with sterile core samplers from a 1200 m depth in Sagami Bay (35°00' N, 139°13'E) and at 1977 m in Suruga Bay (34°43'N, 138°36'E) with a manned submersible vessel, the Shinkai 2000. The samples were transferred to sterile sample bottles and stored at 4°C for 1 week to several months until sample treatment. For yeast growth, YPG medium, which contained 10 g of yeast extract (Difco), 20 g of Bacto-peptone (Difco), and 10 g of glucose per liter of distilled water; YPTG medium, which contained 0.5 g of yeast extract, 4 g of bacto-peptone, 1 g of trypticase peptone (BBL), 0.5 g of glucose, 3 g of $(NH_4)_2SO_4$, 0.1 g of K_2HPO_4, 0.5 g of $Mg_2SO_4{\cdot}7H_2O$, 1 g of $CaCl_2{\cdot}2H_2O$, and 30 g of NaCl per liter of distilled water; and yeast nitrogen base (YNB) medium (Difco) were used. For YPTG agar medium, agar was added at a concentration of 2.0%. (w/v) to YPTG medium. When needed, both penicillin G potassium (Wako, Osaka, Japan) and streptomycin sulfate (Nacalai Tesque, Kyoto, Japan) were added to these media at 0.010% each, and the pH was adjusted to 2.5.

For isolation of yeasts that could tolerate organic solvents, the deep-sea sediment

Table 16. Tolerance of organic solvents by yeast from collections

Strain	Tolerance								
	n-Dodecane 7.0	*n*-Decane 6.0	*n*-Nonane 5.5	Hexyl ether 5.1	*n*-Octane 4.9	Iso-octane 4.8	Cyclo-octane 4.5	Diphenyl ether 4.2	*n*-Hexane 3.9
Baker's yeast (Oriental Yeast Co.)	+	+	+	–	–	–	–	–	–
Ambrosiozyma platypodis IFO 1471	+	+	–	+	–	–	–	+	–
Brettanomyces bruxellensis IFO 1589	–	–	–	–	–	–	–	–	–
Candida glabrata ATCC 15126	+	+	+	–	–	–	–	–	–
C. glabrata ATCC 36909	+	+	+	–	–	–	–	–	–
C. glabrata IFO 0622	+	+	+	–	–	–	–	–	–
C. maltosa IFO 1977	+	+	+	+	–	+	–	–	–
C. saka IFO 1634	+	+	+	+	+	+	–	+	–
C. tropicalis IFO 1400	+	+	+	+	–	+	–	–	–
C. utilis IFO 1086	+	+	–	+	–	–	–	–	–
Cryptococcus albidus IFO 0378	+	–	–	–	–	–	–	–	–
Cr. lumicolus IFO 0760	+	–	–	–	–	–	–	–	–
Cr. laurentii IFO 0609	+	–	–	–	–	–	–	–	–
Debaryomyces hansenii IFO 0083	+	+	–	+	–	–	–	–	–
D. polymorphus IFO 1166	+	+	+	+	–	+	–	–	–
Hansenula anomala ATCC 36904	+	–	–	–	–	–	–	–	–
Issatchenkia scutulata IFO 1895	+	+	–	+	–	–	–	–	–
Kloeckera corticis IFO 0631	+	–	–	–	–	–	–	–	–
Kluyveromyces lactis IFO 1090	+	–	–	–	–	–	–	–	–
K. lactis IFO 1267	+	+	+	–	–	–	–	–	–
K. marxianus ATCC 36906	+	+	–	+	–	–	–	–	–
K. marxianus IFO 10005	+	–	–	–	–	–	–	–	–
K. thermotolerans IFO 1985	+	+	–	+	–	–	–	–	–
Pichia anomala IFO 10213	+	–	–	–	–	–	–	–	–
P. anomala IFO 0119	+	+	+	+	–	–	–	–	–
P. angusta IFO 1476	+	–	–	+	–	–	–	+	–
P. bimundalis IFO 1366	+	+	–	+	–	+	–	–	–
P. canadensis IFO 0973	+	–	–	+	–	–	–	+	–
P. canadensis IFO 0976	+	–	+	+	–	–	–	–	–
P. carsonii IFO 0795	–	–	–	–	–	–	–	–	–
P. ciferrii IFO 0793	+	–	–	–	–	–	–	–	–
P. fabianii IFO 1370	+	+	–	–	–	+	–	–	–
P. henricii IFO 1477	+	+	–	+	–	–	–	+	–
P. jadinii IFO 0987	+	+	–	+	–	–	–	–	–
P. membranaefaciens ATCC 36908	+	+	–	+	–	–	–	–	–
P. petersonii IFO 1372	+	+	+	+	–	+	–	+	–
Rhodosporidium toruloides IFO 8766	+	–	–	–	–	–	–	–	–
Rhodotorula rubra IFO 1539	+	+	–	+	–	–	–	–	–
Saccharomyces bayanus IFO 1127	+	–	–	–	–	–	–	–	–
Sacch. capeusis ATCC 36899	+	+	+	–	–	–	–	–	–
Sacch. cerevisiae ATCC 4098	+	–	–	–	–	–	–	–	–
Sacch. cerevisiae ATCC 60782	+	–	–	–	–	–	–	–	–
Sacch. cerevisiae IFO 0565	+	–	–	–	–	–	–	–	–
Sacch. cerevisiae IFO 1333	+	–	–	–	–	–	–	–	–
Sacch. cerevisiae IFO 1334	+	–	–	–	–	–	–	–	–
Sacch. cerevisiae IFO 2347	+	–	–	–	–	–	–	–	–
Sacch. cerevisiae IFO 10159	+	–	–	–	–	–	–	–	–
Sacch. diastaticus ATCC 36902	+	+	+	–	–	–	–	–	–
Sacch. pastorianus IFO 2027	+	–	–	–	–	–	–	–	–
Schizosaccharomyces pombe IFO 1608	+	–	–	–	–	–	–	–	–
Sporobolomyces antarcticus IFO 10260	+	+	–	+	–	–	–	–	–
Willopsis beijerinckii IFO 0941	+	+	–	+	–	–	–	–	–
W. mrakii ATCC 10743	+	+	+	+	–	–	–	–	–
W. mrakii IFO 0897	+	+	+	+	–	–	–	–	–
W. saturnus IFO 0117	+	+	+	+	–	–	–	–	–
Yarrowia lipolytica IFO 1548	+	–	–	–	–	–	–	–	–
Zygosaccharomyces fermentati IFO 0479	–	–	–	–	–	–	–	–	–
Zygosacch. rouxii IFO 1914	+	–	–	–	–	–	–	–	–

Each strain was incubated in YPG medium with 10% organic solvant at 30°C. After overnight incubation, yeast growth was evaluated by measurement of the OD_{660} of the midium.
+, grew; –, did not grow.
The number accompanying each solvent is the log *p*.

Table 17. Tolerance of organic solvents by marine yeast isolated from deep-sea sediments

Strain	Tolerance								
	n-Dodecane 7.0	*n*-Decane 6.0	*n*-Nonane 5.5	Hexyl ether 5.1	*n*-Octane 4.9	Iso-octane 4.8	Cyclo-octane 4.5	Diphenyl ether 4.2	*n*-Hexane 3.9
Y-1	+	+	–	+	–	–	–	–	–
Y-2	+	+	–	–	–	–	–	–	–
Y-3	+	+	+	+	–	–	–	+	–
Y-4	+	+	–	+	–	–	–	+	–
Y-5	+	+	–	+	–	–	–	–	–
Y-6	+	+	–	+	–	–	–	–	–
Y-7	+	+	–	+	–	–	–	–	–
Y-8	+	+	–	+	–	–	–	–	–
Y-9	+	+	–	–	–	–	–	–	–
Y-10	+	+	–	+	–	–	–	+	–
Y-11	+	+	+	+	–	–	–	–	–
Y-12	+	+	+	+	–	–	–	–	–
Y-13	+	+	–	+	–	–	–	–	–
Y-14	+	+	–	+	–	–	–	–	–
Y-15	+	–	–	+	–	+	–	–	–
Y-16	+	+	–	+	–	–	–	–	–
Y-17	+	+	+	+	–	–	–	+	–
Y-18	+	+	+	+	–	–	–	–	–
Y-19	+	–	–	–	–	–	–	–	–
Y-20	+	+	–	+	–	–	–	–	–
Y-21	+	+	+	+	–	–	–	–	–
Y-22	+	+	+	+	–	–	–	–	–
Y-23	+	+	–	+	–	–	–	–	–
Y-24	–	–	–	–	–	–	–	–	–
Y-25	+	–	–	–	–	–	–	–	–
Y-26	+	–	–	+	–	–	–	–	–
Y-27	+	+	–	+	–	–	–	–	–
Y-28	+	–	–	–	–	–	–	–	–
Y-29	+	–	–	–	–	–	–	–	–
Y-30	+	+	–	+	–	–	–	–	–
Y-31	+	–	–	+	–	–	–	–	–
Y-32	+	–	–	–	–	–	–	–	–
Y-33	+	–	–	–	–	–	–	–	–
Y-34	+	+	+	+	–	–	–	–	–
Y-35	+	+	+	+	–	+	–	+	–
Y-36	+	+	+	+	–	+	–	+	–
Y-37	+	+	–	+	–	–	–	–	–
Y-38	+	+	–	+	–	+	–	–	–
Y-39	+	+	–	+	–	–	–	–	–
Y-40	+	+	+	+	+	+	+	+	–
Y-41	+	–	+	+	–	–	–	–	–
Y-42	+	+	–	+	–	–	–	–	–
Y-43	+	+	+	+	–	–	–	+	–

Each strain was incubated in YPG medium with 10% organic solvant at 30°C. After overnight incubation, yeast growth was evaluated by measurement of the OD_{660} of the midium.
+, grew; –, did not grow.
The number accompanying each solvent is the log R.

samples were treated with 50% (v/v) *n*-hexane for 2 days by a slight modification of the described elsewhere method (Moriya and Horikoshi 1993a) Some organic solvent was spread directly on YPTG agar medium with penicillin and streptomycin. Incubation was at 20°C for 1 week. After incubation, yeast-like colonies growing on the medium were picked up and purified.

Tolerance of organic solvents by yeast from collections is shown in Table 16. These yeast-type strains were purchased from the Institute for Fermentation (Osaka), the American Type Culture Collection, and Oriental Yeast Co.

Altogether 43 yeast strains were isolated and purified from deep-sea sediment sample treated with *n*-hexane. Tolerance of organic solvents was examined on YPG medium. *n*-Dodecane, *n*-decane, *n*-nonane, hexyl ether, *n*-octane, isooctane, cyclooctane, diphenyl ether, *n*-hexane, and kerosene were the organic solvents used. A 10-μl aliquot of a 1-day culture was used to inoculate 4.5 ml of YPG medium in test tubes. Then 0 ml (control), 0.5 ml (10% v/v), 1.5 ml (25%), or 4.5 ml (50%) of an organic solvent was overlaid on the medium. Incubation was at 30°C for 2 days with shaking at 260 rpm. At the end of the incubation the concentration of viable cells was measured by conventional dilution-plate counting. Yeast strains that had the same concentration of viable. Tolerance of organic solvents by these strains is shown in Table 17. Cells with and without the solvent were regarded as tolerating the solvent.

8.6.2 Identification and Characteristics of Marine Yeast Strain Y-40

One of the deep-sea isolates, strain Y-40, tolerated the aliphatic hydrocarbons *n*-dodecane, *n*-decane, *n*-nonane, *n*-octane, isooctane, and cyclooctane and the ethers hexyl ether and diphenyl ether. This strain could grow even in the presence of 50%

Table 18. Tolerance of organic solvents by marine yeast strain Y-40

Solvent	Concentration (%)	Cell growth (OD_{660})
None	–	21.5
n-Octane	10	16.3
	25	9.4
	50	20.1
Isooctane	10	22.2
	25	19.1
	50	17.4
Cyclooctane	10	12.7
	25	7.9
	50	7.0
Kerosene	10	20.5
	25	14.1
	50	17.8

After incubation at 30°C for 24 h, yeast growth was evaluated by measurement of the OD_{660}.

(v/v) organic solvents (Table 18). Some other deep-sea isolates and some type strains grew in the presence of diphenyl ether, which had the lowest log p value of the solvents used; but they could not grow in cyclooctane. Accordingly, we selected strain Y-40 as a possible hydrocarbon degrader.

Characteristics of strain Y-40 are summarized in Table 19. The yeast cells were short ovals and formed budding cells during growth in YPG medium. The strain

Table 19. Characteristics of marine yeast strain Y-40

Form	3.0 – 5.0 × 4.0 – 9.0 μm, short, oval
Fermentation	
D-Glucose	+
Maltose	–
Galactose	–
Lactose	–
Sucrose	–
Assimilation of carbon compounds	
Galactose	–
Maltose	–
D-Xylose	–
D-Ribose	–
Trehalose	+
Cellobiose	–
Erythritol	–
D-Mannitol	+
D-Glucitol	+
Succinic acid	–
Methanol	+
Raffinose	–
Sucrose	–
L-Arabinose	–
Lactose	–
L-Rhamnose	+
Soluble starch	–
Ribitol	+
Inositol	–
Glycerol	+
Citric acid	–
Ethanol	+
Assimilation of nitrate	+
Growth in vitamin-free medium	+
Splitting of arbutin	–
Tolerance to NaCl	7%
Tolerance to cycloheximide	–
Growth at 37°C	+
DNase	–
Urease	–
Lipase	–

–, not found; +, found.

fermented only D-glucose of the sugars tested. It grew in YNB medium, which contained trehalose, L-rhamnose, ribitol, D-mannitol, D-glucitol, or glycerol as the sole carbon source. Based on its morphological and biochemical characteristics it seemed to be a variant strain of *Candida*. Strain Y-40 could not grow in YNB medium containing any of the organic solvents tested as the sole carbon source (data not shown).

8.6.3 Degradation of *n*-Alkanes in Kerosenes

The ability of yeasts to assimilate *n*-alkanes and organic solvents was examined with 5 ml of YNB medium to which *n*-undecane, *n*-tridecane, *n*-tetradecane, *n*-pentadecane, *n*-hexadecane, or one of the organic solvents listed above was added at 0.5% as the sole carbon source. Incubation was at 30°C. Strains that grew within a week were regarded as being hydrocarbon assimilators.

Degradation of *n*-alkanes in kerosene was examined in YPG medium under a mixture of organic solvents. A 10 µl aliquot from a 1-day culture was used to inoculate 4 ml of YPG medium in a test tube; 1 ml of an organic solvent that contained 10% kerosene was then overlaid on the medium. *n*-Octane, isooctane, and cyclooctane were the organic solvents used. The control with no organic solvent and the other control without inoculation were used. Incubation was for 2 days at 30°C with shaking at 200 rpm. At the end of the incubation, the *n*-alkanes remaining in the kerosene were extracted and analyzed. The amount of degraded *n*-alkanes was recorded as the difference between one of the controls and the inoculated sample or control.

n-Hexane (8 ml) containing 0.2 mg *n*-tetracosane as the internal standard was

Table 20. Degradation of *n*-alkanes in organic solvents by yeasts

n-Alkane	*n*-Alkane degradation (mg)					
	Marine yeast Y-40		*Candida tropicalis* IFO 1400		*Yarrowia lipolytica* IFO 1548	
	None	*n*-Octane	None	*n*-Octane	None	*n*-Octane
n-Nonane	0.59	1.04	1.20	ND	0.68	NG
n-Decane	1.15	1.92	2.40	ND	1.28	NG
n-Undecane	1.24	2.38	2.80	ND	1.34	NG
n-Dodecane	1.29	2.04	2.38	ND	0.85	NG
n-Tridecane	1.11	1.73	1.94	ND	0.94	NG
n-Tetradecane	0.65	1.02	1.12	ND	0.78	NG
n-Pentadecane	0.40	0.48	0.53	ND	0.30	NG
n-Hexadecane	0.12	0.16	0.24	ND	0.17	NG

Yeast was incubated in 4 ml of YPG with 1 ml (20%) octane containing 2% kerosene at 30°C for 3 days. The control was incubated without *n*-octane.

ND, no degradation; NG, no growth.

added to the incubated cultures. The *n*-alkanes remaining were extracted twice with *n*-hexane. Water in the *n*-hexane extracts was removed with anhydrous sodium sulfate. The anhydrous extracts were transferred to a sample vial and stored at -20°C until tested by gas chromatography, which was done with a Shimadzu model GC-14B gas chromatograph equipped with a flame ionization detector. A glass column (3.2 mm i.d. × 2.1 m) packed with 10% silicone OV-101 on 60/80 mesh Chromosorb was used. Nitrogen as the carrier gas was run through the column at the rate of 40 ml/min. The temperature was programmed to rise from 35° to 280°C at the rate of 13°C/min. Peak areas were measured. Column efficiency and detector response were measured using a mixture of *n*-alkanes with carbon numbers C_9–C_{16}.

For a comparison of the hydrocarbon-degrading ability of strain Y-40 with those of two strains that assimilate *n*-alkane (*Candida tropicalis* IFO 1400 and *Yarrowia lipolytica* IFO 1548), kerosene was used because it contains a mixture of *n*-alkanes. The results of degradation of *n*-alkanes are shown in Table 20. The three strains tested degraded *n*-alkanes in the absence of an organic solvent. Of the three strains, IFO 1400 degraded *n*-alkanes in kerosene without an organic solvent most. Neither IFO 1400 nor IFO 1548 degraded *n*-alkanes in the presence of any of the three organic solvents tested. Y-40 degraded *n*-alkanes in the presence of *n*-octane. The amount of *n*-alkanes in kerosene with *n*-octane degraded by strain Y-40 was greater than that without *n*-octane. In the presence of isooctane or cyclooctane, strain Y-40 grew but did not degrade *n*-alkanes (data not shown).

8.6.4 Discussion

In an earlier paper (Moriya and Horikoshi 1993a), benzene (log p 2.0) was used for sample treatment, but its toxicity was too great for yeast. Accordingly, we used *n*-hexane (log p 3.9) here because of its lower toxicity. Most of the marine yeasts isolated did not grow in the presence of *n*-hexane but did survive in its presence, although the type strains did not survive (data not shown). These results showed that treatment with an organic solvent was useful for isolating yeasts that tolerated such solvents.

Some *Candida* species (*C. maltosa, C. sake, C. tropicalis*) had tolerance to some organic solvents (e.g., *n*-octane and isooctane) that have a low log p value. Species in the genera *Saccharomyces* and *Williopsis* had no tolerance to *n*-octane or isooctane. Yeast usually takes a Gram strain, so the isolation of a *Candida* sp. that tolerated organic solvents was of interest.

A more recent study showed that the log p value is not always related to OS tolerance (Moriya and Horikoshi 1993a). Many yeast strains tolerated hexyl ether (log p 5.1), isooctane (log p 4.8), and diphenyl ether (log p 4.2), but few strains grew in the presence of *n*-nonane (log p 5.5), *n*-octane (log p 4.9), or cyclooctane (log p 4.5). There was no relationship between the log p value and tolerance of organic solvents. None of the marine yeasts isolated grew in the presence of *n*-nonane, *n*-octane, or *n*-hexane; but they could survive. These findings show that the yeasts have enzymes that degrade organic OS-solvents, and that the influence of OS-emulsifying and OS-solubilizing substances and the difference of cell wall structure are relevant. The results suggest that these organic solvents were related to the metabolism of the yeast cells.

Although *C. tropicalis* IFO 1400 and *Y. lipolytica* IFO 1548 degraded *n*-alkanes in kerosene without other organic solvents, neither strain degraded *n*-alkanes in the presence of 20% (v/v) *n*-octane, isooctane, or cyclooctane, which have highly toxic effects on yeast growth. The degradation of *n*-alkanes in kerosene by strain Y-40 increased from 20% to 75% when *n*-octane was added at a concentration of 20% (v/v). The results showed that the metabolic system for *n*-alkanes in strain Y-40 was different from that in strains IFO 1400 and IFO 1548. The increase in *n*-alkane degradation by Y-40 might be due to induction of enzymes that degrade *n*-alkanes after the addition of *n*-octane. Strain Y-40 could grow in the presence of 20% (v/v) isooctane or cyclooctane but did not degrade *n*-alkanes in the presence of either organic solvent. This characteristic would be useful for the microbial conversion of water-insoluble compounds by fermentation in a two-phase system of organic solvent and aqueous medium.

References

Abe A, Inoue A, Usami R, Moriya K, Horikoshi K (1995) Properties of newly isolated marine bacterium that can degrade polyaromatic hydrocarbons in the presence of organic solvents. J Mar Biotechnol 2:182-186

Aono R, Aibe K, Inoue A, Horikoshi K (1991) Preparation of organic solvent- tolerant mutants from *Escherichia coli* K- 1 2. Agric Biol Chem 55:1935-1938

Aono R, Ito M, Inoue A, Horikoshi K (1992) Isolation of novel toluene-tolerant strain of *Psendomonas aeruginosa*. Biosci Biotechnol Biochem 56:145-146

Bocren S, Laane C (1987) Steroid conversions by *Flavobacterium dehydrogenans* in two-liquid-phase systems. Biotechnol Bioeng 29:305-309

Flygare S, Larsson P (1987) Steroid transformation using magnetically immobilized *Mycobacterium* sp. Enzyme Microb Technol 9:494-499

Fukumaki T, Inoue A, Moriya K, Horikoshi K (1994) Isolation of a marine yeast that degrades hydrocarbon in the presence of organic solvent. Biosci Biotech Biochem 58:1784-1788

Hansch C, Fujita T (1964) A method for the correlation of biological activity and chemical structure. J Am Chem Soc 86:1616-1626

Hansch C, Muir MR, Fujita T, Maloney PP, Geiger F, Streich M (1963) The correlation of biological activity of plant growth regulators and chloromycetin derivatives with hammett constants and partition coefficients. J Am Chem Soc 85:2817-2824

Inoue A, Horikoshi K (1989) A *Pseudomonas* thrives in high concentration of toluene. Nature 338:264-266

Inoue A, Horikoshi K (1991a) Estimation of solvent-tolerance of bacteria by the solvent parameter log P. J Ferment Bioeng 71:194-196

Inoue A, Horikoshi K (1991b) *Pseudomonas putida* which can grow in the presence of toluene. Appl Environ Microbiol 57:1560-1562

Jones D, Collins DM (1984) Irregular, nonsporing gram-positive rods. In: Bergey's manual of systematic bacteriology, vol 2. Williams & Wilkins, Baltimore, pp 1261-1434

Kitagawa M (1956) Studies on the oxidation mechanism of methyl-group. J Biochem 43:553-564

Krieg NR, Holt JG (1984) Gram-negative aerobic rods and cocci. In: Murray RGE, Breuner DJ (eds) Bergey's manual of systematic bacteriology, vol. 1. Williams & Wilkins, Baltimore, p 140

Moriya K, Horikoshi K (1993a) Isolation of a benzene-tolerant bacterium and its hydrocarbon degradation. J Ferment Bioeng 76:168-173

Moriya K, Horikoshi K (1993b) A benzene-tolerant bacterium utilizing sulfur compounds isolated from deep-sea. J Ferment Bioeng 76:397-399

Moriya K, Yanagitani S, Usami R, Horikoshi K (1995) Isolation and some properties of an organic-solvent tolerant marine bacterium degrading cholesterol. J Mar Biotechnol 2:131-133

Nakajima H, Kobayashi H, Aono R, Horikoshi K (1992) Effective isolation and identification of toluene-tolerant *Pseudomonas* strains. Biosci Biotechnol Biochem 56:1872-1873

Raymond RL, Jamison VW, Hudson JO (1967) Microbial hydrocarbon cooxidation. I. Oxidation of mono- and dicyclic hydrocarbons by soil isolates of the genus *Nocardia*. Appl Microbiol 15:857-865

Tamaoka J, Komagata K (1984) Determination of DNA base composition by reversed-phase high-performance liquid chromatography. FEMS Microbiol Lett 25:125-128

Walker JD, Colwell RR (1974) Microbial petroleum degradation: use of mixed hydrocarbon substrates. Appl Environ Microbiol 27:1053-1060

Walker JD, Colwell RR, Hamming MC, Ford HT (1975) Extraction of petroleum hydrocarbons from oil-contamrinated sediments. Bull Environ Contam Toxicol 13:245-248

Worsey MJ, Williams PA (1975) Metabolism of toluene and xylene by *Pseudomonas putida* (arvilla) mt-2: evidence for a new function of the Tol plasmid. J Bacteriol 124:7-15

9 Microbial Processes Associated with a Methane Seep Tubeworm

Takeshi Naganuma

9.1 Introduction

Recent findings of the deep-sea vestimentiferan tubeworms that live on sulfur oxidation gave a Copernican change to the view of life. The worms have a unique strategy of life, wherein they lack digestive organs but harbor chemosynthetic (mainly sulfide oxidizing, or thiotrophic) bacteria in a specialized bag-like tissue, the trophosome (Childress et al. 1987; Fisher 1990).

The tubeworm, centimeters to meters long and centimeters wide, has no mouth, no gut, and no anus; it lives on sulfide oxidation by symbiotic bacteria. Tubeworms inhabit chemically reducing environments, such as hydrothermal vents and methane seeps; the latter supplies abundant methane but little sulfide directly. Thus the source of sulfide for the seep-dwelling tubeworms has been questioned. The clue lies in the sea floor. Sediment microflora produces sulfide from sulfate using methane, coupled with in situ formation of calcite/dolomite. The seep tubeworms may take in interstitial sulfide from sediment but not from ambient water. To this, we have hypothesized that the process may occur symbiotically in the tubeworm's body. This hypothesis claims the interface between sulfur reduction and sulfur oxidation (and the associated O_2 interface), which is most likely formed and stabilized by a microbial consortium. The microbial consortium is fueled by methane and recycles sulfer, in possible association with calcification and pyritization. The microbial consortium is potentially applicable for structuring the microspatial O_2 gradient for nano-fabricated minerals.

Key words. Chemoautotroph, Symbiosis, Vestimentiferan tubeworm, *Lamellibrachia*, Methane seep, 16S rDNA, in situ hybridization, ε–Proteobacteria, *Campylobacter*, *Arcobacter*, Microaerophilic, Multiformity, Electron probe Microanalysis, Microbial consortium

Fig. 1. Tubeworms (Sagami Bay 1160 m deep). Without a month of digestive tract, this creature was found to be and entirely new biological form in the deep-sea beyond the reach of sunlight, using hydrogen sulfide as its source of enegy

This chapter focuses on tubeworms (*Lamellibrachia* sp.) (Fig. 1) inhabiting the methane seep in Sagami Bay, central Japan (Masuzawa et al. 1992). The tubeworms were collected at the off-Hatsushima Island methane seep, 1170-1200 m deep, using the submersible Shinkai 2000 and the remotely operated vehicle, Dolphin 3K, operated by the Japan Marine Science and Technology Center (JAMSTEC).

9.2 Vestimentiferan Tubeworms

9.2.1 Biological Background: What Do They Look Like?

The *tubeworms* identified in the reducing habitats so far include polychaetous annelids, pogonophorans, and vestimentiferans. There is a possibility that hemichordate tubeworms may be included because a novel nontubulous hemichordate "spaghetti worm" was reported from a hydrothermal area (Woodwick and Sensenbaugh 1985). A close phylogenetic relationship among these worms has been suggested (Williams et al. 1993).

Vestimentiferan tubeworms typically inhabit sulfide- and methane-rich environments such as deep-sea hydrothermal vents and cold seeps. The worms have a unique strategy of life wherein they lack digestive organs but harbor chemoautotrophic bacteria in a trophosome. Reportedly more than 3×10^9 bacteria were packed in 1 g (wet) of trophosome of the hot vent tubeworm *Riftia pachyptila* (Cavanaugh et al. 1981). The trophosome mass, in turn, accounts for 40%–60% of total worm body mass (Felbeck and Childress 1988), which makes the worm a chemoautotrophic animal.

Microscopic observations suggested more than one endosymbiont morphotype for vestimentiferan worms (Cavanaugh et al. 1981; de Burgh et al. 1989; Fisher 1990) and a gutless annelid (Dubilier et al. 1995). However, morphotypes were often ascribed to pleomorphism of a single species. Ribosomal RNA analyses suggested that the trophosomal endosymbionts are monospecific, or at least 90% homogeneous, and related to chemoautotrophic sulfur-oxidizing bacteria (Stahl et al. 1984; Distel et al. 1988). More than one 16S rDNA sequence was identified and located in the trophosome content of the methane seep tubeworm *Lamellibrachia* sp. (Naganuma et al. 1997a). Some of the 16S rDNA sequences were closely related to those of *Arcobacter*, a member of the ε–subdivision of Proteobacteria (ε–Proteobacteria).

9.2.2 Ecological Background: Where Do They Live?

Vestimentiferan tubeworms inhabit the submarine environments that provide sulfides, methane, or both such as:

1. Methane-rich cold-seeps in subduction zones (e.g., Kulm et al. 1986; Hashimoto et al. 1989)
2. Hydrothermal vents (reviewed by Lutz and Kennish 1993) and submarine volcanic vents (Hashimoto et al. 1993)
3. Hydrocarbon seeps and a related brine-filled pockmark (e.g., Kennicutt et al. 1985)
4. Seeps at an escarpment base and on the continental slopes (e.g., Paull et al. 1984)
5. Floors of stagnant fjords (e.g., Schmaljohann et al. 1990)
6. Organic-rich shipwrecked cargoes (Dando et al. 1992)
7. Whale carcasses (Smith et al. 1989; Naganuma et al. 1996).

A whale carcass is now known to harbor a vestimentifera (*Escarpia spicata*) (Feldman et al. 1997). The mechanism that tubeworms inhabit the whale carcass is discussed below.

9.2.3 Whale Carcasses: Stepping Stones for Vestimentiferan Dispersal

Whale Carcass as a Deep-Sea Oasis

A whale skeleton was discovered on the 1200-m deep sea floor of the Santa Catalina Basin (SCB). The SCB carcass is assumed to be that of a blue whale *(Balaenoptera musculus)* or a fin whale *(B. physalis)*. The whale skeleton was colonized by vent/seep-type biological fauna (Smith et al. 1989), and this discovery added a new site of deep-sea oases for chemosynthesis-based fauna. However, the hypothesis proposed by Smith et al. (1989) that whale carcasses serve as "stepping stones" for the dispersal of some vent organisms was controversial, because: (1)the carcass-associated fauna was not necessarily vent-specific; and (2)large whales and vent/seep fauna appeared at different geological/evolutionary times. However, the co-occurrence of fossil whale bones and fossil *Calyptogena-Lucina* clams (Hachiya 1993) suggests that the whale bones are chemosynthetic habitats for those bivalves, whose modern species are specifically adapted to vents, seeps, and other reducing environments.

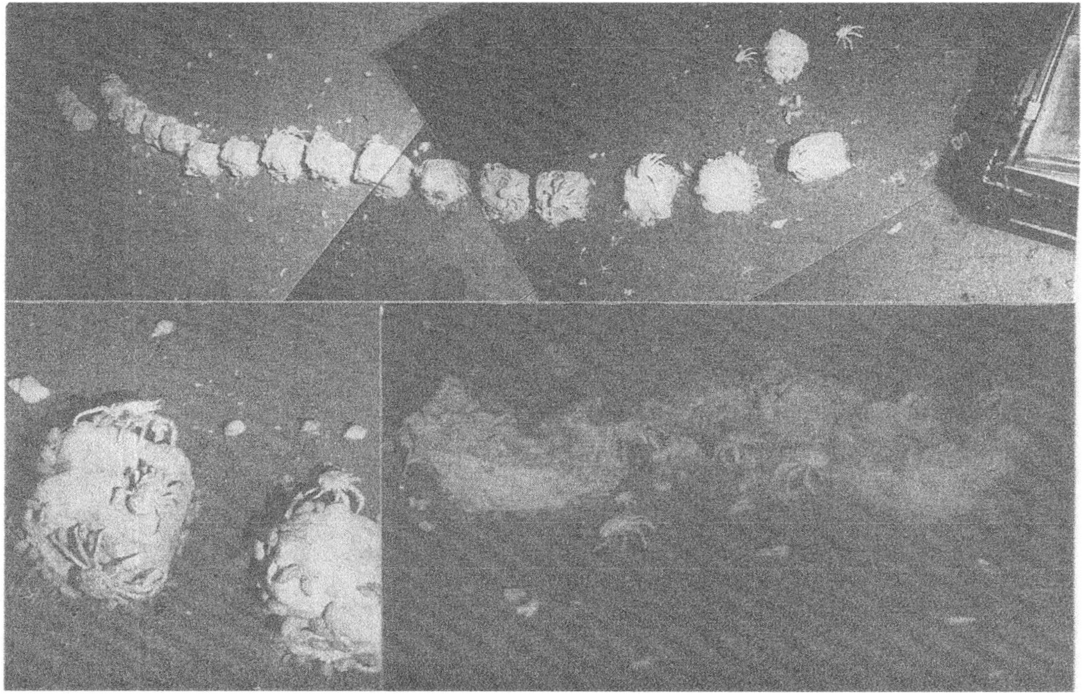

Fig. 2. **Whale skeleton and associated fauna at the 4037-m deep Torishima Seamount near the Izu-Bonin Trench, Pacific Ocean.** *Top* View of the vertabrae, ca. 6 m long. *Bottom left* galatheid crabs, *Munidopsis* sp. on vertebrae colonized by *Idasola* mussels, with buccinid snails crawling on the nearby sediment. *Bottom right* View of the 1.5-m wide jawbones with associated *Munidopsis* crabs, *Idasola* mussels, buccinid snails, actiniid anemones, and spionid polychaetes

Whale Skeleton at the Torishima Seamount

A whale skeleton was found near the eastern summit of the Torishima Seamount (TSM, 4037 m deep) in the western Pacific Ocean during a dive by the DSV Shinkai 6500 (Fig. 2). The skeleton consisted of 22 vertebrae and an assemblage of jawbones. Based on the ear tympanic bone morphology, the whale was identified as *Balaenoptera edeni* (according to Dr. H. Kato, National Institute of Far Seas Fisheries, Japan). The vertebrae were colonized by mytilid mussels and harbored galatheid crabs. A year after the discovery in 1992 the TSM whale skeleton was revisited, and the bone-associated sediment cores were collected (Naganuma et al. 1996).

Chemical and Microbial Gradients Associated with the Whale Skeleton

The original body mass of the TSM whale skeleton is thought to be more than 10 m long and to weigh more than 10 tons. Degradation of such a huge biomass must have influenced the ambient and in situ environment, for example, by producing anaerobic, reducing conditions. The typical products of anaerobic protein degradation are sulfides. Sulfides may also be produced by sulfate-reducing bacteria (SRB). SRB may utilize hydrogen produced via anaerobic oxidation of lipids; whale bones are generally abundant in lipids (Allison et al. 1991). Methane might possibly be co-produced with hydrogen and utilized by methane-oxidizing bacteria. In addition, microbial sulfate reduction using methane was geochemically implicated (Masuzawa et al. 1992). The co-occurrence of sulfate-reducing bacteria and methane-oxidizing bacteria in the skeleton-associated sediment was indicated by the biomarker fatty acids (Naganuma et al. 1996).

Distributions of rhodanese activity, bacterial viable counts (both autotrophic and heterotrophic), and total fatty acid contents were generally similar to that of the sulfide concentration. Linear correlations of these variables and the sulfide concentration were demonstrated (Naganuma et al. 1996). It can thus be concluded that microbial processes in the whale skeleton system were affected by the concentrations of sulfides derived from the whale carcass degradation.

Microbial Populations Associated with Whale Skeleton

Rhodanese is assumed to be primarily an autotrophy-related enzyme; however, the good correlation with heterotrophic viable counts (and the poor correlation with autotrophic viable counts) was discrepant. Rhodanese is known to be not limited to autotrophs but is widespread among heterotrophs (Westley 1973). Thus, some of the rhodanese activity may be due to heterotrophic bacteria. An alternative explanation is that rhodanese activity was mostly by autotrophs, most of which were uncultured.

The autotrophic/heterotrophic viable count ratios were generally high, reaching 0.58 in the upper part of the sediment and 0.2 m from the skeleton. These ratios compare favorably with those observed in hydrothermal plumes (up to 0.3) and contrast with the ratio of less than 0.001 in normal seawater (Naganuma et al, 1989) and with the low numbers of thiosulfate oxidizers at hydrothermal plume boundaries

(Naganuma and Seki 1994). Thus, the high of autotrophic/heterotrophic bacterial count ratio strongly indicates a similarity between hydrothermal plumes and the whale skeleton-associated sediment in terms of microfloral chemosynthetic potential and abundance.

9.3 Methane-Rich Seeps: Geochemical Processes Mediated by Microbial Consortia

9.3.1 Past Seeps: Fossil Records

Worm tubes are thus involved in the reduction/oxidation and mineral deposition of sulfur and other elements. Involved in such geochemical processes are not only present-day tubeworms but also fossil tubeworms (briefly reviewed by Naganuma et al. 1995). A recent discovery of fossil tubeworms was reported from the Miocene Hayama Group, central Miura Peninsula, Japan, along with a massive assemblage of *Calyptogena-Acharax* clam fossils (Kanie et al. 1992).

Fossil Tubeworms, Pyrite (FeS_2), and Life

Pyrite is often localized at the fossil worm tubes in the carbonate matrix when the tubes are present. Naganuma et al. (1995) observed by X-ray microanalysis (XRMA) that iron (Fe) and sulfur (S) co-precipitated in fossil worm tubes 15 million years old. The Fe-S co-occurrence inside the tubes is presumed to be pyrite based on preliminary X-ray diffraction studies. Pyritization is often mediated microbially. It can be said that pyritization in the ancient worm tubes was possibly associated with energy metabolism, at least in part, coupled with microbial S oxidation.

Tubeworm as the Site of Controlled Calcification and Pyritization

Elemental mapping by XRMA was also performed on the body content of modern, living tubeworms (Naganuma et al. 1997b). In contrast to the fossil worm tube, pyritization in the living worm body must be slow or not occur, as Fe and S were not co-accumulated significantly in the worm's body content. Also, co-precipitation of Ca and Mg was not visible, against the anticipation of high-Mg calcite formation in the worm's body. These observations suggest that the formation of high-Mg calcite and pyrite be regulated through certain physiological mechanisms in the living worm's body, probably controlled by the symbiotic microflora.

9.3.2 Modern Seeps: Dowsed by Fatty Acid Analysis

Sediment fatty acid analysis is a tool for detecting methane seepage on the sea floor. Methane seepage with variable activity should be reflected in the microfloral composition of the seep-associated sediment, and the microfloral 'signature' can be seen in the species and composition of fatty acids.

An example is a newly found biological community in the Northern Okushiri Ridge (NOR) in the northeastern Japan Sea (Naganuma et al. 1999). The short-chain ($C_{9\text{-}20}$) sediment fatty acids were analyzed for amounts and compositions and then were compared with those from the sediments and tissues of the *Lamellibrachia* worms obtained from the off-Hatsushima methane seep. As a result, a close relationship among the NOR sediments, the seep sediments, and *Lamellibrachia* tissues was found. This finding represents the first discovery of methane seepage and associated biological communities in the Japan Sea.

Among the fatty acids, 3-hydroxylated fatty acids (C10:0 3-OH and C17:0 3-OH) were characteristically found in the trophosome tissue of the *Lamellibrachia* worm. Generally, 3-OH fatty acids are regarded as biomarkers for *Campylobacter*, a major group of ε-Proteobacteria (Goodwin et al. 1989). Trophosome-specific occurrence of these 3-OH fatty acids may be a reflection of endosymbiotic occurrence of ε-Proteobacteria.

9.3.3 Mystery of a Methane Seep

Tubeworms take up sulfide from the ambient water via a specialized gill-like organ at the tip-opening of the tube; they live on sulfide oxidation. Tubeworms inhabit methane seeps, but sulfide is sometimes missing in the seep seawater at lower-than-requirement levels. From where does sulfide in the worm's body derive? A clue lies in the sea floor sediment: sulfate reduction using methane or methane oxidation using sulfate (Masuzawa et al. 1992).

The seep-dwelling tubeworm anchors its posterior end to the high-Mg calcite, which is a product of sulfate reduction using methane (Naganuma 1998). This observation leads to the idea that the worms may take up sediment interstitial sulfide from their posterior body (MacDonald and Fisher 1996). To this thought, an alternative idea is that tubeworms acquire sulfide from inside their tissue, not from any outside source. Sulfate reduction may occur inside of the worm's body, along with sulfide oxidation.

9.4 Microbial Consortium: Hypothesis

9.4.1 Multiformity of Endosymbiotic(?) Microorganisms

It has been suggested that the populations of trophosomal endosymbiotic bacteria are monospecific, or at least 90% homogeneous, based on the 5S and 16S rRNA sequence analyses (Stahl et al. 1984; Distel et al. 1988). On the other hand, more than one bacterial morphotype was observed with the bacterial populations in vestimentiferan worms (Cavanaugh et al. 1981; de Burgh et al. 1989; Fisher 1990) and a gutless annelid (Dubilier et al. 1995). However, those morphotypes were considerately ascribed to pleomorphism of a single species. Naganuma et al. (1997b) reported the microbial morphotypes observed in the endosymbiotic population of a vestimentiferan tubeworms, along with various elemental accumulations among the bacterial cells revealed by X-ray microanalysis.

Morphotypes of the Trophosome Microflora

At least three morphotypes of spindle-shaped, spiral-shaped, and sphere-shaped (coccoid) bodies have been observed in the trophosome homogenate (Naganuma et al. 1997b). Spindle-shaped bodies were larger than the darkened particles, with dimensions of 10-20 μm length and 4-8 μm width. The size and shape suggests their relation to scuticociliate protozoans, which are frequently found as endosymbionts in many marine bivalves, echinoids, and polychaete worms (Capriulo 1990). A nucleus was not clearly visible. An alternate possibility is that the bodies are large symbiotic bacteria, as shown in other host animals (e.g., Angert et al. 1993).

Spiral-shaped filaments were 50 μm or more long and 0.5 μm or more wide. They characteristically appeared bundled in barrel-like assemblages more than 20 μm wide consisting of more than 100 filaments. Their morphology and habitat suggest that they may be related to *Campylobacter* spp., many of which inhabit animals' oral cavities and gastric tracts (Penner 1988). Some *Campylobacter* spp. are known to be viable but nonculturable (VBNC) species in natural aquatic environments (Rollins and Colwell 1986). This may help explain the difficulty culturing the tubeworms' symbionts.

The spherical bodies had an average diameter of about 2.5 μm. Some of the spherical particles distinctively autofluoresced when illuminated with blue light (460-490 nm). This seemed to be similar to F420 autofluorescence, which is known for methanogens and sulfur reducers.

Populations of the Trophosome Microflora

The population density of the spindle-shaped particles was in the range of 10^5-10^6/ml trophosome content. Assuming an average spindle volume of 420 μm^3 (as a cylinder 15 μm long and 6 μm wide) and a specific gravity of 1, the total mass of the spindle-shaped particles is estimated to be 4.2×10^{-4} to 4.2×10^{-5} g/ml trophosome content.

The population density of the spiral-shaped filaments was in the range of 10^7-10^8 bundles/ml trophosome content. Each bundle consisted of more than 100 filaments. The total mass of the filaments is estimated to be 1×10^{-1} to 1×10^{-2} g/ml trophosome content, assuming an average filament volume of 10 μm^3 (as a cylinder 50 μm long and 0.5 μm wide), a specific gravity of 1, and 100 filaments in a bundle.

The population density of the spherical particles was in the range of 10^9-10^{10}/ml trophosome content. Assuming the average spindle volume of 8 μm^3 (as a sphere of 2.5 μm wide) and a specific gravity of 1, the total mass of the spherical particles is estimated to be 8×10^{-2} to 8×10^{-3} g/ml trophosome content. Thus, the whole mass of the spindles, filaments, and spheres is calculated to be in the range of 18–180 mg/ml trophosome content. This suggests that a considerable fraction of the trophosome mass is from the trophosome microflora.

X-ray Microanalysis of the Trophosomal Cells and Particles

The accumulation of elements in the trophosomal cells or particles were surveyed by XRMA. Sulfur was found to be accumulated in some, not all, spherical cells and particles. Iron, phosphorus, and potassium were co-accumulated. Calcium and zinc were not visibly accumulated (data not shown). As a consequence, XRMA revealed that at least two types of spherical cell or particle were present in the tubeworm trophosome. This supports the multiformity view regarding possible heterogeneity of the trophosomal microflora.

9.4.2 Intracellular Occurrence of ε–Proteobacteria

Analysis of 16S rDNA Sequence

Partial sequences of 16S rDNA were amplified and sequenced from the trophosome content of the seep-dwelling *Lamellibrachia* tubeworm (Naganuma et al. 1997a). Of the total 20 clones, 2 were successfully sequenced and were found to be related to a novel bacterial group in association with vestimentiferan tubeworms. The other cloned sequences are being analyzed, but they are unlikely to be closely related to the two clones; the sequences of the other clones will be reported elsewhere.

Two 16S rDNA sequences in intracellular association with a vestimentifera were sequenced and designated TW-1 and TW-2 (accession numbers D83060 and D83061,

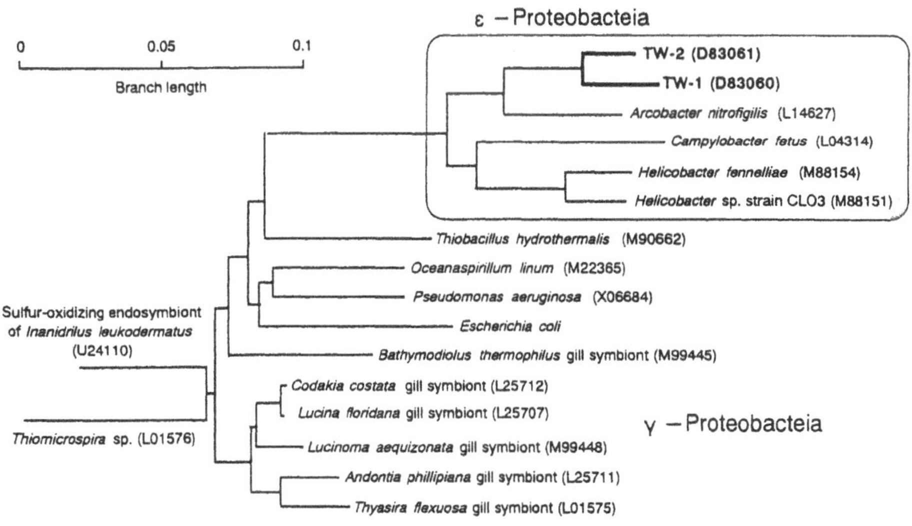

Fig. 3. Unrooted dendrogram showing the phylogenetic positions of the 16S rDNA sequences TW-1 and TW-2, retrieved from the tissure of the vestimentiferan tubeworm *Lamellibrachia* sp. The source of other bacterial 16S rDNA sequences are shown in parentheses (EMBL/GenBank accession numbers)

respectively). The matching between TW-1 and TW-2 was 95%, which suggests that the two belong to different species but the same genus. Both TW-1 and TW-2 were most closely related to *Arcobacter nitrofigilis* (87%-88% matching) and to *A. cryaerophilus* and *A. butzlerii* (86% to 88% matching), all of which fell in the ε-subgroup of the class Proteobacteria, or ε-Proteobacteria (Vandamme et al. 1991). An unrooted dendrogram based on the neighbor-joining method was constructed to position TW-1 and TW-2 in comparison with the *Arcobacter-Campylobacter-Helicobacter* cluster and other chemosynthetic symbionts (Fig. 3).

In Situ Localization

Intracellular localization of the ε-Proteobacterial 16S rDNA sequences TW-1 and TW-2 was demonstrated by in situ hybridization. The vestimentiferan trophosome has a lobular intrastructure. Positive hybridization of the probes TW-1R and TW-2R (TTTCCAGGCCCGAAGGC and AGAAGCTTTAGTAAACTA, respectively) were detected in the periphery of the lobules. The TW-1R hybrid was localized in inner lobules, and the TW-2R hybrid was detected in inner and outer lobules. The periph-

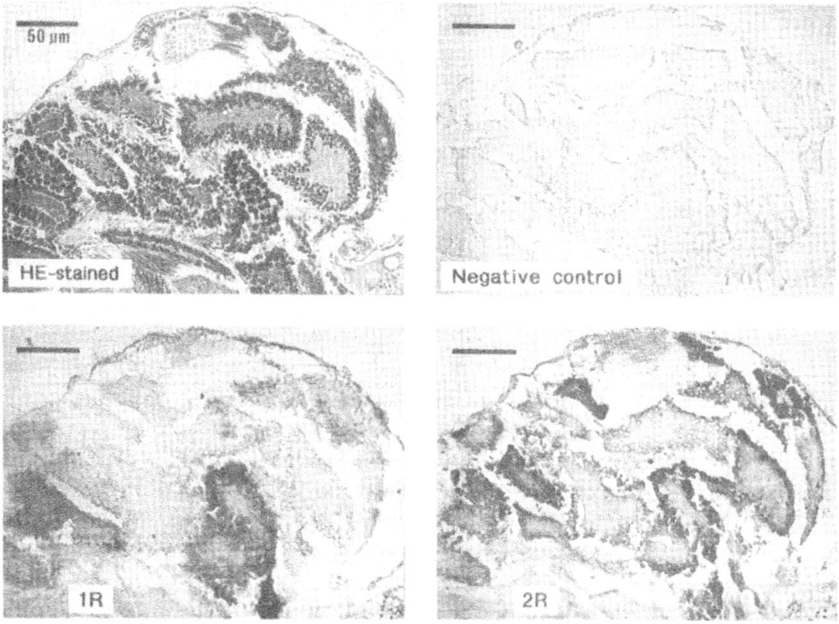

Fig. 4. In situ hybridazation to localize symbiotic 16S rDNA in the cross-sections of the vestimentiferan tubeworm *Lamellibrachia* sp. *Top left* Cross section of trophosome stained with hematoxylin-eosin. Lobular intrastructure of the trophosome is shown. *Top right* Negative control with an antisense probe from the human prostate-specific antigen gene. *Bottom left* Hybrids of the symbiont-specific probe TW-1R localized in periphery of inner lobules. *Bottom right* Hybrids of the symbiont-specific probe TW-2R localized in the periphery of the inner and outer lobules. Bars 50µm

eral localization of symbiont-specific hybrids was previously reported with the vestimentiferan worm *Riftia piscesae* (Cary et al. 1993). In the hemotoxylin-eosin (HE)-stained section several morphotypes (globules, ovoids, rods, threads) (Naganuma et al. 1997b) were observed, and the positive hybridization was restricted to globules and ovoids. TW-1R hybrid was detected in slightly ovoid particles, and TW-2R hybrid was localized in association with the smaller globules (4–5 μm) (Fig. 4). Separate localization of hybrids to different morphotypes suggests that TW-1 and TW-2 were from different bacterial species though they were closely related to each other.

ε–Proteobacteria and Tubeworms

ε–Proteobacterial 16S rDNAs are known to predominate among the epibiotic microflora associated with rocks and animals at hydrothermal vents (Haddad et al. 1995; Moyer et al. 1995; Polz and Cavanaugh, 1995). Thus, it is not surprising that ε–Proteobacteria would be in intracellular association with a vestimentiferan worm.

Vestimentiferan worms are known to bear sulfur-oxidizing endosymbionts that are presumed to belong to the γ–Proteobacteria (Distel et al. 1988). Sulfur oxidation by ε–Proteobacteria is known for free-living *Thiovulum* sp. (Wirsen and Jannasch 1978), and some other ε–Proteobacterial species were reported to be capable of sulfur and sulfate reduction (Vandamme et al. 1991). Intracellular association of ε–Proteobacterial 16S rDNAs with the vestimentiferan trophosome suggests that ε–Proteobacteria are involved in multiple metabolic pathways regarding sulfur oxidation, reduction or both. These multiple metabolic actions may support intracellular microbial production, as shown for a pogonophoran tubeworm and a vent mussel (Schmaljohann et al. 1990; Distel et al. 1995).

9.4.3 Endosymbiotic Microbial Consortium?

Sulfate reduction using methane was hinted at with the question about the source of the tubeworm's sulfide. The hypothesis is that the in-sediment process is present in the tubeworm's body. If so, tubeworms must have a microbial consortium composed of, at least, sulfate-reducing and sulfide oxidizing bacteria. sulfur eduction is anaerobic, sulfur oxidation is aerobic, and they cannot coexist in terms of O_2 requirement. Thus, certain bacteria (probably those that are microaerobic) may also be needed to stabilize the O_2 gradient in the worm's body.

It has been suggested that the tubeworm's symbiotic bacteria are composed of a single species of sulfur oxidizing ε–Proteobacteria (monospecific view). On the other hand, more than one bacterial morphotype and intracellular occurrence of ε–Proteobacteria were observed in the trophosome of the seep-dwelling *Lamellibrachia* worms.

The symbiotic ε–Proteobacteria have not been cultured. Inferred from cultured ε–Proteobacterial species, the symbiotic ε–Proteobacteria could be microaerobic, requiring low levels of O_2 (microaerobic). They probably inhabit and stabilize the microspatial O_2 gradient inside the tubeworm's body. Internal occurrence of ε–Proteobacterial suggests that they are involved in multiple metabolic pathways regarding sulfur oxidation, reduction or both. This multiple metabolic responsibility may support intracellular microbial production, as shown for a pogonophoran tubeworm (Schmaljohann et al. 1990).

References

Allison PA, Smith CR, Kukert H, Deming JW, Bennett BA (1991) Deep-water taphonomy of vertebrate carcasses: a whale skeleton in the bathyal Santa Catalina Basin. Paleobiology 17:78-89

Angert ER, Clements KD, Pace NR (1993) The largest bacterium. Nature 362:239-241

Capriulo GM (ed) (1990) Ecology of marine protozoa. Oxford University Press, New York

Cary SC, Warren W, Anderson E. Giovannoni SJ (1993) Identification and localization of bacterial endosymbionts in hydrothermal vent taxa with species-specific PCR amplification and in situ hybridization techniques. Mar Molec Biol Biotechnol 2:51-62

Cavanaugh CM, Gardiner SL, Jones ML, Jannasch HW, Waterbury JB (1981) Prokaryotic cells in the hydrothermal vent tube worm, *Riftia pachyptila*: possible chemoautotrophic symbionts. Science 213:340-342

Childress JJ, Felbeck H, Somero GN (1987) Symbiosis in the deep sea. Sci Am 255:114-120

Dando PR, Southward AJ, Southward EC, Dixon DR, Crawford A, Crawford M (1992) Shipwrecked tube worms. Nature 356:667

De Burgh ME, Juniper SK, Singla CL (1989) Bacterial symbionts in Northeast Pacific vestimentifera: a TEM study. Mar Biol 101:97-105

Distel DL, Lane DJ, Olsen GJ, Giovannoni SJ, Pace B, Pace NR, Stahl DA, Felbeck H (1988) Sulfur-oxidizing bacterial endosymbionts: analysis of phylogeny and specificity by 16S rRNA sequences. J Bacteriol 170:2506-2510

Distel DL, Lee HK-W, Cavanaugh CM (1995) Intracellular coexistence of methano- and thioautotrophic bacteria in a hydrothermal vent mussel. Proc Natl Acad Sci USA 92:9598-9602

Dubilier N, Giere O, Distel DL, Cavanaugh CM (1995) Characterization of chemoautotrophic bacterial symbionts in a gutless marine worm (Oligochaeta, Annelida) by phylogenetic 16S rDNA sequence analysis and in situ hybridization. Appl Environ Microbiol 61:2346-2350

Felbeck H, Childress JJ (1988) *Riftia pachyptila*: a highly integrated symbiosis. Ocean Acta 8:131-138

Feldman RA, Black MB, Cary CS, Lutz RA, Vrijenhoek RC (1997) Molecular phylogenetics of bacterial endosymbionts and their vestimentiferan hosts. Mol Mar Biol Biotechnol 6:268-277

Fisher CR (1990) Chemoautotrophic and methanotrophic symbioses in marine invertebrates. Rev Aquat Sci 2:399-436.

Goodwin CS, McConnell W, McCulloch RK, McCullough C, Hill R, Bronsdon MA, Kasper G (1989) Cellular fatty acid composition of *Campylobacter pylori* from primates and ferrets compared with those of other campylobacters. J Clin Microbiol 27:938-943

Hachiya K (1993) Mammalia (in Japanese with English abstract). In: Morosaki-so-gun no kaseki [Fossils from the Morosaki Group]. Tokai Kaseki Kenkyukai [Tokai Fossil Association, Nagoya, Japan], pp 263-274

Haddad A, Camacho F, Durand P, Cary SC (1995) Phylogenetic characterization of the epibiotic bacteria associated with the hydrothermal vent polychaete *Alivinella pompejana.* Appl Environ Microbiol 61:1679-1687

Hashimoto J, Miura T, Fujikura K, Ossaka J (1993) Discovery of vestimentiferan tube-worms in the euphotic zone. Zool Sci 10:1063-1067

Hashimoto J, Ohta S, Tanaka T, Hotta H, Matsuzawa S, Sakai S (1989) Deep-sea communities dominated by the giant clam, *Calyptogena soyoae,* along the slope foot of Hatsushima Island, Sagami Bay, Japan. Paleogeogr Plaeoclimatol Paleoecol 71:179-192

Kanie Y, Asami S, Okada H, Watanabe M (1992) White clam community discovered from claystone of the Miocene Hayama Group, Miura Peninsula, south-central Japan (in Japanese with English abstract). Sci Rep Yokosuka City Mus 40:31-35

Kennicutt MC II, Brooks JM, Bidigare RR, Fay RR, Wade TL, McDonald TJ (1985) Vent-type taxa in a hydrocarbon seep region on the Louisiana slope. Nature 317:351-353

Kulm LD, Suess E, Moore JC, Carson B, Lewis BT, Ritger SD, Kadgo DC, Thornburg TM, Embly RW, Rugh WD, Massoth GJ, Langseth MG, Cochrane GR, Scamman RL (1986) Oregon subduction zone: venting, fauna, and carbonates. Science 231:561-566

Lutz RA, Kennish MJ (1993) Ecology of deep-sea hydrothermal vent communities: a review. Rev Geophys 31:211-242

MacDonald IR, Fisher C (1996) Life without light. Natl Geogr 190(4):86-97

Masuzawa, T, Handa N, Kitagawa H, Kusakabe M (1992) Sulfate reduction using methane in sediments beneath a bathyal "cold seep" giant clam community off Hatsushima Island, Sagami Bay, Japan. Earth Planet Sci Lett 110:39-50

Moyer CL, Dobbs FC, Karl DM (1995) Phylogenetic diversity of the bacterial community from a microbial mat an active, hydrothermal vent system, Loihi seamount, Hawaii. Appl Environ Microbiol 61:1555-1562

Naganuma T (1998) A hypothetical microbial consortium that stabilizes oxygen-gradient in chemosynthetic microenvironment. Supramolec Sci 5:439-443

Naganuma T, Seki H (1994) Microbial populations of hydrothermal fluid and plumes in the North Fiji Basin with reference to chemosynthesis. Mar Geol 116:243-253

Naganuma T, Kato C, Hirayama H, Moriyama N, Hashimoto J, Horikoshi K (1997a) Intracellular occurrence of ε–proteobacterial 16S rDNA sequences in the vestimentiferan trophosome. J Oceanogr 53:193-197

Naganuma T, Meisel CJ, Wada H, Kato Y, Takeuchi A, Fujikura K, Naka J, Fujioka K (1999) sea floor fissures, biological communities and sediment fatty acids of the Northern Okushiri Ridge, Japan Sea: implications for possible methane seepage. Island Arc 8 (in press)

Naganuma, T, Naka J, Okayama Y, Minami A, Horikoshi K (1997b) Multiformity of the microbial population in a vestimentiferan tubeworm. J Mar Biotechnol 5:119-123

Naganuma T, Okayama Y, Hattori M, Kanie Y (1995) Fossil worm tubes from the presumed cold-seep carbonates of the Miocene Hayama group, central Miura Peninsula, Japan. Island Arc 4:199-208

Naganuma T, Otsuki A, Seki H (1989) Abundance and growth rate of bacterioplankton community in hydrothermal vent plumes of the North Fiji Basin. Deep Sea Res 36:1379-1390

Naganuma T, Wada H, Fujioka K (1996) Biological community and sediment fatty acids associated with a deep-sea whale skeleton at the Torishima Seamount. J Oceanogr 52:1-15

Paull CK, Hecker B, Commeau R, Freeman-Lynde RP, Neumann C, Corso WP, Golubics S, Hook JE, Sikes E, Curray J (1984) Biological communities at the Florida Escarpment resemble hydrothermal vent taxa. Science 226:965-967

Penner JL (1988) The genus *Campylobacter*: a decade of progress. Clin Microbiol Rev 1:157-172

Polz MF, Cavanaugh CM (1995) Dominance of one bacterial phylotype at a Mid-Atlantic Ridge hydrothermal vent site. Proc Natl Acad Sci USA 92:7232-7236

Rollins DM, Colwell RR (1986) Viable but nonculturable stage of *Campylobacter jejuni* and its role in survival in the natural aquatic environment. Appl Environ Microbiol 52:531-538

Schmaljohann R, Faber E, Whiticar MJ, Dando PR (1990) Co-existence of methane- and sulphur-based endosymbioses between bacteria and invertebrates at a site in the Skagerrak. Mar Ecol Prog Ser 61:119-124

Smith CR, Kukert H, Wheatcroft RA, Jumars PA, Deming JW (1989) Vent fauna on whale remains. Nature 341:27-28

Stahl DA, Lane DJ, Olsen GJ, Pace NR (1984) Analysis of hydrothermal vent-associated symbionts by ribosomal RNA sequences. Science 224:409-411

Vandamme P, Falsen E, Rossau R, Hoste B, Segers P, Tytgat R, De Ley J (1991) Revision of *Campylobacter, Helicobacter, and Wolinella* taxonomy: emendation of genetic descriptions and proposal of *Arcobacter* gen. nov. Int J Syst Bacteriol 41:88-103

Westley J (1973) Rhodanese. Adv Enzymol 39:327-368

Williams NA, Dixon DR, Southward EC, Holland PWH (1993) Molecular evolution and diversification of the vestimentiferan tube worms. J Mar Biol Assoc UK 73:437-452

Wirsen CO, Jannasch HW (1978) Physiological and morphological observations on *Thiovulum* sp. J Bacteriol 136:765-774

Woodwick KH, Sensenbaugh T (1985) *Saxipendium coronatum*, new genus, new species (Hemichordata: Enteropneusta): the unusual spaghetti worms of the Galapagos Rift hydrothermal vents. Proc Biol Soc Wash 98:351-365

3 New Approaches and Future Scope for Deep-Sea Microbiology

10 Barophysiology (Piezophysiology)

Fumiyoshi Abe and Chiaki Kato

10.1 Perspectives on Barophysiology (Piezophysiology)

Hydrostatic pressure is a thermodynamic parameter that has recently received further consideration in various experimental fields. This parameter acts to decrease the total volume of a system at equilibrium in the case of liquids and solutions. The pressure effects in biological systems have been analyzed from two perspectives: (1) the physiology of deep-sea organisms and (2) biochemical reactions as a function of pressure. Although the physicochemical basis of pressure effects is well established (Heremans 1982; Balny et al. 1989), the pressure-induced phenomena that occur in living microorganisms have not been fully defined. Many bacteria adapted to deep-sea environments, called "barophiles," have been reported; and these organisms can grow under high hydrostatic pressure conditions below 100 MPa, the pressure at the deepest point in the ocean. Gene expression under elevated hydrostatic pressure conditions has been explored extensively in barophilic bacteria in recent studies (Bartlett et al. 1995; Kato and Horikoshi 1995; Kato and Bartlett 1997).

We have been analyzing the effects of elevated hydrostatic pressure, focusing on cell growth, cellular metabolism, gene expression, the functions of organelles, and chemical reactions with large volume changes (ΔV), and we propose to establish a new research field of barophysiology (piezophysiology) in the area of high-pressure

---

Key words. Hydrostatic pressure, Barophysiology (Piezophysiology), *Saccharomyces cerevisiae*, Fluorescent analysis, Vacuolar acidification, Cytoplasmic pH homeostasis, Volume change (ΔV), Flow cytometry, Efficient detection, Gene Expression, Transcription, Pressure-regulation, *Escherichia coli*, *lac* Promoter, *tac* Promoter, *malB* operon, Lambda phage, *lamB* bioscience and biotechnology. Research in barophysiology (piezophysiology) investigates the physiological effects of elevated hydrostatic pressure on living organisms to answer such questions as: How do organisms respond to changes in hydrostatic pressure? In this chapter we describe the effects of hydrostatic pressure on intracellular metabolic changes in *Saccharomyces cerevisiae*, as an example of eucaryote, and on gene expression in *Escherichia coli* as an example of prokaryote.

10.2 Barophysiology (Piezophysiology) in Eucaryotes

10.2.1 Effects of Hydrostatic Pressure in Yeast

In the yeast *Saccharomyces cerevisiae*, high hydrostatic pressure (> 100 MPa) induces cytoplasmic petite mutation (Rosin and Zimmerman 1977) and tetraploid or homozygous diploid forms (Hamada et al. 1992), or it disrupts the ultrastructure of the cells (Kobori et al. 1995). Based on the latter observation, high hydrostatic pressure has been applied to food processing to sterilize food materials without significant loss of nutrients or flavor. A short duration of heat shock induces barotolerance, allowing cells to survive at 150 MPa (Iwahashi et al. 1991). Intracellular trehalose and Hsp104 have a role in heat shock–induced barotolerance (Fujii et al. 1996; Iwahashi et al. 1997).

Application of hydrostatic pressure to a logarithmic–phase culture of strain IFO 2347 (Sake yeast, kyokai no. 7) caused a delay in cell growth (Fig. 1). The delay in growth was dependent on the applied pressure, and a pressure above 40 MPa completely inhibited cell division. There was no considerable change in relative colony-forming units (CFU) when hydrostatic pressure was applied for 1 h, indicating that a

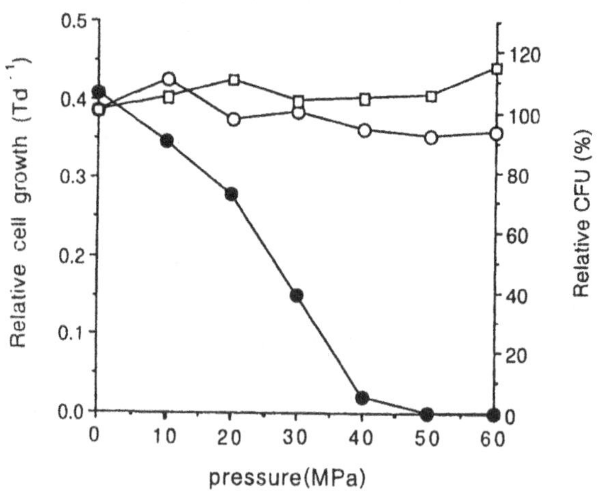

Fig. 1. Growth rate and relative colony–forming units (*CFU*) of strain IFO 2347 at various pressures. Growth rate is indicated as the doubling time(Td)$^{-1}$ (h^{-1}). *Filled circles*, growth rate; relative CFU after high pressure treatments for 1 h in YPD broth (*open circles*) or YPD broth pH 3.0 (*squares*)

high pressure of 60 MPa and below did not affect the viability of strain 2347 (Fig. 1). Pressure-induced repression of cell growth may be ascribed to: inhibition of DNA replication (Murakami and Zimmerman 1973), alteration in protein synthesis (Jaenicke et al. 1988), or deactivation of several enzymes (Schmid et al. 1975).

Hydrostatic Pressure Promotes Acidification of Yeast Vacuoles

The yeast vacuole is an acidic organelle and plays an important role in (1) degradation of cellular proteins, (2) storage of amino acids and carbohydrates, and (3) storage of cytoplasmic calcium. Acidic vacuolar pH (pH 6.0) is maintained through the function of vacuolar H^+-ATPase (V-H^+-ATPase) on the vacuolar membrane (Kakinuma et al. 1981; Nelson and Taiz 1989). The V-H^+-ATPase is a DCCD-sensitive electrogenic proton pump that generates a proton motive force of 180 mV, inside positive and acidic, in vacuolar membrane vesicles. The low vacuolar pH is particularly important for all cellular functions mentioned above. A number of biochemical and genetic studies of vacuolar functions have been reported, and have been reviewed by Anraku et al. (1992a, 1992b). Exposure to a pressure of 40-60 MPa for 1 h reduces the vacuolar pH [determined using the pH-sensitive fluorescent dye carboxyfluorescein (CF) (Preston et al. 1989)] from pH 6.05 to pH 5.88 in strain IFO2347 (Abe and Horikoshi 1995a) and in other *Saccharomyces* strains (Abe and Horikoshi 1995b, 1996). The pressure-induced acidification of vacuoles has been confirmed by measuring vacuolar accumulation of the weak base quinacrine, which was known to accumulate in acidic organelles by a mechanism dependent on the pH gradient across the membrane. Bafilomycin A_1 a specific inhibitor of V-H^+-ATPase, markedly inhibited the pressure-induced acidification, indicating that the process is mediated by V-H^+-ATPase (Abe and Horikoshi 1995a).

Fluorescence Analysis Under High Hydrostatic Pressure: Transient Acidification of Vacuoles

Our next challenge was the analysis of vacuolar pH under elevated hydrostatic pressure conditions, that is, the vacuolar pH of CF-labeled cells was determined when high hydrostatic pressure was applied to the cells. Two items of equipment were required: (1) a hydrostatic chamber with transparent windows originally designed by Morita (1957) (Fig. 2A), which was slightly modified (Abe and Horikoshi 1997a, 1997b) by connecting an optical cable to one of the windows to detect the fluorescence emission, and (2) a hydrostatic chamber with three transparent windows (Fig. 2B) (Abe 1998).

Fluorescence was emitted through the windows at a right angle. The early time course of changes in vacuolar pH in response to elevated hydrostatic pressure was examined. The vacuolar pH was transiently reduced by approximately 0.33 within 4 min after a pressure of 40-60 MPa was applied (Fig. 3) (Abe and Horikoshi 1997a). After decompression, the pH gradually recovered to its original level. The *vma3* mutant lacks subunit c of the V-H^+-ATPase, and this deficiency results in a vacuolar acidification defect (Umemoto et al. 1990; Tanida et al. 1995). Consequently, the mutant cells never show vacuolar acidification when hydrostatic pressure is applied

(Abe and Horikoshi 1997a, 1997b). This observation indicates that transient vacuolar acidification is mediated through the function of the V-H^+-ATPase and is not due to simple changes in the equilibrium of the vacuolar medium.

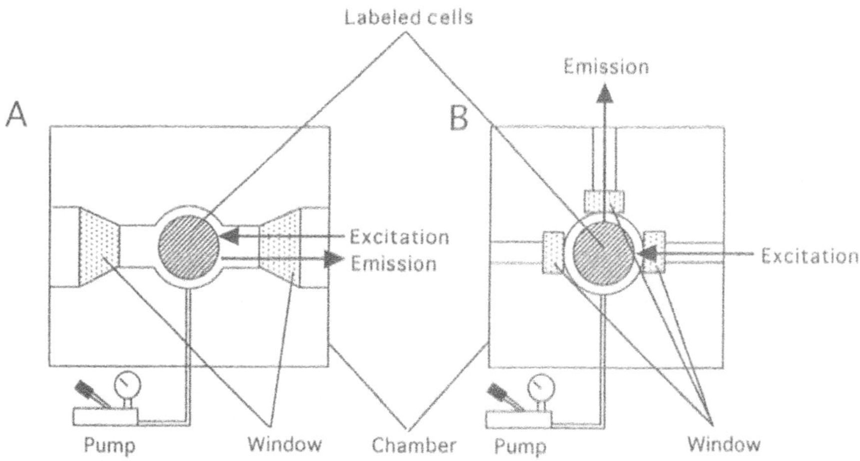

Fig. 2. Hydrostatic chambers for fluorescence analysis under elevated hydrostatic pressure conditions. Labeled cells are placed in the hydrostatic chamber, and fluorescence is emitted through the window at the same angle (**A**) or at a right angle (**B**)

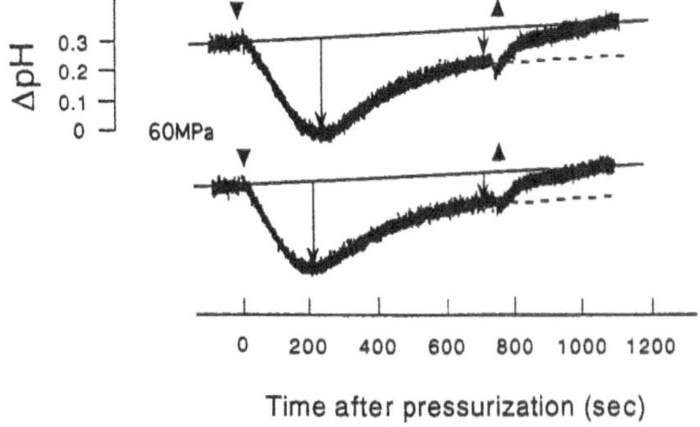

Fig. 3. Time course of vacuolar acidification after application of hydrostatic pressure. Cells of strain X2180 were exposed to hydrostatic pressures of 10, 20, 40, and 60 MPa at *P*, and subsequently removed at *R*. ΔpH-max, maximal change in vacuolar acidification; ΔpHst, steady level of acidification

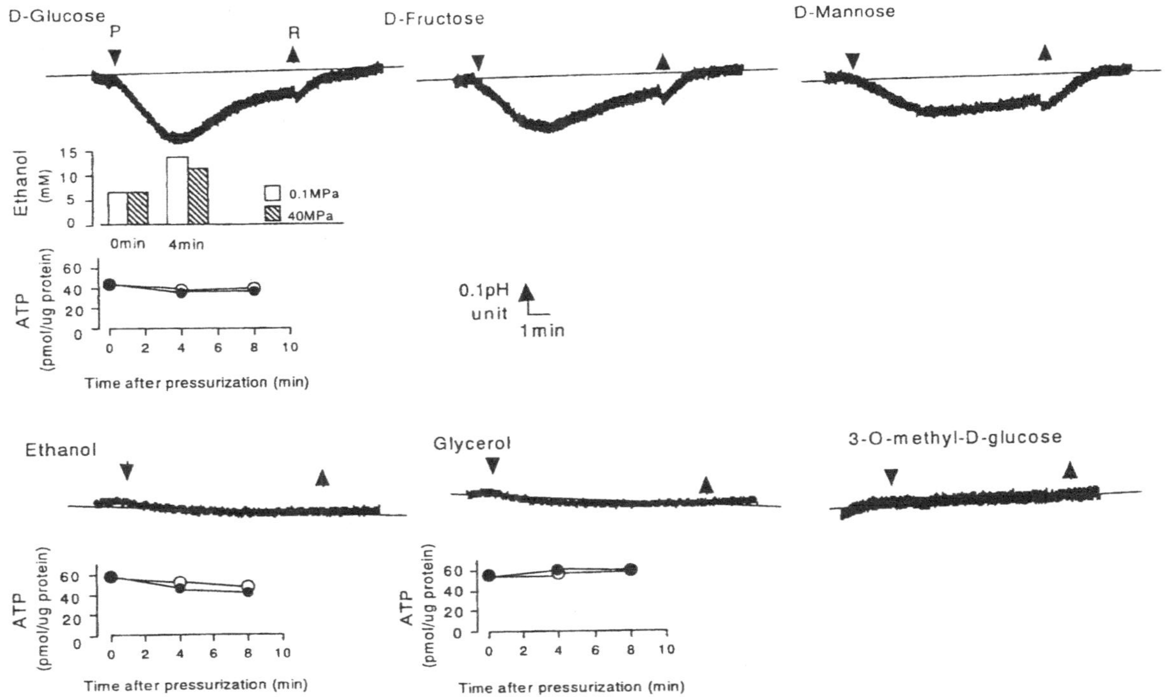

Fig. 4. Time course of vacuolar acidification, concentration of ethanol and amount of cellular ATP in the presence of various carbohydrates (100 mM) after application of hydrostatic pressure. Cells were exposed to a pressure of 40 MPa at *P* and subsequently removed at *R*. *Open circles*, cellular ATP at atmospheric pressure; *filled circles*, cellular ATP at 40 MPa

What Primary Factors Influence Vacuolar Acidification Induced by Elevated Hydrostatic Pressure?

Transient vacuolar acidification was found to occur in the presence of fermentable sugars such as glucose, fructose, or mannose but never when ethanol or glycerol was supplied as the carbon source (Fig. 4) (Abe and Horikoshi 1997a, 1997b). Pressure-induced vacuolar acidification was examined in a glycolysis-defective mutant, ATCC90946 (DFY568; *hxk1*, *hxk2*, *GLK*) (Vojtek and Fraenkel 1990), in the presence of various hexoses. Glucose and fructose are known to be transported through the plasma membrane by the same hexose transporter. The first step in metabolism of these hexoses is phosphorylation catalyzed by a hexokinase or glucokinase. Because DFY568 cells completely lack both hexokinases PI and PII, the mutant cannot metabolize fructose.

As a result, this mutant displays vacuolar acidification in the presence of glucose but never in the presence of fructose when hydrostatic pressure is applied (Fig. 5) (Abe and Horikoshi 1998). These results indicate that glycolysis or ethanol fermentation is involved in the process. CO_2 and glycolytic intermediates such as glucose-6-phosphate or fructose-6-phosphate are weak acids and possibly release protons in solution. Alterations in production or accumulation of these weak acids might have a key role in the mechanism by which hydrostatic pressure induces vacuolar acidification. CO_2 generation in the presence of glucose at 40 MPa was approximately 75% of that produced at atmospheric pressure (Fig. 4) (Abe and Horikoshi 1997a). This means that a considerable amount of CO_2 would be generated in the cytoplasm at 40 MPa. CO_2 is quite soluble in water; at atmospheric pressure more that 99% of aqueous CO_2 as the dissolved gas and less than 1% as carbonic acid, H_2CO_3, which

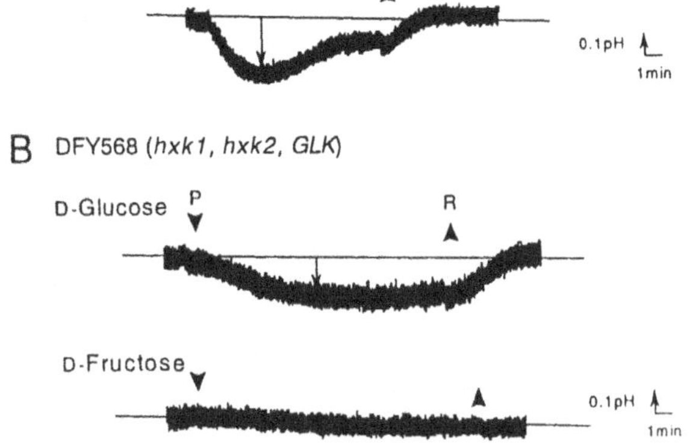

Fig. 5. Time course of vacuolar acidification in a glycolysis-defective mutant after application of hydrostatic pressure. Cells were exposed to a pressure of 40 MPa at *P* and subsequently removed at *R*. **A** Parental strain DFY1. **B** Glycolysis-defective mutant DFY568

partly dissociates to give H^+, HCO_3^-, and CO_3^{2-}. The reaction volume of the reaction $H_2CO_3 \rightarrow H^+ + HCO_3^-$ is negative (-26.0 ml/mol), which means that the dissociation of the weak acid is facilitated by elevated hydrostatic pressure. Therefore large numbers of protons are likely to accumulate in the cytoplasm under elevated hydrostatic pressure conditions. It is possible that the yeast vacuole takes up these protons from the cytoplasm, which results in vacuolar acidification.

Hydrostatic Pressure Inhibition of Glucose-Induced Cytoplasmic Alkalinization

The plasma membrane ATPase (Pma1) is a P-type ATPase and a major essential protein required for establishment of cellular membrane potential. It plays a central role in the regulation of cytoplasmic pH (Serrano et al. 1986). This ATPase is essential for cell viability and functions to maintain a suitable cytoplasmic pH (Eraso and Gancedo 1987; Serrano 1993; Van der Rest et al. 1995). When starved cells are challenged with glucose, the external medium is rapidly acidified and simultaneously cytoplasmic alkalization occurs (Haworth et al. 1991; Eraso and Portillo 1994). The efflux of protons is mediated through the direct activation of plasma membrane H^+-ATPase. Hydrostatic pressure reduces the rate of proton extrusion to an extent dependent on the magnitude of applied pressure (Abe and Horikoshi 1995b). The cytoplasmic pH, determined using the pH-sensitive fluorescent dye carboxy SNARF-1, has been analyzed under elevated hydrostatic pressure conditions. At atmospheric pressure, glucose rapidly induces cytoplasmic alkalinization (Fig. 6A). However, a pressure of 25 or 50 MPa considerably slows the process of cytoplasmic alkalinization (Fig. 6B, C) (Abe and Horikoshi 1998). This may be ascribed to inactivation of the plasma membrane H^+-ATPase. Meanwhile, vacuolar acidification is markedly facilitated by elevated hydrostatic pressure up to 50 MPa (Fig. 6A-C). This means

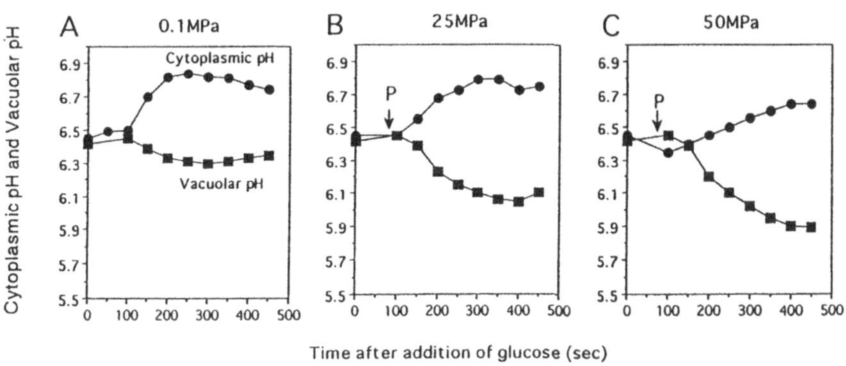

Fig. 6. Time course of cytoplasmic alkalinization and vacuolar acidification after addition of glucose at atmospheric pressure (**A**) or under elevated hydrostatic pressure (**B, C**). Cells were exposed to a pressure of 25 or 50 MPa at *P*. Changes in cytoplasmic pH (*circles*) and vacuolar pH (*squares*) were analyzed under hydrostatic pressure

that the vacuoles are acidified when cytoplasmic alkalinization is inhibited, and the V-H^+-ATPase functions when the plasma membrane H^+-ATPase is inactivated by elevated hydrostatic pressure.

Conclusion: Yeast Vacuole May Serve as Proton Sequestrant Under Hydrostatic Pressure

Cytoplasmic pH tends to decrease in response to elevated hydrostatic pressure for the following reasons: (1) Some of the intracellular compounds present in large amounts, such as inorganic phosphates, carbohydrate-phosphates or H_2CO_3 (CO_2 + H_2O), may release protons because negative volume changes ($\Delta V < 0$) accompany the ionizing reactions. (2) The plasma membrane H^+-ATPase seems be inactivated by elevated hydrostatic pressure. Low cytoplasmic pH could slow glycolysis because pH changes readily affect the activity of a key enzyme of the glycolytic pathway, phosphofructokinase (Hofmann and Kopperschlager 1982).

Figure 7 shows a model proposed to explain the intracellular events under elevated hydrostatic pressure conditions. To maintain a favorable cytoplasmic pH, the yeast vacuoles may serve as proton sequestrants under hydrostatic pressure. It is still unclear why the plasma membrane H^+-ATPase seems to be more sensitive than the V-H^+-ATPase to elevated hydrostatic pressure. Hydrostatic pressure is known to effect the lipid–protein interactions in P-type ATPases, such as Na^+, K^+-ATPase (De Smedt et al. 1979) or Ca^{2+}-ATPase (Heremans and Wuytack 1980), decreasing the activities. It is worthwhile analyzing the pressure dependence of the transition tem-

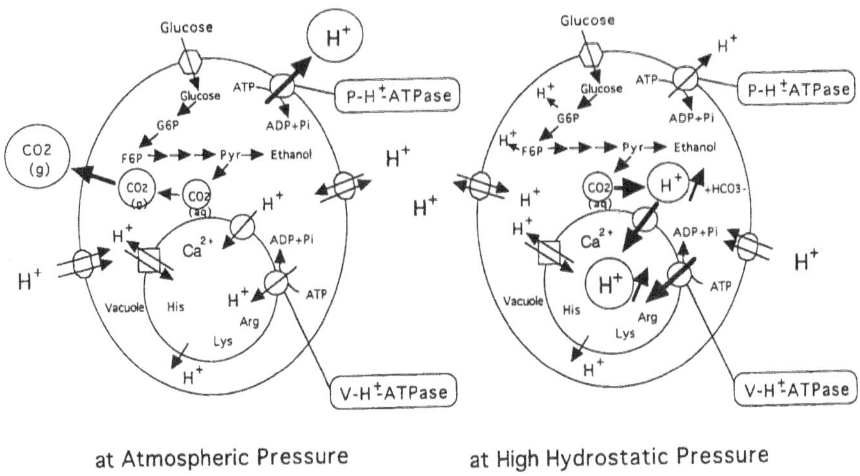

Fig. 7. Model proposed to explain the intracellular events under elevated hydrostatic pressure conditions. Hydration and dissociation of CO_2 are facilitated by elevated hydrostatic pressure, which leads to accumulation of protons in the cytoplasm. To maintain a favorable cytoplasmic pH, the yeast vacuoles may serve as proton sequestrants under elevated hydrostatic pressure conditions

perature on the activities of the plasma membrane H^+-ATPase or the V-H^+-ATPase. Although further experimentation is required to explain fully the significance of the vacuolar contribution to regulation of cytoplasmic pH under elevated hydrostatic pressure conditions, the findings so far suggest that chemical reactions involving intracellular low molecular weight compounds should be noted to elucidate the physiological responses of living organisms to elevated hydrostatic pressure.

10.2.2 New Application of Moderate-Pressure Treatment for Flow Cytometry

It is widely recognized that fluorescence labeling of microorganisms is an effective means to determine total cell numbers or the number of viable cells in a sample. Flow cytometry combined with fluorescence staining is a powerful tool for analyzing heterogeneous microbial populations (Troussellier et al. 1993; Davey and Kell 1996). Fluorescein diacetate (FDA) and its derivatives are nonfluorescent molecules that defuse into cells and are hydrolyzed by intracellular nonspecific esterases to give fluorescent products. The fluorescent products accumulate only in those cells that have intact cell membranes, therefore, dead cells with leaky membranes are not stained. It is known that the fluorescence intensity of labeled yeast varies considerably among strains, probably because of differences in intracellular esterase activities. This may cause inaccurate detection of yeast cells in heterogeneous populations by flow cytometry. We have found that accumulation of CF or carboxy-

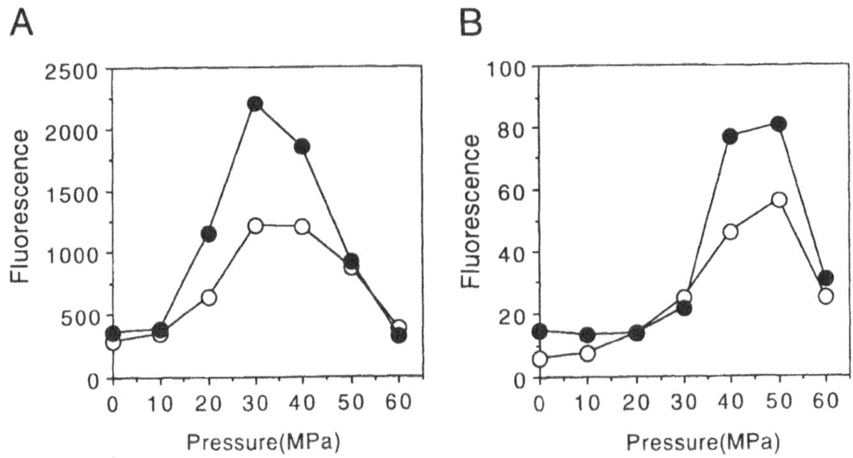

Fig. 8. Pressure-induced accumulation of carboxyfluorescein (CF) or carboxydichlorofluorescein (CDCF) in IFO2347 (**A**) and IFO10159 (**B**). Cells were labeled with 10 μM carboxyfluorescein diacetate (CFDA) or 10 μM carboxydichlorofluorescein diacetate (CDCFDA) under several hydrostatic pressures for 1 h, and the labeled cells were analyzed using CytoFluor 2350. *Open circles*, CF-labeled cells; *filled circles*, CDCF-labeled cells

dichlorofluorescein (CDCF) was facilitated by nonlethal hydrostatic pressures. In this section we discuss a new application of moderate hydrostatic pressure, which may contribute to the efficient detection of microorganisms by flow cytometry (Abe 1998).

Effect of Hydrostatic Pressure on Accumulation of Fluorescent Dye

Carboxyfluorescein diacetate (CFDA) and carboxydichlorofluorescein diacetate (CDCFDA) are known to be hydrolyzed by intracellular nonspecific esterases; and the fluorescent products CF (pH-sensitive) and CDCF (pH-insensitive) accumulate in acidic compartments such as vacuoles (Preston et al. 1989). Because the variability of staining different strains is known, two strains, the Sake yeast IFO2347 (strongly labeled strain) and IFO10159 (weakly labeled strain, which is less readily detectable by flow cytometry), were used in this study. Application of hydrostatic pressure markedly promoted the accumulation of CF and CDCF in IFO 2347, peaking at 40

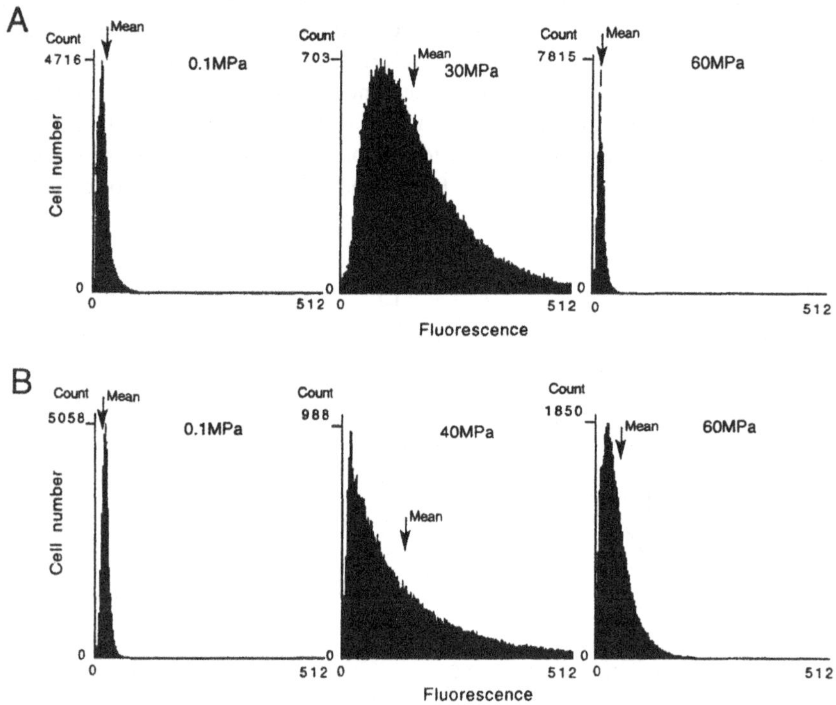

Fig. 9. Histograms of populations of labeled cells. Labeled cells (approximately 100 000 cells) were analyzed using the Bryte-HS flow cytometry system. **A** Cells of IFO2347 were subjected to hydrostatic pressures of 0.1, 30, and 60 MPa in the presence of 10 μM CDCFDA for 1 h. **B** Cells of IFO10159 were subjected to hydrostatic pressures of 0.1, 40, and 60 MPa in the presence of 10 μM CDCFDA for 1 h

and 30 MPa, respectively (Fig. 8A). A similar result was obtained for IFO10159, even though enhanced dye accumulation was observed at a higher pressure (50 MPa) (Fig. 8B). Although there are no data to explain that the five- to tenfold-enhancement would have a significant impact on detection limits in natural samples, application of moderate hydrostatic pressure could be a potentially new procedure for detecting living yeasts.

Hydrostatic pressure above 40 MPa markedly inhibited cell growth, although cell survival (determined as relative number of CFUs) was not considerably affected by application of hydrostatic pressure for 1 h (Abe and Horikoshi 1995a). It is evident that hydrostatic pressure can enhance CF or CDCF staining in the absence of cell proliferation or significant loss of viability, which is necessary for precise determination of the viable cell number in a sample.

Flow Cytometry Analysis

Figure 9A shows the histograms of populations of IFO2347 cells labeled with CDCFDA under several hydrostatic pressures and analyzed at atmospheric pressure. When the cells were subjected to a pressure of 30 MPa for 1 h, the accumulation of CDCF increased four- to six fold. Similar results were obtained with IFO10159 (Fig. 9B). These results of flow cytometry analysis are mostly consistent with the results obtained by ordinary fluorescence analysis shown in Fig. 8.

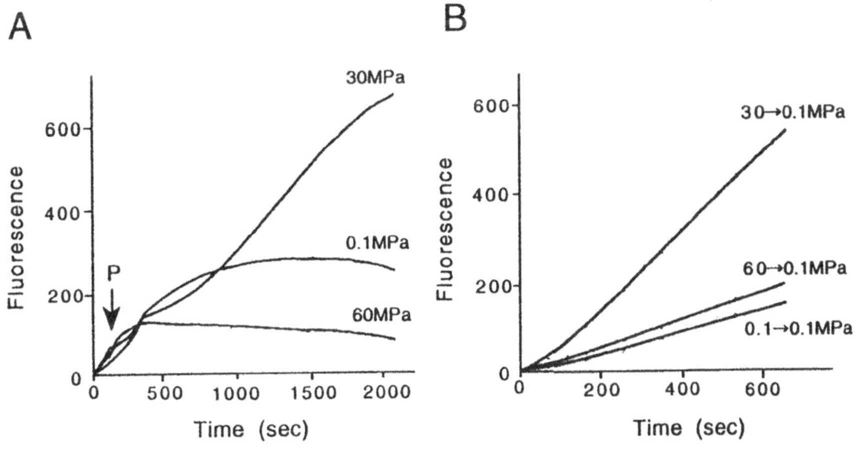

Fig. 10. Effect of hydrostatic pressure on the hydrolysis of CDCFDA in IFO2347. Fluorescence was emitted in a hydrostatic chamber with transparent windows and analyzed using a RF5300PC spectrofluorometer. **A** Hydrolysis of CDCFDA under elevated hydrostatic pressures. A cuvette containing the cells was placed in the chamber, and hydrostatic pressure was applied at *P* after addition of 50 μM CDCFDA. **B** Hydrolysis of CDCFDA at atmospheric pressure after preincubation at several hydrostatic pressures. A cuvette containing the pressure-adapted cells was placed in the chamber, then 50 μM CDCFDA was added

Effect of Hydrostatic Pressure on Hydrolysis of CDCFDA

Fluorescence of IFO2347 was measured in a hydrostatic chamber after addition of CDCFDA. The fluorescence intensity increased to 300 and 130 arbitrary units (AU) when hydrostatic pressures of 0.1 and 60 MPa were applied, respectively (Fig. 10A). The fluorescence increased considerably when a pressure of 30 MPa was applied, compared with 0.1 and 60 MPa. Cells were preincubated at several hydrostatic pressures for 1 h in YPD (pH 5.0), and the hydrolysis activity was subsequently measured at atmospheric pressure. The cells adapted to 30 MPa hydrolyzed CDCFDA at a rate approximately three times greater than observed in the case of cells adapted to 0.1 or 60 MPa (Fig. 10B). These results indicate that preincubation of the cells at 30 MPa induced CDCFDA hydrolysis activity, and it was maintained after decompression.

Dependence on Glucose Metabolism

Glucose was found to be required for the pressure-induced accumulation of CDCF. The glucose concentration required for half-maximal CDCF accumulation was approximately 10 mM (Fig. 11A). 2-Deoxyglucose considerably inhibited the pressure-induced accumulation of CDCF (Fig. 11B). The concentration at which half-maximal inhibition occurred was approximately 12 mM, close to the concentration of glucose required for half-maximal CDCF accumulation. Although both values

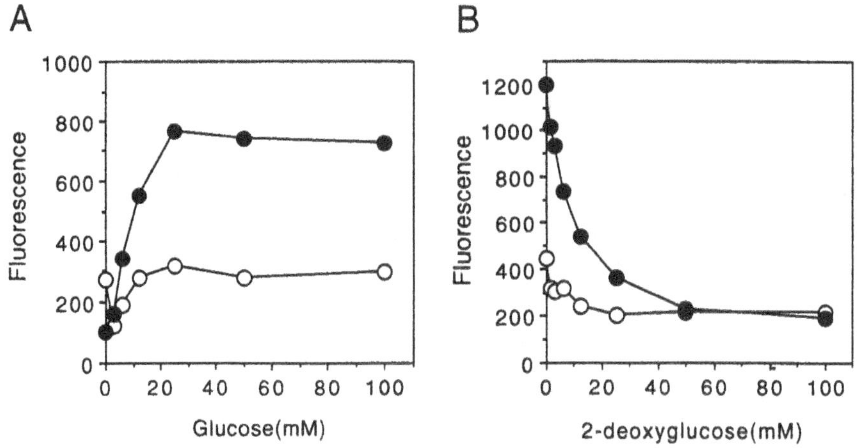

Fig. 11. Effects of D-glucose concentration and 2-deoxyglucose on the pressure-induced accumulation of CDCF in strain IFO2347. **A** Cells were incubated with 10 μM CDCFDA at 0.1 MPa (*open circles*) or 30 MPa (*filled circles*) in the presence of various concentrations of D-glucose in MBA buffer for 1h. **B** Cells were incubated with 10 μM CDCFDA at 0.1 MPa (*open circles*) or 30 MPa (*filled circles*) in the presence of 100 mM D-glucose and various concentrations of 2-deoxyglucose in MBA buffer for 1 h. Labeled cells were analyzed using CytoFluor 2350

were slightly lower than the Michaelis-Menten constant (K_m) of the low affinity site of the hexose transporter (20 mM) (Bisson and Fraenkel 1983) and much greater than the K_m for sugar kinases (<<1 mM), the results suggest that glucose metabolism or ATP production has a role in the process of CDCF accumulation.

Conclusion: Possible Intracellular Events Induced by Elevated Hydrostatic Pressure

The dependence of the kinetics of a simple chemical reaction on pressure yields a direct measurement of the volume change associated with the formation of the activation state of the reaction. Preliminary results suggest that the hydrolysis of CFDA and CDCFDA in both buffer and cell extract is simply facilitated by elevated hydrostatic pressure, which means the chemical reaction of dye hydrolysis accompanies negative volume changes ($\Delta V^{\ddagger} < 0$).

Preincubation of cells at a pressure of 30 MPa promoted the accumulation of CDCF, suggesting that hydrostatic pressure may (1) induce the synthesis of esterases, (2) promote the hydrolytic activity of esterases, (3) promote the passive diffusion of prefluorochrome CDCFDA through the cell membrane, or (4) stimulate the glucose metabolism required for dye hydrolysis at 30 MPa. A pressure of 60 MPa did not effectively promote accumulation of dye in the cells, although preliminary results suggested that dye hydrolysis activity in the cell extract increased linearly in response to elevated pressures up to 60 MPa. How hydrostatic pressure might affect the processes of items (1)-(5) must await experimentation on the subject. As we mentioned previously, an increase in hydrostatic pressure to 40-60 MPa promotes acidification of vacuoles. Such an increase in hydrostatic pressure also induces acidification of the cytoplasm. It is likely that a reduction of vacuolar pH by 0.2-0.3 affects the activity of some esterases. It would be worthwhile in future studies to analyze whether a pressure of 30-50 MPa affects the availability of yeast esterases through stimulation of transcription or protein synthesism, through stabilization of mRNA, or through inhibition of the degradation of esterases. Although the precise course of the induction pathway is still unclear, these findings may contribute to improve methods of analysis by flow cytometry and reveal the need to investigate intracellular metabolic events in living organisms under hydrostatic pressure during experiments into a new research field: barophysiology (piezophysiology).

10.3 Barophysiology (Piezophysiology) in Prokaryotes

We are investigating the molecular mechanisms of gene regulation in microorganisms isolated from deep-sea sediment to determine how they have become adapted to high hydrostatic pressure conditions. We have already described the gene regulation by high pressure in deep-sea barophilic (piezophilic) bacteria (see section 5.2). As a simple model to examine these mechanisms, we have also studied the relationship between gene expression and pressure conditions using the *Escherichia coli* system. In this section, we describe the gene expression controlled by pressure in the typical prokaryote *E. coli*.

10.3.1 Gene Expression Under High Pressure in Atmospheric Adapted Prokaryote *E. coli*

Positive Regulation by High Pressure

Because of the genetic similarity between *E. coli* and deep-sea bacteria, we hypothesized that *E. coli* could share with many deep-sea bacteria common mechanisms for regulating gene expression at high pressure. To test this possibility, we used promoters from *E. coli*, encoded on plasmids, to analyze the effect of pressure on expression of a reporter gene, the chloramphenicol acetyltransferase (CAT) gene (Close and Rodriguez 1982). The results are summarized in Table 1.

Gene expression initiated from the *lac* promoter region, on plasmid pUC13, and from the *tac* promoter region, on plasmid pKK223-3, was enhanced greatly by growth at 30 and 50 MPa, respectively, in the absence of the inducer isopropyl-β-D-thiogalactopyranoside (IPTG) (Table 1) (Kato et al. 1994; Sato et al. 1995). In contrast, promoters on plasmid pBR322 and pACYC184 were unaffected at high pressure, although elevated pressure did increase the plasmid copy number (Kato et al. 1994). The *lac* and *tac* promoters are activated in the presence of IPTG, whereas the other promoters, which were unaffected by high pressure, function in the absence of interaction with a allosteric effector molecule. IPTG enhances expression of the *lac* and *tac* genes by binding to the repressor protein LacI and releasing it from the repressor binding site on the DNA just downstream of the transcription starting point (Ogata and Gilbert 1979). The transcription start point of *lac* and *tac* promoters induced by high pressure or IPTG are the same (Fig. 12). Therefore, induction of gene expression by pressure must be related to induction by IPTG.

Table 1. Comparison of CAT activity encoded by various plasmids in *E. coli* JM109 grown at 0.1, 30, and 50 MPa

Plasmid	Vector	Gene	Enzyme activity (unit/mg)			Ratio [a]	
			0.1 MPa	30 MPa	50 MPa	30/0.1	50/0.1
pTS1	pBR322	*antitet*-CAT	645	1481	2498	2.3	3.8
pTS2	pBR322	*tet*-CAT	253	512	1294	2.0	5.1
pTS3	pBR322	*amp*-CAT	183	231	399	1.3	2.2
pTS4	pUC13	*lac*-CAT	127	11944	471	**94.0**	3.7
pTS5	pKK223-3	*tac*-CAT	121	587	10973	4.8	**90.5**
pACYC184	pACYC184	*cat*-CAT	12103	7303	13529	0.6	1.1

[a] The ratio of the specific enzyme activity (unit/mg of protein) at 30 MPa compared with 0.1 MPa (30/0.1) and at 50 MPa compared with 0.1 MPa (50/0.1).

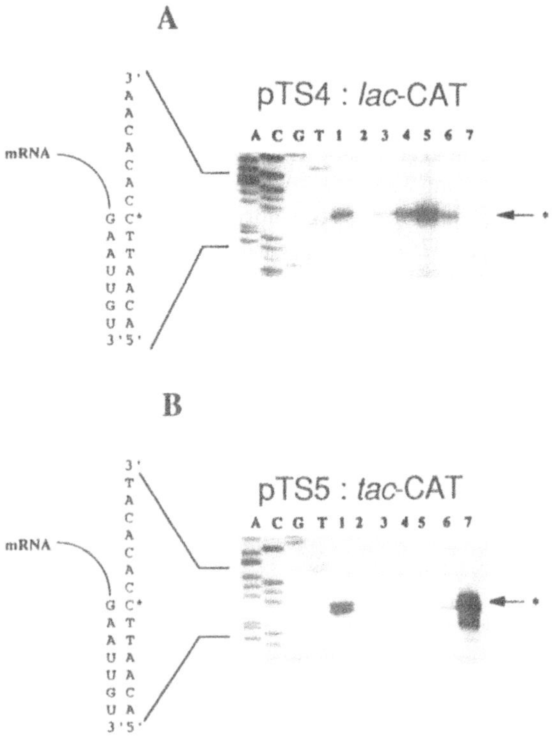

Fig. 12. Primer extension analysis to determine the transcriptional starting point for pressure regulated gene expression by the *lac* (**A**) and *tac* (**B**) promoters in *E. coli* transformants, and the level of mRNA derived from the promoter in response to elevated pressure. The nucleotide sequence corresponding to the ladder is written on the left. The transcriptional starting point is indicated by an *arrow and an asterisk*, and the 5' end of the mRNA is shown. Lanes: *1*, RNA from the *E. coli* transformant grown at 0.1 MPa with 1 mM isopropyl-β-D-thiogalactopyranoside (IPTG); *2*, at 0.1 MPa without IPTG; *3*, at 10 MPa without IPTG; *4* at 20 MPa without IPTG; *5*, at 30 MPa without IPTG; *6*, at 40 MPa without IPTG; *7*, at 50 MPa without IPTG

High pressure may release the repressor protein bound to the DNA by causing a structural change in the DNA or the repressor protein itself. Royer (1990) reported that a pressure of around 200 MPa induced the repressor protein, LacI, to change from a tetramer to a dimer *in vitro*. The LacI protein binds its operator DNA target with high affinity as a tetramer but displays poor DNA binding characteristics when present in the dimer form. As shown in Table 1, gene expression regulated by the *lac* and *tac* promoters was maximum at 30 and 50 MPa, respectively; and these pressure levels were low compared with those in Royer's report. These differences could be explained by differences in the concentration of LacI protein. Two alternative interpretations exist. One is that the observed effects relate entirely to a change in plasmid supercoiling that incidentally interferes with the LacI binding site. Another is that the *lacI* promoter may be inactive at these high pressures. Recently, we observed that detectable amounts of stable *lacI* transcript and LacI protein were present in *E. coli* cells grown under several pressure conditions (unpublished results). Therefore, the first hypothesis may be correct.

Negative Regulation by High Pressure

In other work we have demonstrated that the formation of plaques by λ phage in *E. coli* is prevented at 30 MPa hydrostatic pressure (Fig. 13) (Sato et al. 1996).

When λ phage infects *E. coli*, the phage binds to an outer membrane protein, LamB (Emr et al. 1980). The λ receptor, LamB, is expressed from the *mal*B region, which is composed of two operons, *mal*EFG and *mal*K*lam*B, transcribed divergently from an interoperon regulatory interval (Hofnung 1974; Raibaud et al. 1979; Shilhavy et al. 1979). Both operons of the *mal*B region are positively controlled by the *mal*T gene product through its interaction with an inducer, maltose (Debarbouille et al. 1978). Using promoter fragments derived from the *mal*B region, we showed that gene expression initiated from both promoters (*pmal*K-*lam*B and *pmal*EFG) is re-

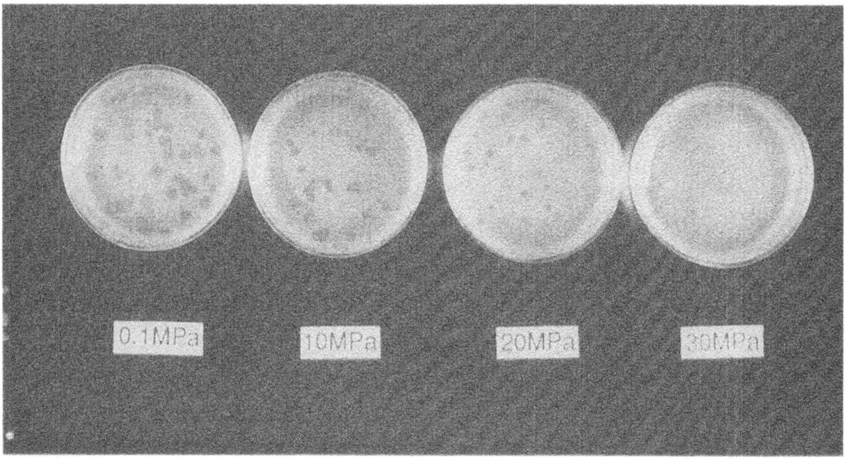

Fig. 13. Ability of the λ phage, ZAP II, to infect *E. coli* JM109. Incubation pressures are indicated

Table 2. Effect of pressure on gene expression in both directions from the *malB* operon in *E. coli*

Promoter	Inducer [a]	Enzyme activity (unit/mg)				
		0.1 MPa	10 MPa	20 MPa	30 MPa	40 MPa
pmalKlamB	+	173.9	64.8	26.7	10.9	10.8
	–	20.5	14.8	15.9	25.1	9.3
pmalEFG	+	326.4	205.1	57.2	17.9	18.3
	–	27.5	22.4	27.9	37.9	16.9

Data are from Sato et al. 1996.

[a] +, inducer present at 0.2% for maltose; –, no inducer.

pressed by elevated hydrostatic pressure (Table 2). High pressure repressed gene expression directed by the *mal*B regulatory interval even in the presence of inducer, thereby preventing synthesis of the λ receptor protein.

Nakashima et al. (1995) reported that in *E. coli* the amounts of outer membrane proteins OmpF and OmpC are reduced under high pressure, and that this is not the result of effects on the amounts of the regulatory proteins EnvZ and OmpR. The extent of expression of another OM protein, designated OmpX, was also found to be

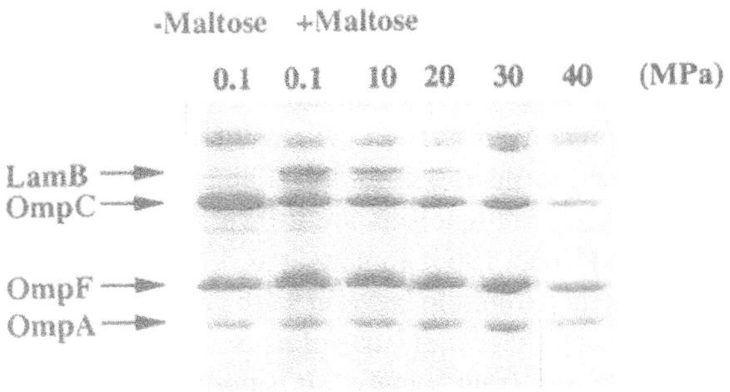

Fig. 14. Expression of LamB protein in *E. coli* JM109 with or without maltose under elevated pressures. Incubation pressures are indicated

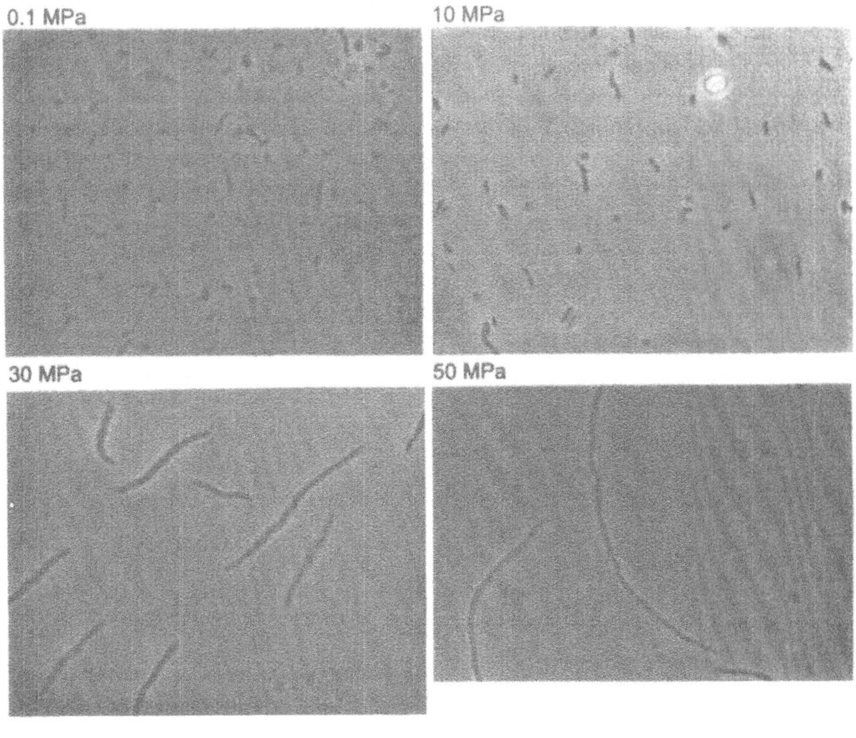

Fig. 15. Morphological change of *E. coli* on growth under elevated hydrostatic pressure

reduced under high pressure culture conditions. Because the N-terminal amino acid sequence of OmpX is identical to that of the λ phage receptor protein, LamB, it seems likely that OmpX is LamB (Fig. 14). These observations led us to conclude that prevention of plaque formation by phage λ under high pressure conditions is due to a paucity of the λ receptor protein, LamB.

Our findings suggest that high pressure affects gene expression directed by the *mal*B regulatory interval, and this causes a decrease in the quantity of λ receptor protein, LamB. In a previous study by Marquis and Keller (1975), binding of the Lac repressor of *E. coli* to β-galactosides was pressure-sensitive in vivo. Thus, it is possible that binding of the positive regulatory protein, MalT, to maltose also may be pressure-sensitive. This might be one of the mechanisms by which high pressure acts to repress expression of the *mal*B region.

Filament formation in *E. coli* adapted to growth at 1 atm pressure has been reported when the organisms are grown at increased hydrostatic pressure (Fig. 15) (Zobel and Cobet 1963; Tamura et al. 1992). We have observed filament formation when strictly barophilic bacteria were grown at low hydrostatic pressure, though no change in cell shape indicative of a pressure effect was detected in moderately barophilic or barotolerant bacteria. It is possible that high pressure affects cell division and influences the composition of the bacterial cell wall. Aspartate β-D-semialdehyde dehydrogenase (ASD) is a key enzyme in the biosynthetic pathway of lysine, threonine, methionine, diaminopimelic acid (DAP), and isoleucine. DAP is an essential component present in the peptidoglycan of all gram-negative and some gram-positive bacteria (Schleifer and Kandler 1972). Therefore, expression of ASD in strictly and moderately barophilic deep-sea bacteria and in *E. coli* may be influenced by hydrostatic pressure. We have investigated *asd* gene expression in *E. coli* under several pressure conditions. Three transcript sizes were detected at pressures up to 30 MPa, but at 50 MPa they were almost undetectable (Fig. 16) (Kato et al. 1997).

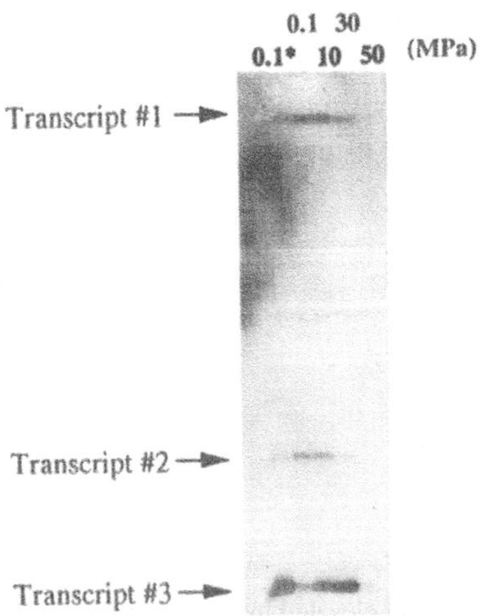

Fig. 16. Primer extension analysis of the *E. coli asd* gene. *Arrows* show transcripts 1, 2, and 3. Lanes: 0.1, RNA from *E. coli* JM109 grown at 0.1 MPa; 10, at 10 MPa, 30; at 30 MPa; and 50, at 50 MPa. * Cell was grown with shaking

Table 3. Effect of high pressure on gene expression in deep-sea adapted bacteria and atmospheric pressure-adapted *Escherichia coli*

Gene	Barophile (DB6705)	Barotolerant (DSS12)	*E. coli*	References
Pressure-regulated operon	+	+	+[a]	Kato et al. 1995 Kato et al. 1996
asd	+	+, ∓	–	Kato et al. 1997
lac (pUC13)			+	Kato et al. 1994 Sato et al. 1995
tac (pKK223-3)			+	Kato et al. 1994 Sato et al. 1995
tet, antitet, amp (pBR322)			∓	Kato et al. 1994
cat (pACYC184)			∓	Kato et al. 1994
ompF, ompC			–	Nakashima et al. 1995
envZ, ompR			∓	Nakashima et al. 1995
malB operon			–	Sato et al. 1996
*lac*I			∓	Unpublished

+, high pressure-inducible gene expression; ∓, pressure-independent gene expression; –, high pressure-repressible gene expression.
[a] Recombinant *E. coli* harboring the pressure-regulated operon from DB6705.

Thus, lack of the ASD enzyme may be one cause of filament formation and growth inhibition at high pressure in *E. coli*.

10.3.2 Conclusion of Gene Regulation by High Pressure in Prokaryote

As shown in Table 3, we have observed three responses to pressure in studies of gene expression in high-pressure-adapted barophilic and barotolerant bacteria and in atmospheric-pressure-adapted *E. coli*. Gene expression in barophilic bacteria was mainly subject to high pressure-inducible regulation, although in barotolerant bacteria both high pressure-inducible and pressure-independent regulatory mechanisms were evident. There are three responses in *E. coli*: high-pressure-inducible, pressure-independent, and high-pressure-repressible. Recently, several groups have proposed that life might have originated in the deep-sea hydrothermal vents (Kandler 1992; Stetter 1993), so it seems possible that the high-pressure-adapted mechanisms of gene expression could represent a feature present during the early stages of life.

References

Abe F (1998) Hydrostatic pressure enhances vital staining with carboxyfluorescein or carboxydichlorofluorescein in *Saccharomyces cerevisiae*: efficient detection of labeled yeasts by flow cytometry. Appl Environ Microbiol 64:1139-1142

Abe F, Horikoshi K (1995a) Hydrostatic pressure promotes the acidification of vacuoles in *Saccharomyces cerevisiae*. FEMS Microbiol Lett 130:307-312

Abe F, Horikoshi K (1995b) Effect of hydrostatic pressure on the yeast vacuole. In: Trzeciakowski WA (ed) High pressure science and technology. World Scientific Publishing, Singapore, pp 875-877

Abe F, Horikoshi K (1996) Vacuolar acidification under high hydrostatic pressure in *Saccharomyces cerevisiae*. High Press Biosci Biotechnol 13:53-58

Abe F, Horikoshi K (1997a) Vacuolar acidification in *Saccharomyces cerevisiae* induced by elevated hydrostatic pressure is transient and is mediated by vacuolar H^+-ATPase. Extremophiles 1:89-93

Abe F, Horikoshi K (1997b) Yeast vacuoles may serve as proton sequestrants under high hydrostatic pressure. In: Heremans K (ed) High pressure research in the biosciences and biotechnology. Leuven University Press, Leuven, Belgium, pp 209-212

Abe F, Horikoshi K (1998) Analysis of intracellular pH in the yeast *Saccharomyces cerevisiae* under elevated hydrostatic pressure: a study in baro-(piezo-) physiology. Extremophiles 2:223-228

Anraku Y, Hirata R, Wada Y, Ohya Y (1992a) Molecular genetics of the yeast vacuolar H^+-ATPase. J Exp Biol 172:67-81

Anraku Y, Umemoto N, Hirata R, Ohya Y (1992b) Genetic and cell biological aspects of the yeast vacuolar H^+-ATPase. J Bioeng Biomem 24:395-405

Balny C, Masson P, Travers F (1989) Some recent aspects of the use of high pressure for protein investigations in solution. High Press Res 2:1-28

Bartlett DH, Kato C, Horikoshi K (1995) High pressure influences on gene and protein expression. Res Microbiol 146:697-706

Bisson LF, Fraenkel DG (1983) Involvement of kinases in glucose and fructose uptake by *Saccharomyces cerevisiae*. Proc Natl Acad Sci USA 80:1730-1734

Close TJ, Rodriguez RJ (1982) Construction and characterization of the chloramphenicol-resistance gene cartridge: a new approach to the transcriptional mapping of extrachromosomal elements. Gene 20:305-316

Davey H, Kell DB (1996) Flow cytometry and cell sorting of heterogeneous microbial populations: the importance of single-cell analyses. Microbiol Rev 60:641-696

Debarbouille M, Shuman HA, Silhavy TJ, Schwartz M (1978) Dominant constitutive mutations in *malT*, the positive regulator gene of the maltose regulation in *Escherichia coli*. J Mol Biol 124:359-371

De Smedt H, Borghgraef R, Ceuterick F, Heremans K (1979) Pressure effects on lipid–protein interactions in (Na^+ + K^+) ATPase. Biochim Biophys Acta 556:479-489

Emr SD, Hedgpeth J, Clement JM, Silhavy TJ, Hofnung M (1980) Sequence analysis of mutations that prevent export of lambda receptor, an *Escherichia coli* outer membrane protein. Nature 285:82-85

Eraso P, Gancedo C (1987) Activation of yeast plasma membrane ATPase by acidic pH during growth. FEBS Lett 224:187-192

Eraso P, Portillo F (1994) Molecular mechanism of regulation of yeast plasma membrane H (+)-ATPase by glucose: interaction between domains and identification of new regulatory sites. J Biol Chem 269:10393-10399

Fujii S, Iwahashi H, Obuchi K, Fujii T, Komatsu Y (1996) Characterization of a barotolerant mutant of the yeast *Saccharomyces cerevisiae*: importance of trehalose content and membrane fluidity. FEMS Microbiol Lett 141:97-101

Hamada K, Nakatomi Y, Shimada S (1992) Direct induction of tetraploids or homozygous diploids in the industrial yeast *Saccharomyces cerevisiae* by hydrostatic pressure. Curr

Genet 22:371-376.

Haworth R, Lemire BD, Crandall D, Cragoe EJ Jr, Fiegel L (1991) Characterization of proton fluxes across the cytoplasmic membrane of the yeast *Saccharomyces cerevisiae*. Biochim Biophys Acta 1098:79-89

Heremans K (1982) High pressure effects on proteins and other biomolecules. Annu Rev Biophys Bioeng 11:1-21

Heremans K, Wuytack F (1980) Pressure effect on the Arrhenius discontinuity in Ca^{2+}-ATPase from sarcoplasmic reticulum: evidence for lipid involvement. FEBS Lett 117:161-163

Hofmann E, Kopperschlager G (1982) Phosphofructokinase from yeast. Methods Enzymol 90:49-60

Hofnung M (1974) Divergent operons and the genetic structure of the maltose B region in *Escherichia coli* K12. Genetics 76:169-184

Iwahashi H, Kaul SC, Obuchi K, Komatsu Y (1991) Induction of barotolerance by heat shock treatment in yeast. FEMS Microbiol Lett 80:325-328.

Iwahashi H, Obuchi K, Fujii S, Komatsu Y (1997) Effect of temperature on the role of Hsp104 and trehalose in barotolerance of *Saccharomyces cerevisiae*. FEBS Lett 416:1-5

Jaenicke R, Bernhardt G, Lüdemann HD, Stetter KO (1988) Pressure-induced alteration in the protein pattern of the thermophilic archaebacterium *Methanococcus thermolithotrophicus*. Appl Environ Microbiol 54:2375-2380

Kakinuma Y, Ohsumi Y, Anraku Y (1981) Properties of H^+-translocating adenosine triphosphatase in vacuolar membranes of *Saccharomyces cerevisiae*. J Biol Chem 256:10859-10863

Kandler O (1992) The archaebacteria: biochemistry and biotechnology. Portland Press, London

Kato C, Bartlett DH (1997) The molecular biology of barophilic bacteria. Extremophiles 1:111-116

Kato C, Horikoshi K (1995) Gene expression under hydrostatic pressure. High Press Biosci Biotechnol 13:59-66

Kato C, Sato T, Smorawinska M, Horikoshi K (1994) High pressure conditions stimulate expression of chloramphenicol acetyltransferase regulated by the *lac* promoter in *Escherichia coli*. FEMS Microbiol Lett 122:91-96

Kato C, Smorawinska M, Sato T, Horikoshi K (1995) Cloning and expression in *Escherichia coli* of a pressure-regulated promoter region from a barophilic bacterium, strain DB6705. J Mar Biotechnol 2:125-129

Kato C, Smorawinska M, Sato T, Horikoshi K (1996) Analysis of a pressure-regulated operon from the barophilic bacterium strain DB6705. Biosci Biotechnol Biochem 60:166-168

Kato C, Smorawinska M, Horikoshi K (1997) Comparison of the gene expression of aspartate β-D-semialdehyde dehydrogenase at elevated hydrostatic pressure in deep-sea bacteria. J Biochem 121:717-723

Kobori H, Sato M, Tameike A, Hamada K, Shimada S, Osumi M (1995) Ultrastructure effects of pressure stress to the nucleus in *Saccharomyces cerevisiae*: a study by immunoelectron microscopy using frozen thin sections. FEMS Microbiol Lett 132:253-258

Marquis RE, Keller DM (1975) Enzymatic adaptation by bacteria under pressure. J Bacteriol 122:575-584

Morita RY (1957) Effect of hydrostatic pressure on succinic formic, and malic dehydrogenases in *Escherichia coli*. J Bacteriol 74:251-255.

Murakami TH, Zimmerman AM (1973) DNA synthesis in *Tetrahymena*: a pressure study. Cytobios 7:171-181

Nakashima K, Horikoshi K, Mizuno T (1995) Effect of hydrostatic pressure on the synthesis of outer membrane proteins in *Escherichia coli*. Biosci Biotechnol Biochem 59:130-132

Nelson N, Taiz L (1989) The evolution of H^+-ATPases. Trends Biochem Sci 14:113-116

Ogata R, Gilbert W (1979) DNA-binding site of *lac* repressor probed by dimethylsulfate methylation of the *lac* operator. J Mol Biol 132:709-728

Preston RA, Murphy RF, Jones EW (1989) Assay of vacuolar pH in yeast and identification of acidification-defective mutants. Proc Natl Acad Sci USA 86:7027-7031

Raibaud O, Roa M, Braun-Breton C, Schwartz M (1979) Structure of the *mal*B region in *Escherichia coli* K12. I: genetic map of the *mal*K-*lam*B operon. Mol Gen Genet 174:241-248

Rosin MP, Zimmerman AM (1977) The induction of cytoplasmic petite mutants of *Saccharomyces cerevisiae* by hydrostatic pressure. J Cell Sci 26:373-385.

Royer CA (1990) Macromolecular binding equilibria in the *lac* repressor system: studies using high-pressure fluorescence spectroscopy. Biochemistry 29:4959-4966

Sato T, Kato C, Horikoshi K (1995) Effect of high pressure on gene expression by *lac* and *tac* promoters in *Escherichia coli*. J Mar Biotechnol 3:89-92

Sato T, Nakamura Y, Nakashima KK, Kato C, Horikoshi K (1996) High pressure represses expression of the *mal*B operon in *Escherichia coli*. FEMS Microbiol Lett 135:111-116

Schleifer KH, Kandler O (1972) Peptideglycan types of bacterial cell walls and their taxonomic implications. Bacteriol Rev 36:407-477

Schmid G, Ludemann HD, Jaenicke R (1975) High pressure effects on the activity of glycolytic enzymes. Biophys Chem 3:90-98

Serrano R (1993) Structure, function and regulation of plasma membrane H(+)-ATPase. FEBS Lett 325:108-111

Serrano R, Kielland-Brandt MC, Fink GR (1986) Yeast plasma membrane ATPase is essential for growth and has homology with (Na^+ + K^+), K^+ and Ca^{2+}-ATPase. Nature 319:689-693

Silhavy TJ, Brickman E, Bassford PJ Jr, Casadaban MJ, Shuman HA, Schwartz V, Guarente L, Schwartz M, Beckwith J (1979) Structure of the *malB* region in *Escherichia coli* K12. II: genetic map of the *mal*E, F, G, operon. Mol Gen Genet 174:249-259

Stetter KO (1993) Frontiers of life. Frontiers, Gif sur Yvette, France

Tamura K, Shimizu T, Kourai H (1992) Effects of ethanol on the growth and elongation of *Escherichia coli* under high pressures up to 40 MPa. FEMS Microbiol Lett 99:321-324

Tanida I, Hasegawa A, Iida H, Ohya Y, Anraku Y (1995) Cooperation of calcineurin and vacuolar H^+-ATPase in intracellular Ca^{2+} homeostasis of yeast cells. J Biol Chem 270:10113-10119

Troussellier M, Courties C, Vaquer A (1993) Recent applications of flow cytometry in aquatic microbial ecology. Biol Cell 78:111-121

Umemoto N, Yoshihisa T, Hirata R, Anraku Y (1990) Roles of the *VMA3* gene product, subunit c of the vacuolar membrane H^+-ATPase on vacuolar acidification and protein transport. J Biol Chem 265:18447-18453

Van der Rest ME, Kamminga AH, Nakano A, Anraku Y, Poolman B, Konings WN (1995) The plasma membrane of *Saccharomyces cerevisiae*: structure, function, and biogenesis. Microbiol Rev 59:304-322

Vojtek AB, Fraenkel DG (1990) Phosphorylation of yeast hexokinases. Eur J Biochem 190:371-375

Zobel CE, Cobet AB (1963) Filament formation by *Escherichia coli* at increased hydrostatic pressures. J Bacteriol 87:710-719

11 Genome Analysis of Facultatively Alkaliphilic *Bacillus halodurans* C-125

HIDETO TAKAMI

Generally, alkaliphilic *Bacillus* strains cannot grow or else grow poorly under neutral pH conditions, but they grow well at a pH higher than 9.5. Since 1969 we have isolated a large number of alkaliphilic *Bacillus* strains from various environments and have purified many alkaline enzymes (Horikoshi 1991). As mentioned above, alkaliphilic bacteria containing *Bacillus* strains have been isolated from even deep-sea sediment collected at depths of 1000–10 000 m depth at a frequency of 0.2×10^2 to 2.3×10^4 per colonies g^{-1} dry sea mud (Takami et al. 1999a).

Over the past two decades our studies have focused on the enzymology, physiology, and molecular genetics of alkaliphilic microorganisms to elucidate their mechanisms of adaptation to alkaline environments (Horikoshi 1991). Industrial applications of these microbes have also been investigated, and some commercial enzymes from alkaliphilic *Bacillus* strains have brought great advantages to industry. Thus, it is clear that alkaliphilic *Bacillus* strains are important and interesting not only academically but also industrially. An alkaliphilic bacterium, strain C-125 (JCM9153), isolated in 1977, was identified as a member of the genus *Bacillus* and reported as a producer of β-galactosidase (Ikura and Horikoshi 1979) and xylanase (Honda et al. 1985). It is the most thoroughly characterized strain, physiologically, biochemically, and genetically, among those in our collection of alkaliphilic *Bacillus* isolates (Hori-

Key words. *Bacillus halodurans*, Genome analysis, Physical map, Shotgun sequencing, Alkaliphile

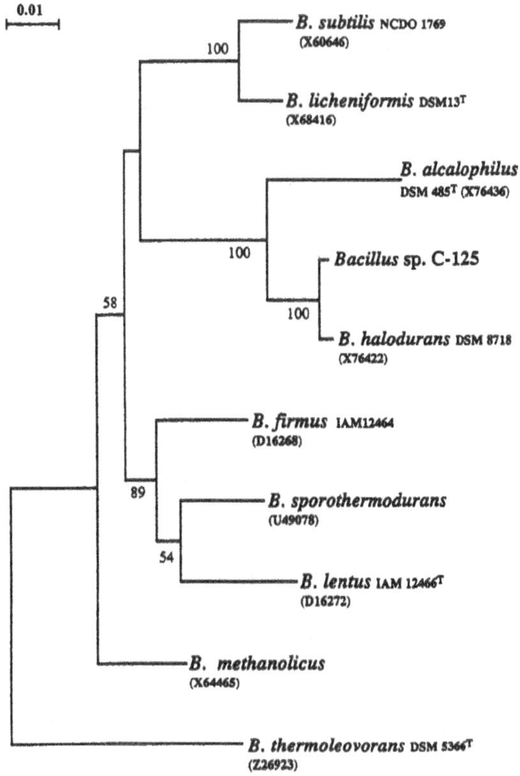

Fig. 1. Unrooted phylogenetic tree showing the relationship of strain C-125 to other alkaliphilic *Bacillus* strains. The numbers indicate the percent of bootstrap samples, derived from 1000 samples, that supported the internal branches (Felsenstein 1985). Bootstrap probability values less than 50% were omitted from this figure. Bar 0.01 Knuc unit

koshi 1991). Recently, this strain was reidentified as *Bacillus halodurans* based on phylogenetic analysis using 16S rDNA sequencing and DNA-DNA hybridization analysis (Takami and Horikoshi 1999). Whole genome analysis of *B. subtilis*, which is taxonomically related to alkaliphilic *B. halodurans* strain C-125 (Fig. 1), except for the alkaliphilic phenotype, has been completed (Kunst et al. 1997). Knowledge of the complete nucleotide sequence of the *B. subtilis* genome will help us analyze *B. halodurans* C-125 genome. From this research background, we initiated genome analysis of alkaliphilic *B. halodurans* C-125.

11.1 Construction of Physical and Genetic Map of the Genome of *B. halodurans* C125

First we attempted to construct a physical and genetic map of the chromosome of strain C-125 to facilitate further genome analysis, and we examined differences in genome structure between strain C-125 and *B. subtilis*. We report a revision of the genome size from 3.7 Mb (Southerland et al. 1993) to 4.25 Mb and a complete *Asc*I/*Sse*8387I physical map of the chromosome of *Bacillus halodurans* C-125 (Takami et al. 1999b).

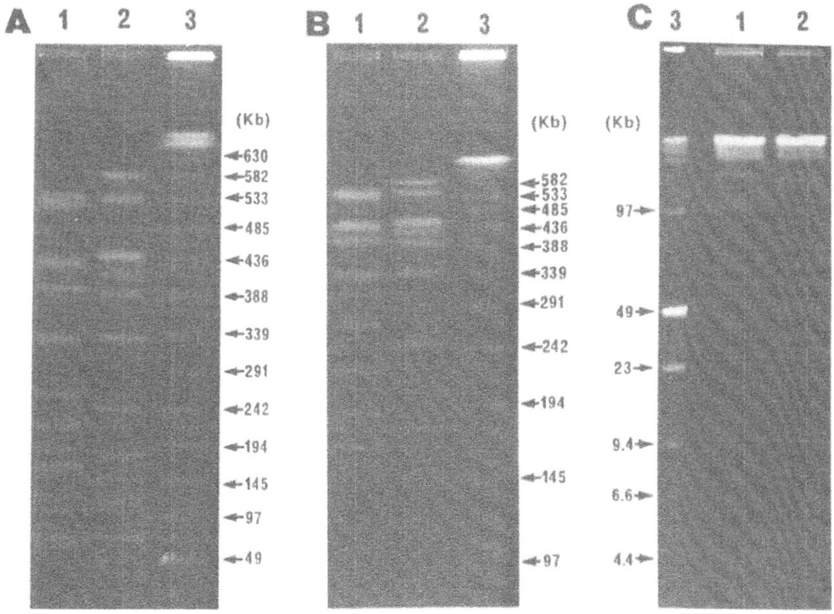

Fig. 2. Digestion patterns of the chromosomal DNA of alkaliphilic *Bacillus halodurans* C-125 obtained with *Asc*I and *Sse*8387I. **A** Separation of fragments ranging in size from 50 to 600 kb using a pulsed time of 20-50 s at 6V cm^{-1} for 26 h. Lanes: *1*, complete *Sse*8387I digestion; *2*, complete *Asc*I digestion; *3*, l DNA ladder. **B** Separation of fragments ranging in size from 50 to 600 kb using a pulsed time of 6.75-54.17 s at 6V cm^{-1} for 28 h Lanes: *1*, complete *Sse*8387I digestion; *2*, complete *Asc*I digestion; *3*, l DNA ladder. **C** Separation of fragments ranging in size from 5 to 75 kb using a pulsed time of 0.22-5.10 s at 6V cm^{-1} for 15 h. Lanes: *1*, complete *Sse*8387I digestion; *2*, complete *Asc*I digestion; *3*, low range PFG marker (2.03–194 kb)

11.1.1 Digestion Pattern of the C-125 Genome with Rare Cut Restriction Endonucleases

Bacillus halodurans C-125 was grown in 100 ml of NII medium for 4-5 h until reaching the mid-logarithmic phase. The cells from 500 μl of culture were harvested by centrifugation, washed once in TE buffer (10 mM Tris-HCl pH 8.0, 1 mM EDTA), resuspended in the same volume of TE buffer (50°C), and mixed with 500 μl of 2% Pulsed Field Certified (PFC) agarose prewarmed to 50°C. The resulting suspension was poured into a mold chamber (BioRad). Solidified blocks were immersed in 10 ml of TE buffer containing 40 mg of lysozyme and incubated at 37°C for 2 h. After washing the blocks in TE buffer twice, they were incubated at 50°C in 10 ml of TE buffer containing 8 mg of proteinase K (GIBCO BRL, MD, USA) overnight. The blocks were washed once in TE buffer and then incubated in TE buffer containing 1 mM phenylmethylsulfonyl fluoride (PMSF) for 1 h at room temperature. They were then washed three times more in TE buffer. Restriction endonucleases that recognize an 8-bp sequence were tested for their ability to digest the chromosome of strain C-125. *Pac*I (5'-TTAAT/TAA-3') and *Pme*I (5'-GTTT/AAAC-3') generated more than 50 fragments each less than 50 kb in size (data not shown). *Not*I (5'-GC/GGCCGCC-

Table 1. Sizes of restriction fragments generated by AscI and *Sse*8387I cleavage of the alkaliphilic *Bacillus* sp. strain C-125 chromosome

Band number	Size (kb) of fragments genarated with		
	*Asc*I	*Sse*8387I	Band number
1A	599	537	1S
2A	549	552	2S
3A [a]	445	433	3S [a]
4A	417	394	4S
5A	388	329	5S
6A	329	273	6S
7A	254	222	7S
8A	174	185	8S
9A	144	162	9S
10A	119	150	10S
11A	112	119	11S
12A [a]	70	104	12S
13A	49	72	13S
14A	33	67	14S
15A	26	61	15S
16A	23	58	16S
17A	3.8	43	17S
18A	0.7	39	18S
		20	19S
Total	4250	4253	

[a] Two bands overlapped at the same position.

3') did not digest the C-125 chromosome as described in a previous paper (Southerland et al. 1993). *Sfi*I (5'-GGCCNNNN/NGGCC-3') generated approximately 20 fragments ranging in size from 20 to 700 kb. The restriction pattern was uneven, however, and smearing made it difficult to identify fragments and their sizes accurately as observed in a previous paper (Southerland et al. 1993).

*Asc*I (5'-GG/CGCGCC-3') and *Sse*8387I (5'-CCTGCA/GG-3') generated 18 and 19 resolvable fragments, respectively (Fig. 2, Table 1). The intensity of each resolvable band was analyzed by means of the *pdi* Desk Top Scanner System (*pdi* Inc., NY, USA), and it was found that two fragments of the same size overlapped at positions 3A, 12A, and 3S (Table 1). The sizes of these fragments were determined by comparison with size standards on a series of PFGE gels (Fig. 2, Table 1). The mean total size of the genome of *Bacillus halodurans* C-125, estimated by totaling the *Asc*I or *Sse*8387I fragments, was 4.25 Mb. It appeared that the genome size had been underestimated previously (Southerland 1993) for the reason described above. Thus the size of the genome was revised from 3.75 Mb to 4.25 Mb in this study. Carlson and Kolsto have determined the sizes of the chromosomes of six strains of *B. cereus* (2.4-5.3 Mb) (Carlson and Kolsto 1994) and one strain of *B. thuringiensis* (5.4-6.3 Mb) (Carlson and Kolsto 1993). The 4.25 Mb size of the chromosome of strain C-125 is almost the same as that of *B. subtilis* (4.21 Mb) (Kunst et al. 1997) and smaller than that of *B. thuringiensis* (5.4-6.3 Mb).

11.1.2 Isolation and Sequencing Analysis of Linking Clone

We decided to construct a physical map of *Asc*I and *Sse*8387I sites on the *Bacillus halodurans* C-125 chromosome, as enzymatic digestion with *Asc*I or *Sse*8387I resulted in 20 resolvable fragments. A linking clone contains DNA, which overlaps two adjacent restriction fragments and can be used as a hybridization probe to establish the identity of these two contiguous fragments (Smith and Condemine 1990). Isolation of *Asc*I- and *Sse*8387I-linking clones was attempted to join adjacent fragments generated by *Asc*I or *Sse*8387I digestion when mapping the chromosome. Linking clone libraries were constructed using *Hin*dIII or *Eco*RI or *Eco*RV as the R enzyme. Plasmid DNA was isolated from individual clones and screened for the presence of two *Asc*I or *Sse*8387I sites and two R enzyme sites. Seventeen possible *Asc*I-linking clones (A–Q) were identified and screened for authenticity by Southern blot analysis of R enzyme-digested *Bacillus halodurans* C-125 chromosomal DNA and partial sequencing of each *Asc*I-linking clone to identify the *Asc*I site in the clone just in case (Table 2). Similarly, 17 possible *Sse*8387I-linking clones (a–q) were identified in the same manner as in the case of the *Asc*I-linking clones.

All *Asc*I-linking clones obtained in this study were sequenced for further characterization. The determined sequence of each linking clone was searched for open reading frames (ORFs), and putative whole or partial ORFs were identified as shown in Fig. 3. Three partial ORFs identified in the linking clones A, C, and D were similar to *B. subtilis resB* (cytochrome C biogenesis protein), *rnh* (ribonuclease H), and *tuaG* (a protein involved in biosynthesis of teichuronic acid), respectively (Fig. 3).

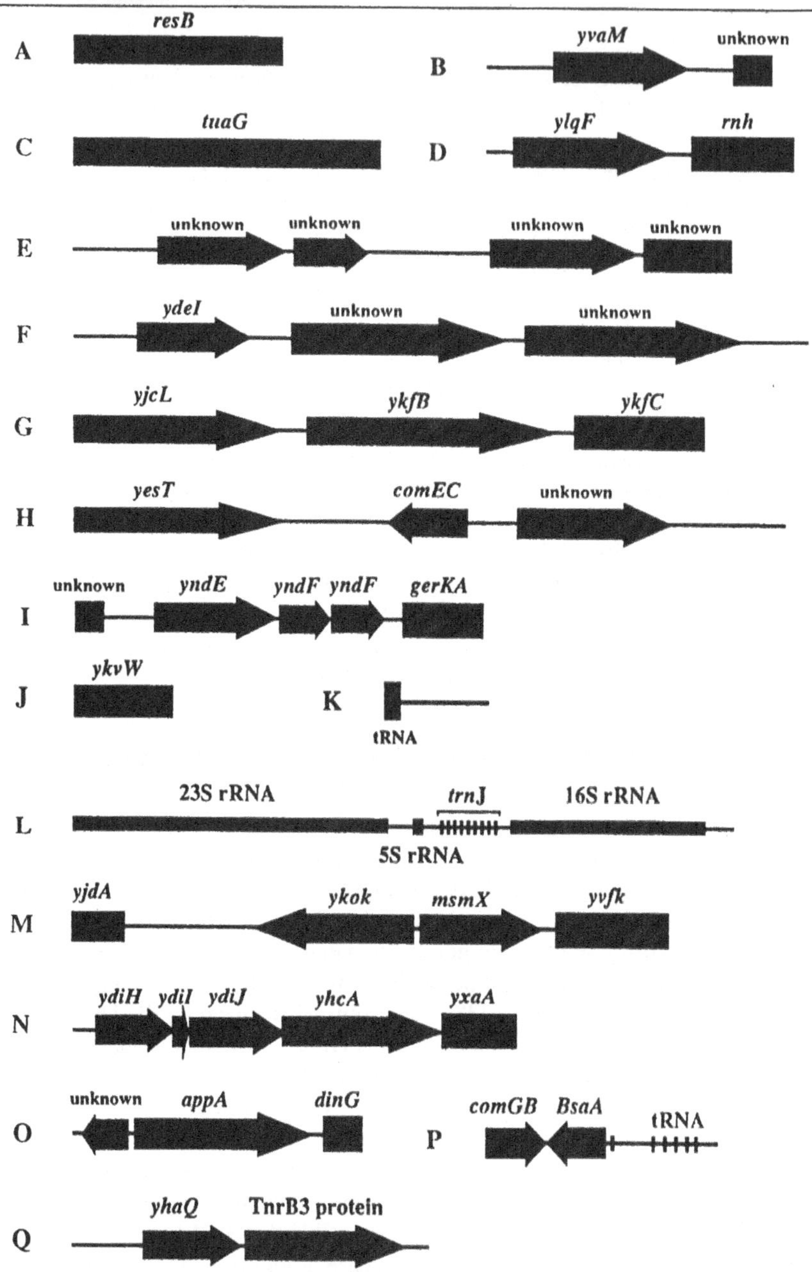

Fig. 3. ORF analysis of *Asc*I-linking clones (**A–Q**). The determined sequence of each linking clone was searched for ORFs consisting of more than 100 codons. The deduced amino acid sequences of these ORFs were used for homology analysis in searches of the BSORF, SWISS-PROT, and PIR protein databases using the FASTA program. As for the ORFs showing similarity to the protein sequences deposited in the databases, the name of the ORF is indicated for each ORF identified. *Arrow* denotes a complete ORF, and *truncated arrow* means a partial one

Table 2. *Asc*I-linking clones sequenced in this study

*Asc*I-linking clones	Length of the nucleotide (bp)	Position of the site(s) for *Asc*I
A	1059	159, 984
B	1445	1258
C	1584	989
D	1688	185
E	2504	2007
F	3638	545
G	2429	24
H	6865	2775
I	3933	380
J	555	417
K	589	76
L	6503	4024
M	5691	1588
N	4208	1244
O	2778	24
P	2207	134
Q	3476	255

Three putative ORFs identified in clones B, D, and F showed similarity to *B. subtilis yvaM*, *ylqF*, and *ydeI*, respectively. In the linking clone H, two ORFs were found to be similar to *B. subtilis yesT* and *comEC* (putative integral membrane protein). In the linking clone L, there was a partial ribosomal operon (Fig. 3). Several whole or partial ORFs were also identified in the linking clones M–Q as shown in Fig. 3.

11.1.3 Constricting an *Asc*I/*Sse*8387I Physical Map

All *Asc*I- and *Sse*8387I-linking clones listed in Table 3 were used as hybridization probes in Southern blot analyses of *Asc*I and *Sse*8387I digests of C-125 chromosomal DNA in PFGE gels. In most cases, the linking clones for a particular enzyme hybridized to two fragments obtained after digestion with the enzyme.

As shown in Fig. 4 and Table 4, *Sse*8387I-linking clone k hybridized to 3S' and 15S, and *Sse*8387I-linking clone l hybridized to 4S and 14S. *Asc*I-linking clone D (5A and 7A) and F (5A and 10A) hybridized to two fragments (Fig. 4, Table 2).

*Asc*I-linking clone A containing two *Asc*I sites in the *Hin*dIII fragment hybridized to three fragments: 1A, 18A, and 3A', respectively. Two DNA fragments that overlapped at the 3A position in the PFGE gel (Fig. 2, Table 1) were distinguished one (3A) from the other (3A') by comparing the hybridization patterns obtained for chromosomal DNA doubly digested with *Asc*I and *Sfi*I when hybridized to a series of *Asc*I-linking clones. It was found that there were at least two *Sfi*I sites in the 3A' fragment (Fig. 5) but no *Sfi*I site in the 3A fragment. Another two DNA fragments that overlapped at the 12A position were distinguished as 12A and 12A' in the same

Table 3. DNA probes used for construction of the physical map

Probe	Size	Characteristic	Source or reference
*Asc*I-A	1.1	C-125 *Asc* I-linking clone	This study
*Asc*I-B	1.5	C-125 *Asc* I-linking clone	This study
*Asc*I-C	1.6	C-125 *Asc* I-linking clone	This study
*Asc*I-D	1.7	C-125 *Asc* I-linking clone	This study
*Asc*I-E	2.5	C-125 *Asc* I-linking clone	This study
*Asc*I-F	3.6	C-125 *Asc* I-linking clone	This study
*Asc*I-G	2.4	C-125 *Asc* I-linking clone	This study
*Asc*I-H	6.9	C-125 *Asc* I-linking clone	This study
*Asc*I-I	3.9	C-125 *Asc* I-linking clone	This study
*Asc*I-J	0.6	C-125 *Asc* I-linking clone	This study
*Asc*I-K	0.6	C-125 *Asc* I-linking clone	This study
*Asc*I-L	6.5	C-125 *Asc* I-linking clone	This study
*Asc*I-M	5.7	C-125 *Asc* I-linking clone	This study
*Asc*I-N	4.2	C-125 *Asc* I-linking clone	This study
*Asc*I-O	2.8	C-125 *Asc* I-linking clone	This study
*Asc*I-P	2.2	C-125 *Asc* I-linking clone	This study
*Asc*I-Q	3.5	C-125 *Asc* I-linking clone	This study
Sse 8387I-a	0.5	C-125 *Sse*8387I-linking clone	This study
Sse 8387I-b	2.2	C-125 *Sse*8387I-linking clone	This study
Sse 8387I-c	4.5	C-125 *Sse*8387I-linking clone	This study
Sse 8387I-d	2.4	C-125 *Sse*8387I-linking clone	This study
Sse 8387I-e	3.2	C-125 *Sse*8387I-linking clone	This study
Sse 8387I-f	6.3	C-125 *Sse*8387I-linking clone	This study
Sse 8387I-g	1.5	C-125 *Sse*8387I-linking clone	This study
Sse 8387I-h	1.3	C-125 *Sse*8387I-linking clone	This study
Sse 8387I-i	2.0	C-125 *Sse*8387I-linking clone	This study
Sse 8387I-j	4.0	C-125 *Sse*8387I-linking clone	This study
Sse 8387I-k	2.8	C-125 *Sse*8387I-linking clone	This study
Sse 8387I-l	2.2	C-125 *Sse*8387I-linking clone	This study
Sse 8387I-m	4.0	C-125 *Sse*8387I-linking clone	This study
Sse 8387I-n	4.0	C-125 *Sse*8387I-linking clone	This study
Sse 8387I-o	2.1	C-125 *Sse*8387I-linking clone	This study
Sse 8387I-p	8.0	C-125 *Sse*8387I-linking clone	This study
Sse 8387I-q	6.0	C-125 *Sse*8387I-linking clone	This study
C-125 *gyr*B	0.4	DNA gyrase subunit B	This study
C-125 *rpo*A	1.0	RNA polymerase a subunit	This study
C-125 *sig*A	1.1	RNA polymerase major sigma factor (s^A)	This study
C-125 *sec*Y	1.3	Preprotein translocase subunit	Kang et al. (1992)
C-125 *hag*	0.5	Flagellin	Sakamoto et al. (1992)
C-125 pALK	2.4	Alkaline resistance	Kudo et al. (1990)
C-125 *xyl* (A)	1.8	Alkaline xylanase	Honda et al. (1985)
C-125 *groEL*	1.5	Heat-shock protein (chaperonin)	Xu et al. (1996)
C-125 *recA*	0.8	Homologous recombination and DNA repair	Takami (unpublished)
B. subtilis trpS	1.0	*B. subtilis* tryptophanyl-tRNA synthetase	Chow and Wong (1988)

manner, except that *Sse*8387I was employed rather than *Sfi*I. *Asc*I-M hybridized to 4A and 16A, *Asc*I-N to 3A and 15A, *Asc*I-K to 3A and 17A, *Asc*I-E to 13A and 17A, *Asc*I-Q to 1A and 13A, *Asc*I-I to 3A' and 10A, *Asc*I-B to 7A and 14A, *Asc*I-O to 12A and 14A, *Asc*I-G to 2A and 8A, *Asc*I-C to 6A and 11A, *Asc*I-H to 6A and 12A', and *Asc*I-J to 9A and 12A' (Fig. 5, Table 4). A series of hybridization experiments resulted in four *Asc*I contiguous fragments as follows: 4A-16A; 15A-3A-17A-13A-1A-18A-3A'-10A-5A-7A-14A-12A; 8A-2A; and 11A-6A-12A'-9A. During mapping, the contiguous fragments were joined by means of hybridization and cross-hybridization patterns obtained on analysis of chromosomal DNA digested with *Asc*I or *Sse*8387I when hybridized to DNA probes specific for *trpS* or *Sse*8387I- or *Asc*I-linking clones (Fig. 5).

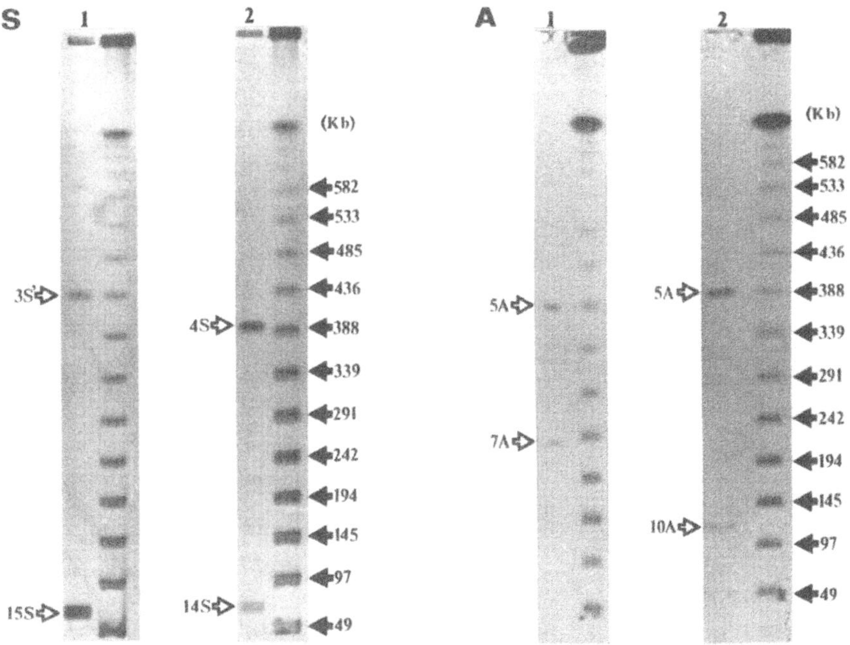

Fig. 4. Southern blot analysis of *Bacillus halodurans* C-125 genomic DNA hybridized to *Asc*I- or *Sse*8387I-linking clones. **S** *Sse*8387I digest. Lanes: *1*, *Sse*8387I-linking clone k; *2*, *Sse*8387I-linking clone l. **A** Lanes: *1*, *Asc*I-linking clone D; 2, *Asc*I-linking clone F. *Numbers on the right* show sizes of fragments in the l DNA ladder. *Arrows on the left* show the bands that hybridized to the linking clone labeled with digoxigenin

Table 4. Hybridization patterns of linking clones and other genes to *Asc*I and *Sse*8387I fragments of *Bacillus* sp. strain C-125 chromosome

Probe	Fragment number hybridized	
	Asc I	*Sse*8387I
Asc I-linking clone		
A	1A/18A/3A'	2S
B	7A/14A	7S
C	6A/11A	1S
D	5A/7A	5S
E	13A/17A	6S
F	5A/10A	18S
G	2A/8A	3S
H	6A/12A'	9S
I	3A/10A	12S
J	9A/12A'	3S'
K	3A/17A	nt
L	nt	nt
M	4A/16A	4S
N	3A/15A	4S
O	12A'/14A	7S
P	nt	nt
Q	1A/13A	6S
*Sse*8387I-linking clone		
a	5A	8S/17S
b	5A	17S/18S
c	3A	6S/13S
d	3A'	10S/12S
e	3A'	2S/13S
f	12A'	3S'/9S
g	2A	1S/16S
h	3A'	11S/19S
i	3A'	14S/19S
j	10A	12S/18S
k	4A	3S'/15S
l	3A	4S/14S
m	3A'	10S/11S
n	6A	1S/9S
o	2A	3S/16S
p	nt	nt
q	nt	nt
Other probe		
gyrB	9A	3S'
rpoA	4A	3S'
sigA	1A	2S
secY	4A	3S'
hag	11A	1S
pALK	1A	2S
xyl (A)	5A	17S
groEL	3A	4S
trpS (B. subtilis)	8A	7S

nt, not tested.

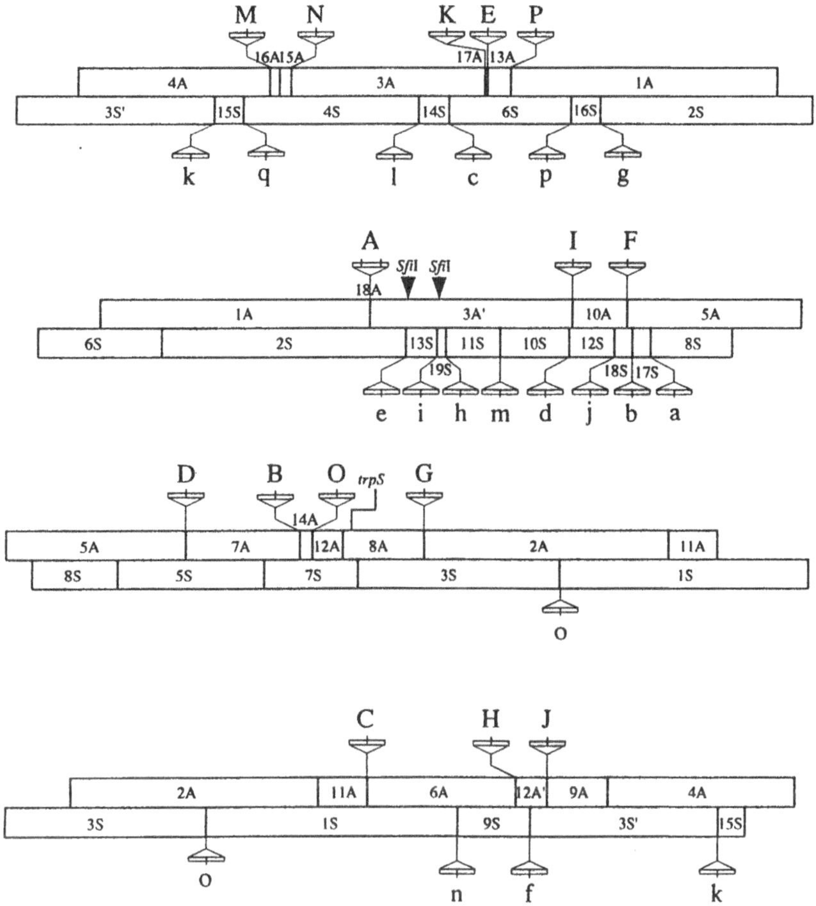

Fig. 5. Linkage map of *Bacillus halodurans* C-125 *Asc*I fragments (A) and corresponding *Sse*8387I (S) fragments. Capital letters indicate *Asc*I-linking clones, and lower-case letters indicate *Sse*8387I-linking clones. Fragment numbers correspond to their assigned band numbers as described in Table 1

11.1.4 Replication Origin Region of the Chromosome of B. halodurans C-125

A 400-bp DNA fragment containing part of the *gyrB* gene was amplified by polymerase chain reaction (PCR) from the chromosomal DNA of strain C-125 using the degenerate primers described (5'-GGNAAYATHGGNYTNATG-3' and 5'-RTTDATNGTYTCDATCAT-3'). A 9.8-kb fragment that hybridized to the PCR fragment was amplified by inverse PCR using the primer set (see legend to Fig. 6). Similarly, a 14 kb fragment was amplified by inverse PCR, and it covered the region upstream of

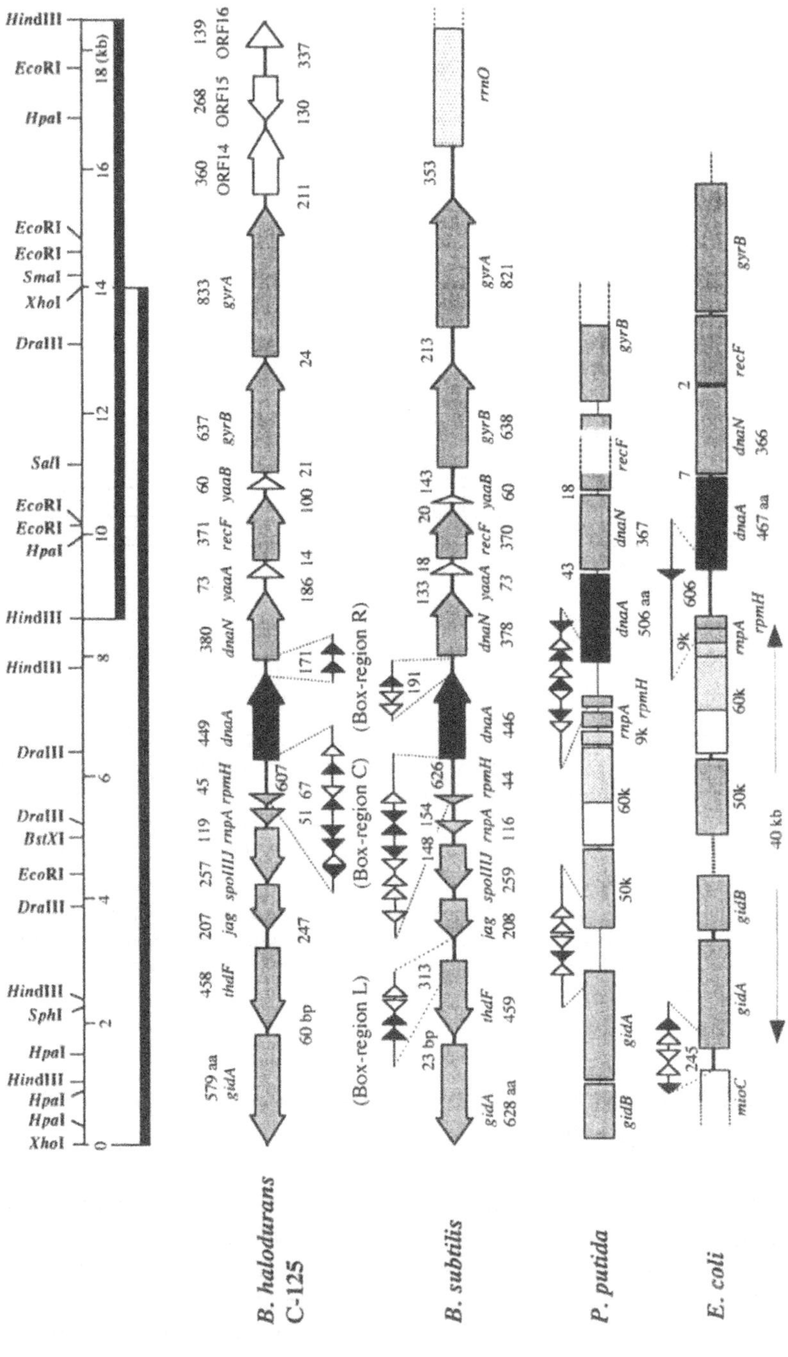
B. halodurans C-125
B. subtilis
P. putida
E. coli
HindIII
EcoRI
HpaI
SmaI
XhoI
DraIII
SalI
BstXI
SphI
18 (kb)
16
14
12
10
8
6
4
2
0
579 aa gidA
458 thdF
207 jag
257 spoIIIJ
119 rnpA
45 rpmH
449 dnaA
380 dnaN
73 yaaA
371 recF
60 yaaB
637 gyrB
833 gyrA
360 ORF14
268 ORF15
139 ORF16
60 bp
247
607
51 67
171
186 14
100 21
24
211
130
337
(Box-region L)
(Box-region C)
(Box-region R)
23 bp
313
148 154
191
626
133 18
20 143
213
353
rrnO
gidA 628 aa
thdF 459
jag 208
spoIIIJ 259
rnpA 116
rpmH 44
dnaA 446
dnaN 378
yaaA 73
recF 370
yaaB 60
gyrB 638
gyrA 821
gidB
gidA
50k
60k
rnpA
9k rpmH
dnaA 506 aa
43
dnaN 367
18
recF
gyrB
mioC
245
gidA
gidB
50k
60k
9k
rnpA
rpmH
606
dnaA 467 aa
7
dnaN 366
2
recF
gyrB
40 kb

the *gyrB* gene, which is located downstream of the *dnaN* gene. Both PCR fragments were sequenced and oligonucleotide primers were designed based on the sequence around the *Hin*dIII or *Xho*I site in each PCR fragment for cloning of the *oriC* region of the chromosome of strain C-125 (Takami et al. 1999c). A 9.8-kb *Hin*dIII fragment and a 14-kb *Xho*I fragment were amplified from the chromosome; a 5.5-kb *Hin*dIII-*Xho*I fragment was common to both of these fragments (Fig. 6). The determined sequence of the *Xho*I fragment and that of the *Hin*dIII fragment were searched for ORFs beginning with the start codon ATG, GTG, or TTG. Sixteen putative ORFs were identified based on the presence of a preceding Shine-Dalgarno (SD) sequence of *B. halodurans* C-125 (Takami et al. 1997) in each instance (see Table 6, below). The initiation codon in three of the ORFs was TTG, six of them had GTG, and the other seven started with ATG (Table 5).

Six ORFs (ORF1–ORF6) were identified in the region upstream of *dnaA* in the chromosome of strain C-125. The ORF organization in this region was found to be fairly well conserved among the eubacteria shown in Fig. 6 and especially resembled that of *B. subtilis* (Moriya et al. 1985). The ORF1 and ORF2 products are similar to the *B. subtilis gidA* and *thdF* products, whose functions are glucose-inhibited division and thiophen/furan oxidation, respectively (Ogasawara and Yoshikawa 1992). The ORF3 product is similar to the *B. subtilis jag* product, a *spoIIIJ*-associated protein. The ORF4 product is similar to the *B. subtilis spoIIIJ* product, which is essential for *sigG* activity in stage III sporulation (Errington et al. 1992). The ORF3 and ORF4 products have 43% and 62% identity to the *jag* and *spoIIIJ* products of *B. subtilis*, respectively. In *B. subtilis*, the *jag* gene appears to overlap the distal *spoIIIJ*

Fig. 6. Gene organization of the 18.5 kb fragment containing *oriC* region of the chromosome of alkaliphilic *Bacillus halodurans* C-125 and its comparison with those of other bacteria. The DNA fragment was amplified by the polymerase chain reaction (PCR) using the two primers above, chromosomal DNA from strain C-125, and Ex Taq polymerase. PCR was performed with a DNA thermal cycler 9700 (Perkin Elmer, CT, USA) under conditions as follows: 25 cycles of 30 s at 94°C, 1 min at 45°C, and 15 s at 72°C. The PCR product was sequenced, and the oligonucleotide primers 5'-TGATCGGCATTTCCAAGCGATTCTC-CCATT-3' and 5'-CCTCCACAATGTAAATCTCACTAATCGA-3' were designed to isolate a large DNA fragment containing the *gyrB* gene by inverse PCR. The chromosomal DNA of strain C-125 was digested with *Hin*dIII or *Xho*I. The digested DNA was properly diluted and circularized using the Takara ligation kit. Inverse PCR was performed with LA Taq polymerase under conditions as follows: 30 cycles of 15 s at 94°C, 8 min at 68°C, and 5 min at 72°C. Open reading frames (ORFs, boxed) and nontranslatable regions (*solid lines*) are indicated. The orientation of ORFs in alkaliphilic *B. halodurans* C-125 and *B. subtilis* (Moriya et al. 1985; Ogasawara et al. 1992) is shown by *arrows*. ORFs homologous among the four bacteria are *shaded* and those conserved in the gram-negative bacteria (Hansen et al. 1982; Fujita et al. 1990) are *dotted*. *Dashed lines* in ORFs indicate regions that are not yet sequenced. The genes of unknown function in strain C-125 are *striped*. The dnaA genes are shown in *black*. DnaA box regions are magnified to show the DnaA boxes. *Black arrowheads* indicate the consensus sequence TTATCCACA, and open *arrowheads* indicate those differing by one base from the consensus sequence. In the case of *E. coli*, in the DnaA-box region L the DnaA boxes that differ by two bases from the consensus sequence are shown as *open arrows* gene, as the GTG start and TGA stop codons overlap in the sequence GTGA (Ogasawara et al. 1994). This pattern is also evident in the case of ORF3 and ORF4 of strain C-125. In addition, the ORF5 and ORF6 products are similar to *B. subtilis rpmH* (Moriya et al. 1985) and *rnpA* (Fujita et al. 1990), whose functions are ribosomal protein L34 and a protein component of RNA-containing ribonuclease P, with 86% and 56% identity, respectively (Table 5).

The ORF7 product showed 78% identity to the *dnaA* gene product (DnaA) of *B. subtilis* (Table 5). DnaA is the protein responsible for initiation of bidirectional chromosome replication in eubacteria. The DnaA protein binds specifically to a 9-mer consensus sequence, the DnaA-box TTAT(C/A)CACA, and closely related sequences (Yoshikawa and Ogasawara 1991). The interaction of DnaA with a cluster of DnaA boxes in the region of the origin forms the initiation complex. DnaA boxes have been identified in the intergenic region flanking *dnaA* in most of the bacteria studied so far. In the chromosome of strain C-125, there are two DnaA-box clusters (box-region C and R) with a consensus sequence on both sides of the *dnaA* gene (Fig. 6). This seems to be a common feature of the chromosomal replication origins in gram-positive bacteria (Ogasawara et al. 1985; Fujita et al. 1989, 1990). However, another DnaA box cluster (box-region L), which is present in the region between *thdF* and *jag* of *B. subtilis*, was not found in the case of strain C-125 (Fig. 6). The chromosome of strain C-125 has 11 DnaA boxes, 7 of which have a complete sequence; in contrast, the chromosome of *B. subtilis* has 15 DnaA boxes, 6 of which have a complete sequence (Yoshikawa and Ogasawara 1991). Another common feature of bacterial replication origins is the presence of AT-rich regions adjacent to the DnaA boxes; these regions were also identified in the case of C-125. This indicates that the replication origin of the chromosome of alkaliphilic *B. halodurans* C-125 is near one of the two clusters of DnaA boxes.

Nine ORFs (ORF8–ORF16) were identified in the region downstream of *dnaA* in the chromosome of strain C-125. The ORF organization in the region from *dnaN* to *gyrB* was known to be well conserved among the eubacteria shown in Fig. 6 and especially resembles that of *B. subtilis* (Moriya et al. 1985). The ORF8 product is similar to the *B. subtilis dnaN* product (Moriya et al. 1985) with 56% identity. The ORF10 product is similar to the *B. subtilis recF* product, whose function is recombination and repair of DNA (Kunst et al. 1997), with 58% identity. The ORF9 and ORF11 products are similar to *B. subtilis yaaA* and *yaaB*, the functions of which are unknown. The ORF12 and ORF13 products show 78% and 76% identity to the *B. subtilis gyrB* and *gyrA* gene products, respectively (Ogasawara et al. 1994). Bacterial DNA gyrase is a topoisomerase of type II involved in DNA replication. As a unique feature, the intergenic region of the gyrase genes of *B. subtilis* consists of 213 nucleotides, and *gyrA* constitutes a single transcriptional unit by itself (Ogasawara et al. 1994). On the other hand, the intergenic region in the case of strain C-125 consists of only 24 nucleotides (Fig. 6), and there is no termination signal-like palindrome in this region. This suggests that the system for transcription of the gyrase gene in strain C-125 differs from that in *B. subtilis*. The last three ORFs (ORF14–ORF16) did not show significant homology to any other protein so far reported. As shown in Fig. 6, there is an *rrnO* gene in the region of the *B. subtilis* chromosome

Table 5. Analysis of ORFs in the 18.5 kb fragment

ORF	Position	Size (aa)	Putative SD sequence	Characteristics	Identity (%)
ORF1	505-2241	579	GGAAAAGGAGGTACAGTCC**atg**	*gidA*	77
ORF2	2302-3675	458	GAAAGaGAGGTGAGCCGAA**atg**	*thdF*	77
ORF3	3923-4543	207	AGGTGGGAGGAAAGAAAAA**gtg**	*jag*	43
ORF4	4543-5313	257	GAAACAAGGAGGAACCGTC**gtg**	*spoIIIJ*	62
ORF5	5365-5721	119	AAGAGGAcGGTGGAACGGA**ttg**	*rnpA*	56
ORF6	5789-5923	45	TCCTCAGGgAGGTGTTATAA**atg**	*rpmH*	86
ORF7	6521-7867	449	GAATAGGAGGGAGCAACGG**ttg**	*dnaA*	78
ORF8	8040-9179	380	ATAAAAGGAGTCTTATTACG**atg**	*dnaN*	58
ORF9	9366-9584	73	CGAAGAAAGcGAGGTCAATC**atg**	*yaaA*	50
ORF10	9599-10711	371	CGGAATAGAGGGTACGAAC**gtg**	*recF*	63
ORF11	10725-10991	89	AAAGAAGGGATAGGTGATAGCTC**gtg**	*yaaB*	29
ORF12	11067-12977	637	GAAGcGaAGGTGAACGAAA**gtg**	*gyrB*	78
ORF13	13002-15500	833	GGGAGATGGAGGTTTAAAA**atg**	*gyrA*	76
ORF14	15700-16779	360	GTGTAAAGGTGAGATTAAC**gtg**	Unknown	—
ORF15	17713-16910	268	ATGCAGAGGAGGAATTCCT**ttg**	Unknown	—
ORF16	18051-18467	139	TAAAATTAGGAGGAATTTAA**atg**	Unknown	—

ORF, open reading frame; SD, Shine-Dalgarno sequence.

The nucleotide numbers corresponding to the beginning and end of each ORF in the 18.5-kb fragment of the chromosome of alkaliphilic *Bacillus halodurans* C-125 and the size of the putative proteins are shown. The putative initiation codons and SD sequences are shown in bold and underlined, respectively. The protein homologous to each ORF is shown, with the identity %.

corresponding to the ORF14–ORF16 region of the C-125 chromosome. This point is a unique feature of the *oriC* region of strain C-125, although the ORF organization from *gidA* to *gyrA* is identical to that of *B. subtilis* (Fig. 6). Compared with the chromosome of *B. subtilis*, we found some differences, such as the length of the intergenic region between the *gyr* genes, and three ORFs of unknown function. We are investigating whether these features are related to some special function required for chromosome replication under alkaline conditions.

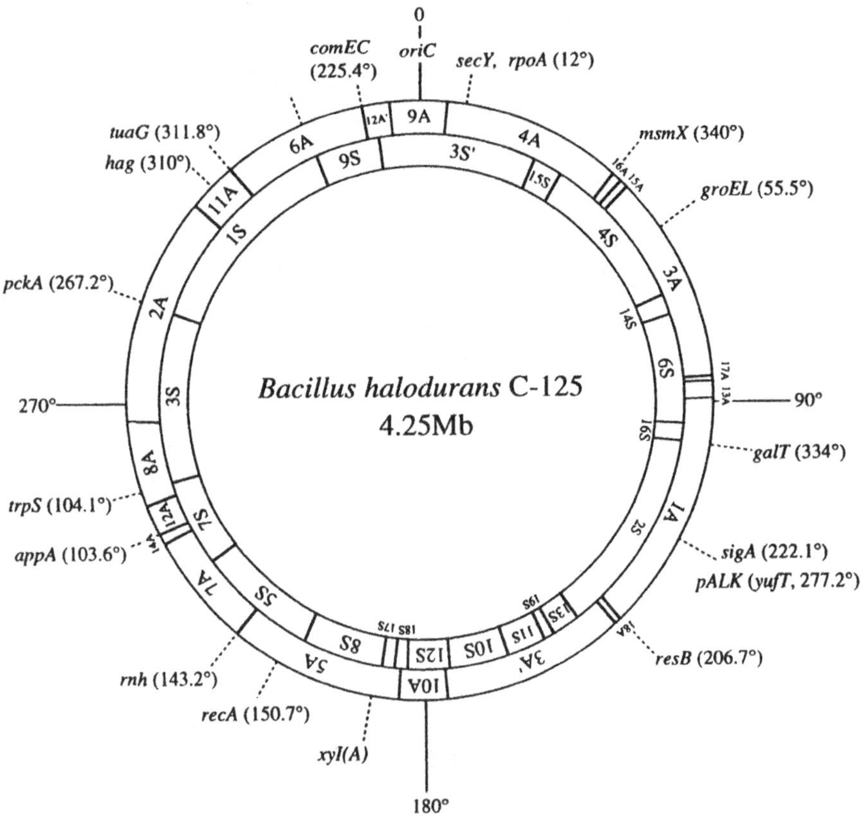

Fig. 7. *AscI/Sse*8387I physical and genetic map of the chromosome of *Bacillus halodurans* C-125. *Outer* and *inner circles* show the *Asc*I and *Sse*8387I physical maps, respectively. The locations of several housekeeping genes are indicated on the map. *Dashed line* indicates the approximate position of the gene

11.1.5 Positioning of the *oriC* Region on the Physical Map

A 9.8-kb *Hin*dIII fragment containing a part of the region extending from the *dnaN* gene to the *gyrA* gene hybridized to the DNA probe and the *Hin*dIII fragment was amplified by inverse PCR (Triglia et al. 1988) using the DNA primer set as described above. In addition, the *Hin*dIII fragment was amplified by long accurate PCR using LA PCR kit Ver. 2 (Takara Shuzo, Japan) from the chromosomal DNA and used as a probe to determine the position of the *oriC* region on the physical map. This DNA probe hybridized to fragments 9A and 3S', and it was found that the *oriC* region was located in the region of overlap between 9A and 3S' (Fig. 7). If the *oriC* region is assigned the position at 12 o'clock as 0 degrees, the 13A fragment is at approximately 90 degrees, the region of overlap between 10A and 12S is at 180 degrees, and the border region between 8A and 2A is at 270 degrees on the map (Fig. 7).

11.1.6 Mapping of Genetic Loci

The DNA probes listed in Table 3 were used for mapping genetic loci on the physical map. The ORFs identified and annotated in the linking clones (Fig. 5) were also used for mapping. The assigned positions of the genes on the physical map of *Bacillus halodurans* C-125 were compared with those on the genetic and physical maps of *B. subtilis* (Itaya and Tanaka 1991; Piggott and Hoch 1993; Kunst et al. 1997). DNA probes for *secY* (Kang et al. 1992) and *rpoA* from strain C-125 hybridized to the same fragments 4A and 3S' (Table 4). The genetic loci on the physical map of strain C-125 were within approximately 12 degrees of their positions on the *B. subtilis* physical map (Fig. 7). It was also found that probes for *groEL* (Xu et al. 1996) and *hag* (Sakamoto et al. 1992) hybridized to 3A and 4S, and 11A and 1S, respectively. Their positions on the map were also close to those of *B. subtilis* (55 degrees for *groEL* and 310 degrees for *hag*) as shown in Fig. 7. On the other hand, it appeared that genes such as *trpS*, *appA*, *rnh*, and *recA* have divergent map positions, being located at 103–150 degrees (Fig. 7) on the *B. subtilis* genetic map and approximately 200–250 degrees on the map of strain C-125. The map positions of three genes, *sigA*, *pALK* (*yuft*), and *resB*, located at 100–140 degrees also differ from the positions 206–277 degrees on the *B. subtilis* map. Thus it appears that the region around 200–250 degrees on the C-125 map may correspond to the region around 100–140 degrees on the genetic map of *B. subtilis*.

11.1.7 Ribosomal Protein Gene Clusters of *B. halodurans* C-125

A 9.3-kb *Sph*I fragment that hybridized to fragment 5 was obtained by means of inverse PCR with a primer set, 4Ss and 4Sas (Fig. 8). This fragment contained the region to *rplV* upstream of *infA*. A 4.7-kb DNA fragment that similarly hybridized to fragment 5 was obtained in the same way using a primer set, 6Ss and 6Sas from a 6.7

kb *Sph*I fragment (Fig. 8). In the adjacent *Sph*I fragment, with a total size of approximately 16 kb, *spc*, the a cluster, and part of the S10 cluster were found. To obtain the region upstream of the S10 cluster, a part of the *rpoB* gene was first amplified by means of PCR with a set of degenerate primers (B1 and B2) from the C-125 chromosome (Takami et al. 1999d). After sequencing the amplified fragment, fragment 1, a 3.5-kb fragment 2 was obtained by inverse PCR using a primer set designed from an internal sequence of fragment 1 (Das and Ds) and chromosomal DNA digested with

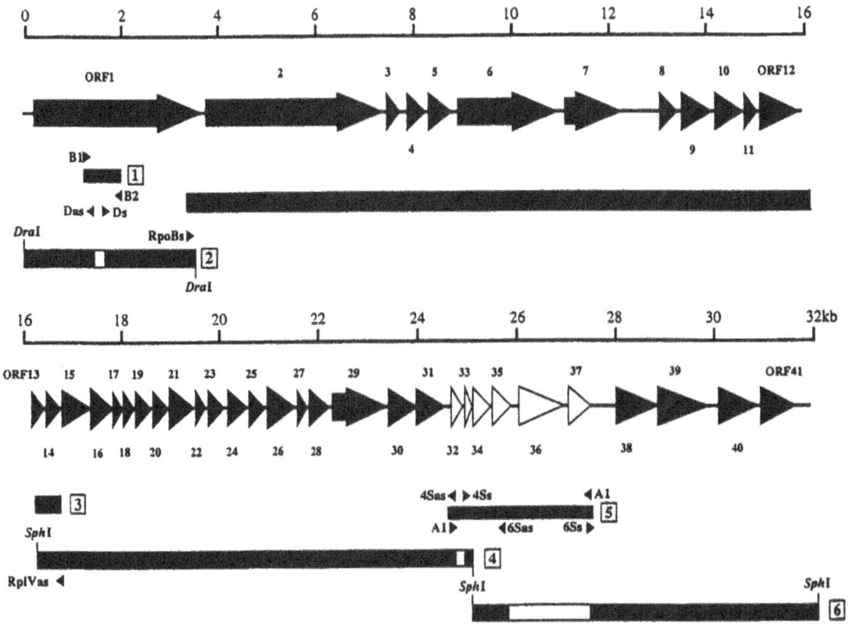

Fig. 8. Cloning strategy and ORF analysis of a 32-kb region of the chromosome of alkaliphilic *Bacillus halodurans* C-125. DNA fragments 1 and 5 were amplified by means of PCR with primer set B1/B2 and A1/A2, respectively, according to the method recommended by the supplier (Perkin Elmer, USA). Fragments 2, 4, and 6 were amplified by means of inverse PCR using the primer set, Das/Ds for 2, 4Sas/4Ss for 4, and 6Sas/6Ss for 6. Fragment 3 was obtained by Logn and Accurate (LA)-PCR using primer set RpoBs/ RplVas according to the method recommended by the supplier (Takara Shuzo, Japan). The PCR primers used in this study were as follows : (**A1**), 5'-AAGCTTAAATGTAGGTGATGAAAA-3'; (**A2**), 5'-AAGCTTTTGAATCGCATCTTGGCCT-3'; (**B1**), 5'-GTGCCGTTCACTACTCYCACTATG-3'; (**B2**), 5'-YTGACGYTGCATGTTCGCWCCCAT-3'; (**5Ss**), 5'-ACTGAAGGAAGGAGAGTAAATTCATGGCCAAGG-3'; (**5Sas**), 5'-AGGTGATTCATGTTTTTTCGCTTC-GCTCGATCG-3'; (**3Ss**), 5'-CATGGCATACGCAAAATTAGGTCGTG-3'; (**3Sas**), 5'-GTGCCTTTACGGC-GGATCACTTTACA-3'; (**Ds**), 5'-CTAGAGATCGTGTGGACTACATGGAC-3'; (**Das**), 5'-GTGTCTCCATAA-AGCCGAACTCGTTG-3'; (**rpoBs**), 5'-AGGTCCTTACTCCCTTGTAACGCAACAGCCATT-3'; (**rplVas**), 5'-AATCGCTTCTCCTACTTGCTTTCCACGGATAAG-3'. The *dark bar* indicates the region amplified in each PCR reaction. *Black arrows* indicate the ORFs identified in this study, and *white arrows* indicate ORFs previously reported (Nakasone et al. 1998)

*Dra*I. The region between the *rpoB* and *rplV* genes was obtained by means of long and accurate PCR (LA-PCR) with a primer set (RpoBs and RplVas) that was designed from an internal sequence of fragment 2 or 4. It was confirmed that part of the *rpoC* gene adjacent to *rpoB* and *rplV* gene was present by sequencing both ends of an amplified 14-kb fragment. All DNA fragments amplified in this study were sequenced by the method described previously (Takami et al. 1999b).

The determined total sequence of the 32-kb fragment was examined for ORFs beginning with an ATG, GTG, or TTG start codon. Forty-one putative ORFs including 6 ORFs (ORF32–ORF37) were identified based on the presence of a preceding Shine-Dalgarno (SD) sequence in each instance. The SD sequence was complementary to one found at the 3' end (UCUUUCCUCCACUAG...) of the 16S rRNA of alkaliphilic *Bacillus halodurans* C-125 (Takami et al. 1997). The initiation codon for 5 of the ORFs was TTG, 2 of them started with GTG, and the other 34 ORFs started with ATG (data not shown). The deduced amino acid sequences of these ORFs were used for homology analysis in searches of the BSORF, SWISS-PROT, and PIR protein databases using the FASTA program. The ORF1 and ORF2 products were similar to RNA polymerase β- and β'-subunits of *B. subtilis,* with 87% and 85% identity, respectively. The gene products of ORF4 and ORF5 were similar to S12 and S7 of *B. subtilis* and *B. stearothermophilus* (Kimura 1991) with 82% and 84% identity, respectively. In addition, two ORFs (ORF6 and ORF7) showed significant identity to elongation factor EF-G (86.9%) and EF-Tu (91.4%). The order of ORF products from ORF4 to ORF7 corresponding to the *str* cluster was same as that in *B. subtilis,* although there was not an ORF corresponding to *ybaC* of *B. subtilis* in this region of the chromosome of strain C-125. The products of 11 ORFs (ORF8–ORF18) downstream from the *str* cluster were similar to the gene products of the S10 cluster in *B. subtilis* (Boylan et al. 1989; Suh et al. 1996; Li et al. 1997) and *B. stearothermophilus* (Kimura et al. 1985; Kromer et al. 1990; Herwig et al. 1992) with high values of percent identity, as shown in Fig. 9. The products of the other 13 ORFs, from ORF19 to ORF31, also showed significant identity to the gene products forming the *spc* cluster in *B. subtilis* (Boylan et al. 1989; Suh et al. 1996; Li et al. 1997) and *B. stearothermophilus* (Ramakrishnan and Gerchman 1991; Herfurth et al. 1994; Kimura and Kimura 1987) as shown in Fig. 9. In the region downstream from the *spc* cluster, six ORFs forming the a cluster previously reported were confirmed to be present, and their products showed significant homology to those of *B. subtilis* and *B. stearothermophilus*. On the other hand, four ORFs of unknown function were identified in the region adjacent to the a cluster, as seen in *B. subtilis*. These ORF products were similar to those in *B. subtilis* with an identity value of more than 50% in each instance (Fig. 9). The nucleotide sequences of the regions of a 32-kb fragment of the chromosome of alkaliphilic *Bacillus halodurans* C-125 containing major ribosomal protein gene clusters, located in the 3S'/4A fragment in the physical map (Takami et al. 1999b), were determined and compared with those of *B. subtilis* and *B. stearothermophilus*. The ORFs and intergenic regions in this fragment were basically homologous in the three *Bacillus* strains compared.

Recently, the *rpoA* gene, encoding the α-subunit of RNA polymerase, was isolated from *Bacillus halodurans* C-125 by PCR (Nakasone et al. 1998). A 3-kb *Hin*dIII

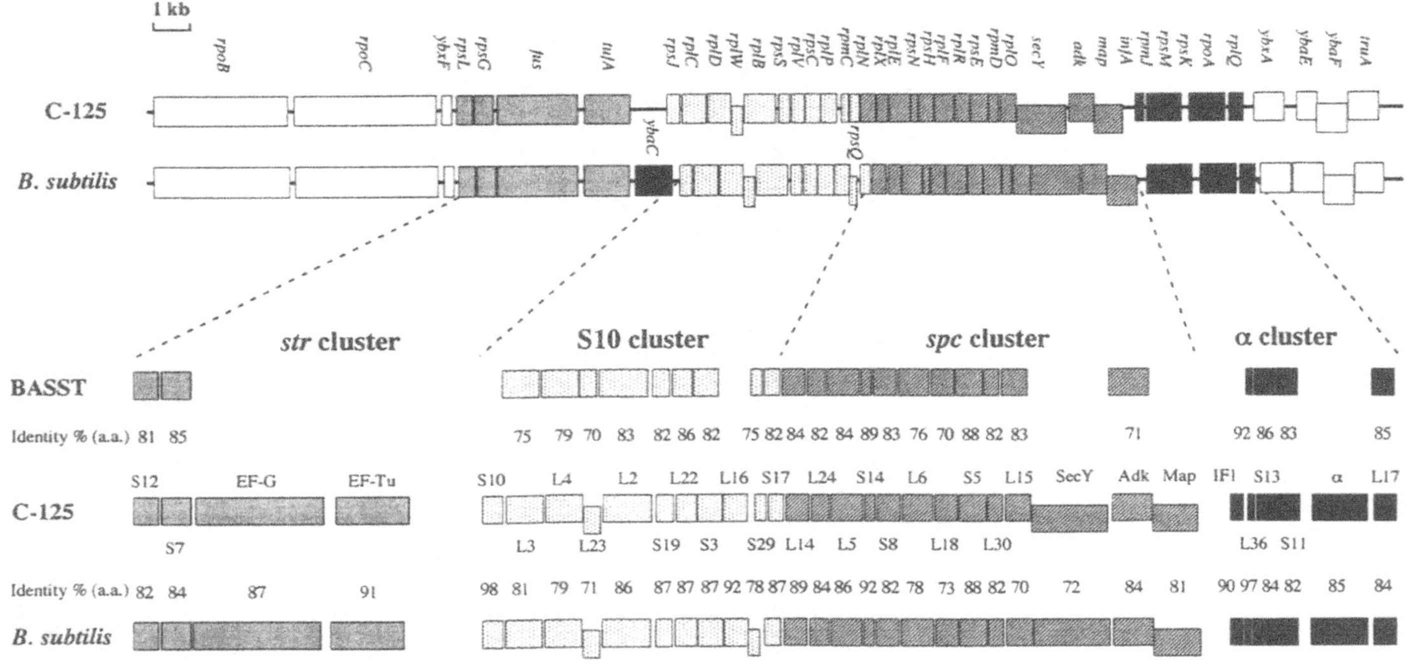

Fig. 9. Comparison of the genetic organization and predicted amino acid sequences of the genes in the *str*-S10-*spc*-a region of the chromosome of alkaliphilic *Bacillus halodurans* C-125 with those of *B. subtilis* and *B. stearothermophilus*. *BASST*, *B. stearothermophilus*. The intercistronic distance is indicated above the map. The identity of amino acid (*a.a.*)sequences between C-125 and the other two strains is shown as a percent (Identity a.a.)

fragment containing the complete *rpoA* gene was cloned and sequenced. The α-subunit gene was found to encode a protein consisting of 314 amino acid residues with a molecular mass of 34 805 daltons. Compared with the amino acid sequences of other known eubacterial RNA polymerase α-subunits, the gene has 84% identity to that of *B. subtilis* while showing 48% and 47% identity to that of *Streptomyces coelicolor* and *Escherichia coli*, respectively. Six conserved regions, which are observed in the case of other eubacteria, were found in the RNA polymerase α-subunits of this strain. Five of them are located in the N-terminal domain involved in assembly of the core enzyme, and one is located in the C-terminal domain, which interacts with several transcriptional factors and a specific DNA element. By means of recombinant plasmids, a hexahistidine-tagged derivative of the RNA polymerase α-subunit of strain C-125 and deletion derivatives (C- and N-terminal domains) this protein were expressed in *E. coli* cells and purified to near homogeneity. Cloning of other *rpo* genes of strain C-125 such as those encoding the β, β', and σ subunits, and construction of expression plasmids is now in progress. To reconstitute the recombinant RNA polymerase of strain C-125 in vitro from the purified subunits, overexpression of these proteins in *E. coli* will be carried out. Our reconstruction system not only will be useful for examining the functional properties of the recombinant RNA polymerase itself but will facilitate identification of promoter elements and *trans*-acting factors that control gene expression in the alkaliphilic strain under alkaline conditions.

11.2 Construction of λ Phage Library from the Genome of *B. halodurans* C-125

As the first step in analysis of the C-125 genome, to compare the gene structure and organization with those of *B. subtilis*, we attempted to construct a λ phage library of the C-125 chromosome for large-scale nucleotide sequencing (Takami et al. 1999e). In this section, we report the sequence determination and analysis of three large fragments (15-20 kb) of the genome of alkaliphilic *Bacillus halodurans* C-125 cloned in a λ phage vector.

11.2.1 Preparation of DNA Libraries from Alkaliphilic *B. halodurans* C-125

Chromosomal DNA was isolated from *Bacillus halodurans* C-125 as described previously (Saito and Miura 1963). DNA fragments partially digested with *Sau*3AI and treated with bacterial alkaline phosphatase (BAP) were ligated to a λDashII/*Bam*HI vector. Three λ clones (λ 3, 4, and 9) were randomly selected for sequencing analysis among the original recombinant phages prepared using the in vitro packaging system from Stratagene (CA, USA). The DNA fragments cloned in λ phage were amplified by LA-PCR using LA PCR Kit Ver. 2 (Takara Shuzo, Japan). A 7.5 μg aliquot of PCR fragments was sonicated for 10–20 s with a Branson Sonifier (Heat Systems, Farmingdale, NY, USA). The sonicated DNA fragments were blunt-ended using a DNA blunting kit (Takara Shuzo, Japan) and fractionated by 1% agarose gel electrophoresis. DNA fragments 1-2 kb in length were excised from the gel and eluted by the freeze-squeeze method (Thuring et al. 1975). The DNA recovered was ligated to the *Sma*I site of vector M13mp19 or pUC18, which had been previously treated with BAP; it was then introduced into competent XL1-Blue cells by the standard method (Sambrook et al. 1989).

11.2.2 DNA Sequencing and Computer Analysis

M13 phage clones isolated were grown in 2 x YT medium for 8 h in the host XL1-Blue. The M13 single-strand DNA was purified by means of a QIAprep 96 M13 kit (Qiagen, CA, USA). The DNA fragment inserted into pUC18 was amplified by PCR using M13-20 and reverse primers. PCR fragments, treated with Exonuclease I and shrimp alkaline phosphatase (Amsersham, OH, USA) to eliminate excess primers in the PCR reaction mixture, were used for sequencing analysis as template DNA. Sequencing was performed with a DNA sequencer ABI PRISM 377 using a Taq Dye Primer and Dye Terminator Cycle Sequencing Kit (Perkin Elmer, CT, USA). Several primers were prepared and used to sequence the DNA covering the gap region.

DNA sequences determined by means of the ABI sequencer were assembled into contigs using a personal computer with AutoAssembler Ver 1.40 (Perkin Elmer). The sequences were analyzed for the locations of possible ORFs using the GeneWorks program (version 2.5.1N) from IntelliGenetics (CA, USA). The deduced amino acid sequences of the identified ORFs were compared with sequences reported previously in a search of the BSORF database (http://bacillus.tokyo-center.genome.ad.jp:8008/) and the nonredundant protein databank using the FASTA and BLAST network service (GenomeNet WWW server, http://www.genome.ad.jp).

11.2.3 ORF Analysis

ORF Analysis of Sequenced Fragment 3

The determined sequence of fragment 3 was searched for ORFs beginning with an ATG, GTG, or TTG start codon and comprising more than 100 codons. Thirteen putative ORFs were identified based on the presence of a preceding Shine-Dalgano (SD) sequence in each instance (Fig. 10-1, Table 6). The SD sequence was complementary to one found at the 3' end (UCUUUCCUCCACUAG...) of the 16S rRNA of alkaliphilic *Bacillus halodurans* C-125 (Takami et al. 1997). The initiation codons for one of the ORFs was TTG, 2 were GTG, and the other 10 started with ATG (Table 6). The deduced amino acid sequences of these ORFs were used for homology analysis in searches of the BSORF, SWISS-PROT, and PIR protein databases

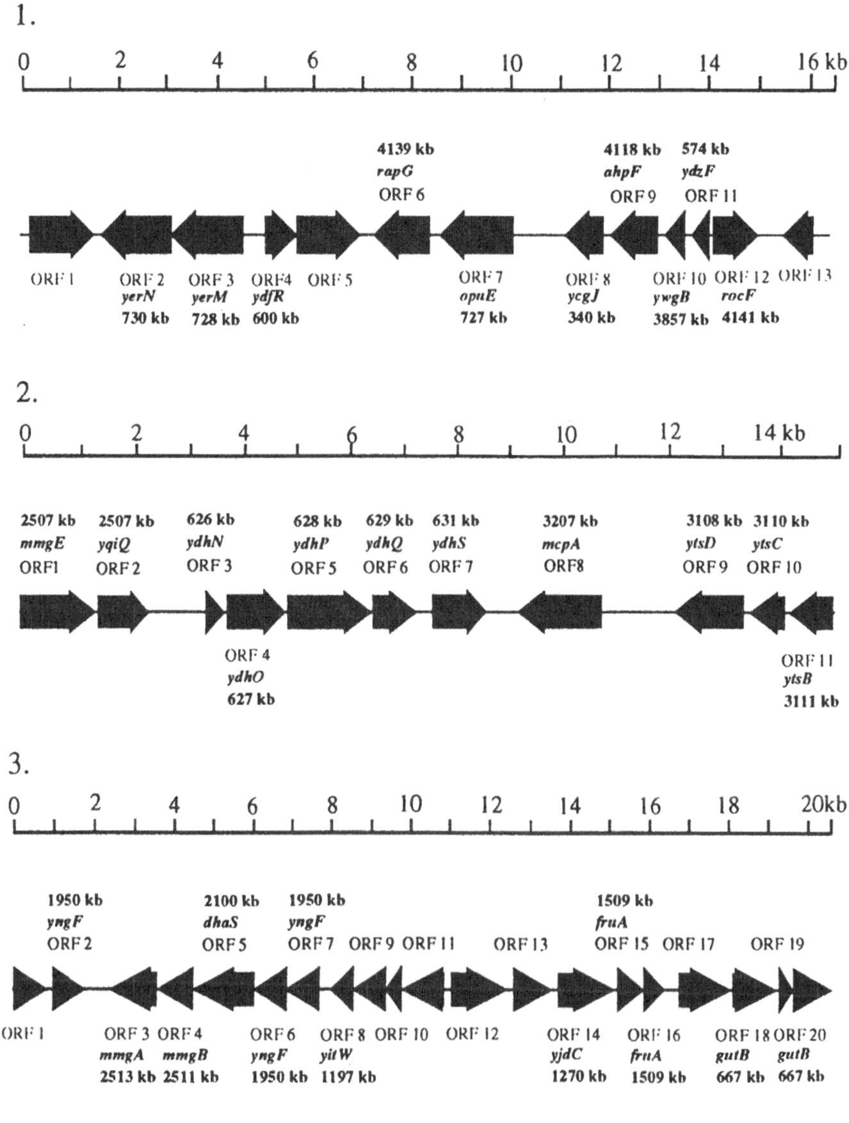

Fig. 10. ORF organization of the genomic fragments of alkaliphilic *Bacillus halodurans* C-125 cloned in l phage. **1** DNA fragment 3. **2** DNA fragment 4. **3** DNA fragment 9. The annotated ORFs are shown with the ORF position in the *B. subtilis* chromosome

using the FASTA program. Seven ORFs showed similarities to unknown *B. subtilis* proteins.

Two ORFs (ORF2 and ORF3) were similar to the products of the *yerN* and *yerM* genes of *B. subtilis*, with 87% and 84% identity, respectively. These genes seemed to constitute an operon with *opuE* at position 727–730 kb of the *B. subtilis* chromo-

Table 6. ORF analysis of 16.8-kb fragment of Bacillus sp. C-125 genome in λ-phage clone 3

ORF	Characteristics	SD/initiation codon
1. (nt 201-1523; length 621; 207 aa)	Similar to *Bordetella parapertussis* transposase for insertion sequence element (26% similarity) (406 aa)	AAAAGGCTCTACAATAA**ATG**
2. (nt 3109-1682; length 1428; 476 aa)	Similar to *B. subtilis yerN* (87% identity) (nt 730192-731445; 417 aa; function unknown)	AGGGGTGAAAGCAG**ATG**
3. (nt 4580-3126; length 1473; 485 aa)	Similar to *B. subtilis yerM* (84% identity) (nt 728544-730001; 485 aa; function unknown)	GGAAcGAGCTTA**ATG**
4. (nt 5118-5726; length 642; 203 aa)	Similar to *B. subtilis ydfR* (40% identity) (nt 600620-599943; 225 aa; function unknown)	AAAGGAGGAACAACCT**TTG**
5. (nt 5743-7014; length 1272; 424 aa)	Similar to hypothetical protein MJ0305 *Methanococcus jannaschii* (25% identity) (395 aa)	AGGATGATAGACGA**ATG**
6. (nt 8418-7306; length 1113; 371 aa)	Similar to *B. subtilis rapG* (33% identity)(nt 4139469-4140566; 365 aa; response regulator aspartate phosphatase)	AAAAGGggAGGGTAAACG**GTG**
7. (nt 10174-8672; length 1503; 501 aa)	Similar to *B. subtilis opuE* (53% identity) (nt 727824-726346; 492 aa; proline transporter)	AGGAGTGAGTGAACT**GTG**
8. (nt 12022-11258; length 765; 252 aa)	Similar to *B. subtilis ycgJ* (35% identity) (nt 340871-340185; 228 aa; function unknown)	AAAAtGGAGGGG**ATG**
9. (nt 13118-12204; length 933; 305 aa)	Similar to *B. subtilis ahpF* (25% similarity) (nt; 4118735 - 4120264; 509 aa; alkyl hydroperoxide reductase (large subunit) NADH dehydrogenase EC 1.6.99.3)	GGAGGTGCTTAAACGT**ATG**
10. (nt 13700-13305; length 396; 132 aa)	Similar to *B. subtilis ywgB* (27% identity) (nt 3857686-3857216; 156 aa; function unknown)	AGGTGAACACTGCATC**ATG**
11. (nt 14212-13874; length 339; 113 aa)	Similar to *B. subtilis ydzF* (54% identity) (nt 574149-573820; 109 aa; function unknown)	AGGAAAGTGTGAAT**ATG**
12. (nt 14319-15230; length 912; 304 aa)	Similar to *B. subtilis rocF* (29% identity) (nt 4141809-4140919; 296 aa; arginase)	AAAGGAGTGTCAATC**ATG**
13. (nt 16339-15737; length 603; 201 aa)	Unknown	AAAGGAGGCGTACATT**ATG**

The beginning and end nucleotide numbers of each ORF in the DNA fragment of the chromosome of alkaliphilic *Bacillus halodurans* C-125 and the size of the putative protein are shown. The putative initiation codons and SD sequences are shown in bold and underlined, respectively. The protein homologous to each ORF is shown with the size of the protein and similarity percent. In the case of the *B. subtilis* protein, the beginning and end nucleotide numbers for each protein are also shown.

Table 7. ORF analysis of 15-kb fragment of Bacillus sp. C-125 genome in λ-phage clone 4

ORF	Characteristics	SD/initiation codon
1. (nt 21-1418; length 1398; 466 aa)	Similar to *B. subtilis mmgE* (74% identity) (nt 2508875-2507457; 472 aa; function unknown)	AGAGAACGACGGATCTT**TTG**
2. (nt 1439-2338; length 900; 300 aa)	Similar to *B. subtilis yqiQ* (73% identity) (nt 2507439-2506534; 301 aa; function unknown)	AGGAGGACACGAAA**ATG**
3. (nt 3458-3784; length 327; 109 aa)	Similar to *B. subtilis ydhN* (59% identity) (nt 626649-626981; 110 aa; function unknown)	AGGAGGACGTTA**ATG**
4. (nt 3813-4955; length 1143; 381 aa)	Similar to *B. subtilis ydhO* (78% identity) (nt 627000-628328; 442 aa; function unknown)	GGAGcGAACGTCCCA**ATG**
5. (nt 4965-6473; length 1509; 503 aa)	Similar to *B. subtilis ydhP* (80% identity) [(nt 628346-629743; 465 aa; β-glucosidase (EC3.2.1.21)]	AAAGGGAT**ATG**
6. (nt 6515-7312; length 798; 266 aa)	Similar to *B. subtilis ydhQ* (70% identity) (nt 629886-630599; 237 aa; function unknown)	GGAGAATA**ATG**
7. (nt 7674-8618; length 945; 315 aa)	Similar to *B. subtilis ydhS* (60% identity) [(nt 631524-632471; 315 aa; mannose-6-phosphate isomerase (EC 5.3.1.8)]	GGAGACTGT**ATG**
8. (nt 10726-9191; length 1536; 512 aa)	Similar to *B. subtilis mcpA* (28% similarity) (nt 3207224-3205239; 661 aa; methyl-accepting chemotaxis protein)	AAAGGATTGATCAA**GTG**
9. (nt 13323-12088; length 1236; 412 aa)	Similar to *B. subtilis ytsD* (64% similarity) (nt 3110418-3108478; 646 aa; function unknown)	AGGTGTGTTAG**GTG**
10. (nt 14063-13458; length 606; 202 aa)	Similar to *B. subtilis ytsC* (80% identity) (nt 3111169-3110408; 253 aa; function unknown)	AGGAGTGAAAATAC**ATG**
11. (nt 14965-14183; length 783; 261 aa)	Similar to *B. subtilis ytsB* (64% similarity) (nt 3112275-3111271; 334 aa; function unknown)	AGGaAtGGGACAAAGAT**TTG**

The beginning and end nucleotide numbers of each ORF in the DNA fragment of the chromosome of alkaliphilic *Bacillus halodurans* C-125 and the size of the putative protein are shown. The putative initiation codons and SD sequences are shown in bold and underlined, respectively. The protein homologous to each ORF is shown with the size of the protein and similarity percent. In the case of the *B. subtilis* protein, the beginning and end nucleotide numbers for each protein are also shown.

Table 8. ORF analysis of 20-kb fragment of Bacillus sp. C-125 genome in λ phage clone 9

ORF	Characteristics	SD/initiation codon
1. (nt 51-836; length 786; 262 aa)	Unknown	AAAGcAGGCGTCATT**ATG**
2. (nt 1001-1777; length 777; 259 aa)	Similar to *B. subtilis yngF* (56% identity) (nt 1951265-1950483; 260 aa; enoyl CoA hydratase)	AGGAGGACAATCG**ATG**
3. (nt 3521-2412; length 1110; 370 aa)	Similar to *B. subtilis mmgA* (38% identity) (nt 2513271 -2512090; 391 aa; acetyl-CoA acetyl transferase)	AGGGGGCTAGTA**ATG**
4. (nt 4387-3539; length 849; 283 aa)	Similar to *B. subtilis mmgB* (42% identity) (nt 2511940-2511206; 244 aa; 3-hydroxybutyryl-CoA dehydrogenase EC 1.1.1.157)	AGGaAGGTTGCT**ATG**
5. (nt 5919-4456; length 1464; 488 aa)	Similar to *B. subtilis dhaS* (44% identity) (nt 2099836-2101323; 495 aa; aldehyde dehydrogenase)	AAAGGAGCGATTCCA**TTG**
6. (nt 6738-5959; length 780; 260 aa)	Similar to *B. subtilis yngF* (32% identity) (nt 1951265-1950483; 260 aa; enoyl CoA hydratase)	AGGAGAGTAAAG**ATG**
7. (nt 7514-6744; length 771; 257 aa)	Similar to *B. subtilis yngF* (39% identity) (nt 1951265-1950483; 260 aa; enoyl CoA hydratase)	AGGAGATGG**ATG**
8. (nt 8336-7833; length 504; 168 aa)	Similar to *B. subtilis yitW* (35% similarity) (nt 1191727-1192035; 102 aa; function unknown)	AGGGGGAGTGAAT**GTG**
9. (nt 9128-8352; length 777; 259 aa)	Unknown	AAGAGGCTAAAACGA**ATG**
10. (nt 9521-9165; length 357; 119 aa)	Unknown	AAAGGAGAAGGATG**ATG**
11. (nt 10528-9548; length 981; 327 aa)	Unknown	AAAAGGAGTTGTCGAT**ATG**

12. (nt 10779-12113; length 1335; 445 aa)	Unknown	AGGAGGGTGCAATTG**ATG**
13. (nt 12302-13195; length 894; 298 aa)	Unknown	AAAcAGGGTGGGATGAATC**TTG**
14. (nt 13404-14735; length 1332; 444 aa)	Similar to *B. subtilis yjdC* (24% similarity) (nt 127012-1272048; 648 aa; function unknown)	AGGgAGGTC**ATG**
15. (nt 14839-15453; length 615; 205 aa)	Similar to *B. subtilis fruA* (33% similarity) [nt 1508661-1510568; 635 aa; phosphotransferase system (PTS) fructose-specific enzyme IIBC component]	AAAAGAacAGGTC**ATG**
16. (nt 15480-15959; length 480; 160 aa)	Similar to *B. subtilis fruA* (29% similarity) (nt 1508661-1510568; 635 aa; phosphotransferase system (PTS) fructose-specific enzyme IIBC component)	AGAGGTGAACTG**ATG**
17. (nt 16333-17589; length 1257; 419 aa)	Similar to *E. coli PTS* system galactitol-specific IIC component (43% identity) (451 aa)	AAAGGGGGAACACC**ATG**
18. (nt 17632-18660; length 1029; 343 aa)	Similar to *B. subtilis gutB* (35% identity) (nt 667183- 668244; 353 aa; sorbitol dehydrogenase EC 1.1.1.14)	AAtGGAGGATCGCATG**ATG**
19. (nt 18775-19074; length 300; 100 aa)	Unknown	AAGAGGCA**TTG**
20. (nt 19119-20036; length 918; 306 aa)	Similar to *B. subtilis gutB* (39% identity) (nt 667183-668244; 353 aa; sorbitol dehydrogenase EC 1.1.1.14)	AGGAGTGACAACCA**ATG**

The beginning and end nucleotide numbers of each ORF in the DNA fragment of the chromosome of alkaliphilic *Bacillus halodurans* C-125 and the size of the putative protein are shown. The putative initiation codons and SD sequences are shown in bold and underlined, respectively. The protein homologous to each ORF is shown with the size of the protein and similarity percent. In the case of the *B. subtilis* protein, the beginning and end nucleotide numbers for each protein are also shown.

some (Fig. 10-1, Table 6). The ORF6 product was similar to the *B. subtilis rapG* product (33% identity), which functions as a response regulator of aspartate phosphatase. In addition, the ORF9 and ORF12 products were similar to the *B. subtilis ahpF* and *rocF* products, which function as alkyl hydroperoxide reductase NADH dehydrogenase and arginase, respectively, as shown in Table 6. ORF5 showed 25% identity to the hypothetical protein of *Methanococcus janaschii* (Bult et al. 1996). The ORF13 product showed no significant similarity to any other proteins so far reported. On the other hand, ORF1 showed 26% similarity to a transposase produced by the gram-negative bacterium *Bordetella parapertussis* (Van der Zee et al. 1993). This finding suggests that this kind of transposable element may be involved in gene transposition in the genome of alkaliphilic *Bacillus halodurans* C-125, considering that the gene organization in this region differs totally from that of *B. subtilis*.

ORF Analysis of the Sequenced Fragment 4

The determined sequence of fragment 4 was searched for ORFs comprising more than 100 codons, and 11 putative ORFs were identified (Fig. 10-2, Table 7). The initiation codons for two of these ORFs were TTG and GTG, and the other seven ORFs started with ATG (Table 7). Seven ORFs from ORF3 to ORF7 showed significant similarities to the products of the *ydhN, ydhO, ydhP, ydhQ* , and *ydhS* of *B. subtilis*, which may constitute a large operon involved in degradation and incorporation of mannan (Sadaie et al. 1997). ORF8 showed comparatively low similarity (28%) to the *mcpA* gene product of *B. subtilis* among the ORFs in this region. Three ORFs, from ORF9 to ORF11 which are similar to the series of *yts* (*ytsD, ytsC, ytsB*) gene products of *B. subtilis* were also conserved in the chromosome of strain C-125. ORF1 was found to be similar to the product of the *mmgE* gene of *B. subtilis*, with 74% identity. ORF2 was similar to the product of *yqiQ* of *B. subtilis*, which is adjacent to *mmgE*.

ORF Analysis of the Sequenced Fragment 9

There were 20 ORFs comprising more than 100 codons in the sequenced fragment 9 as shown in Fig. 10-3 and Table 8. The initiation codons for three of these ORFs were TTG, one was GTG, and the other 16 started with ATG (Table 8). Seven ORFs (ORFs 1, 9, 10, 11, 12, 13, and 20) did not show significant similarity to any protein in the protein databases. Three ORFs (ORFs 2, 6, and 7) were found to be similar to the *yngF* product whose function seems to be enoyl CoA hydratase. The products of ORF3 and ORF4 showed about 40% identity to the products of *mmgA* [acetyl coenzyme A (CoA) acetyltransferase] and *mmgB* (3-hydroxybutyryl-CoA dehydrogenase) of *B. subtilis*, respectively, which should be adjacent to *mmgE* observed in clone 4. (Fig. 10-3). The ORF15 and ORF16 encoded products were similar to the product of *fruA*, the fructose-specific enzyme IIBC, which functions as a component of the phosphotransferase system (PTS). On the other hand, the product of ORF17 showed 43% identity to galactitol-specific IIC, a component of the *E. coli* PTS system, but showed no significant similarity to *B. subtilis* proteins (Fig. 10-3, Table 8). In addition, there were two ORFs (ORFs 18 and 20) at the terminus of fragment 9 (Fig. 10-3, Table 8) that were found to be similar to the product of *gutB* of *B. subtilis* whose function is sorbitol dehydrogenase.

Thus the gene organization of this region of the genome of strain C-125 was different from that of *B. subtilis*, and there were many ORFs of unknown function that are not present in the *B. subtilis* chromosome. We determined the nucleotide sequences of three independent fragments of the chromosome of alkaliphilic *Bacillus halodurans* C-125 and compared them with the sequence of the *B. subtilis* chromosome. The ORFs identified in these fragments were found to be similar to genes in the *B. subtilis* chromosome but were organized in a different order. These results will be helpful for elucidation of the basis of alkaliphily in *Bacillus* and also may contribute to future identification of the function of some *B. subtilis* genes.

11.3 Shotgun Sequencing of the Whole Genome of *B. halodurans* C-125

Systematic sequencing of the whole genome of *Bacillus halodurans* C-125 has been routinized since the beginning of May 1998. More than 90% of the whole genome has been sequenced so far.

11.3.1 Preparation of Whole Genome Shotgun Library

Chromosomal DNA was isolated from *Bacillus halodurans* C-125 as described above. A 20-μg aliquot of chromosomal DNA was sonicated for 5–25 s with a Bioruptor

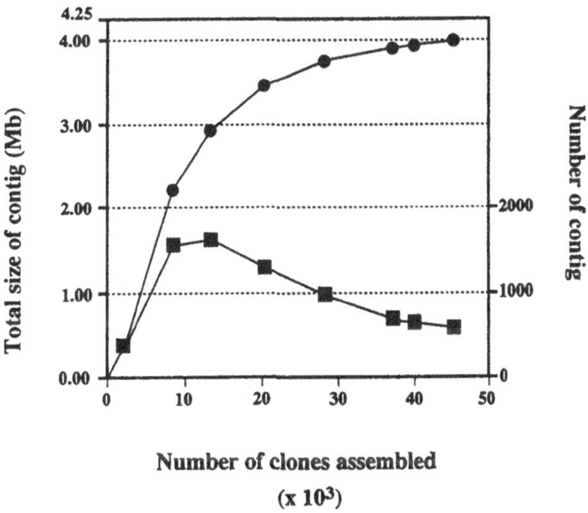

Fig. 11. Assembling summary of shotgun clones from whole genome of alkaliphilic *Bacillus halodurans* C-125. *Circles*, total size of constructed contig; *squares*, number of constructed contig

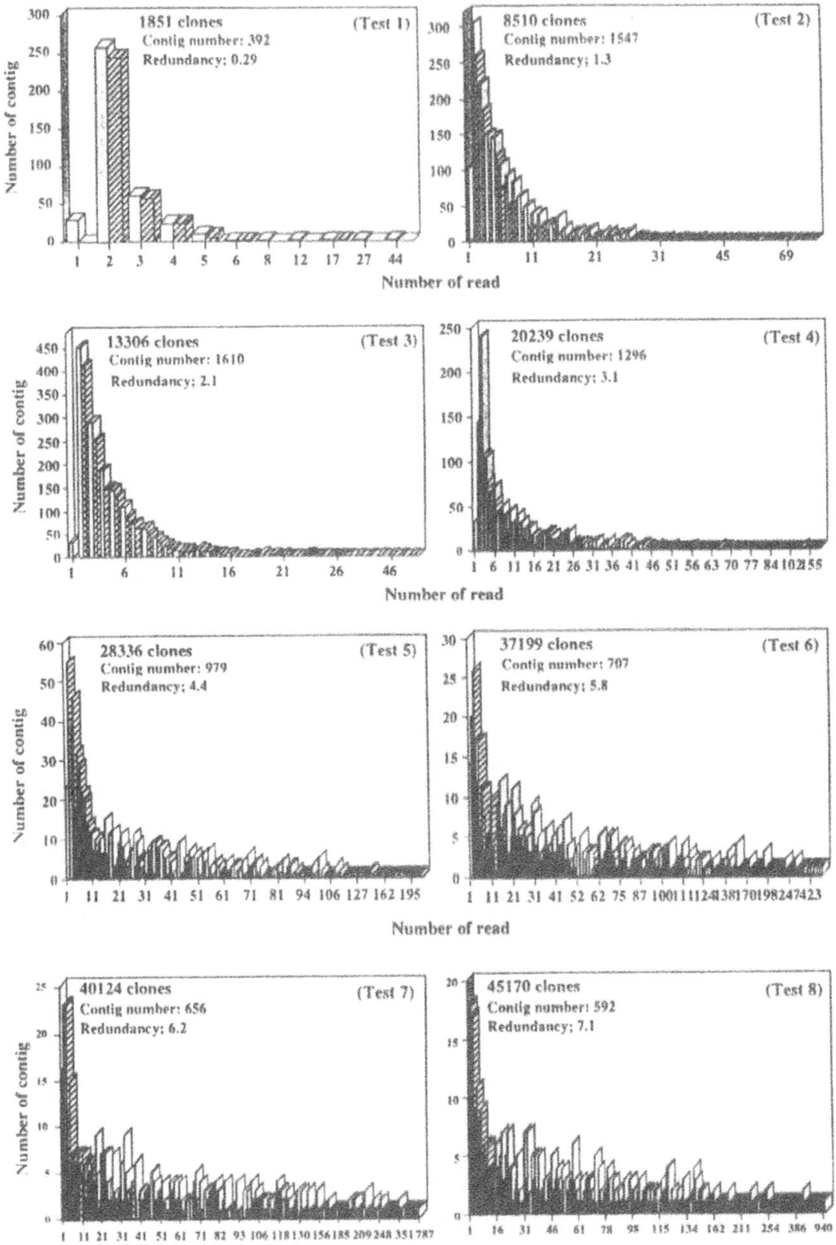

Fig. 12. Assembling pattern of shotgun clones from whole genome of alkaliphilic *Bacillus halodurans* C-125 at each statistical coverage. **1** Number of constructed contig. **2** Total size of contig. *Number of read*, clone number that constructs contig. Total size of contig = number of contig × the length of each contig. *Shaded bars*, regular contig that has some consensus sequence to other contigs; *cross-hatched bars*, isolated contig that has no consensus sequence to any other contig

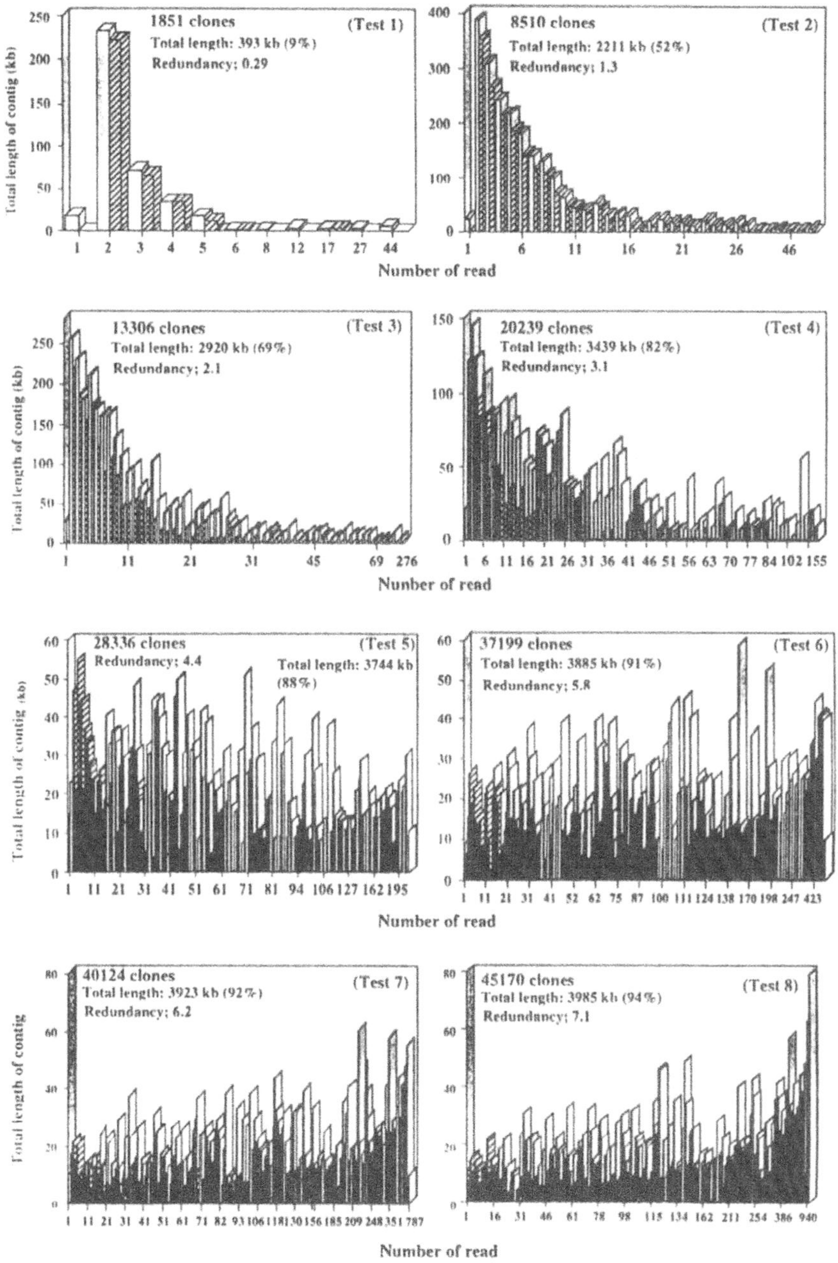

Fig. 12. continued

UCD-200TM (Tosho Denki, Japan). The sonicated DNA fragments were blunt-ended using a DNA blunting kit (Takara Shuzo, Japan) and fractionated by 1% agarose gel electrophoresis. DNA fragments 1-2 kb in length were excised from the gel and eluted by the freeze-squeeze method (Thuring et al. 1975). The DNA recovered was ligated to the *Sma*I site of pUC18, which had been previously treated with bacterial alkaline phosphatase (BAP), and introduced into competent XL1-Blue cells by the standard method (Sambrook et al. 1989). We usually could obtain transformants with a frequency of 5×10^5 to 6×10^5/μg DNA, and they were used for amplification of insert using the colony PCR method.

11.3.2 Sequence Strategy for Whole Genome Shotgun Library

The DNA fragment inserted into pUC18 was amplified by PCR using M13-20 and reverse primers. PCR fragments, treated with Exonuclease I and shrimp alkaline phosphatase (Amsersham) to eliminate excess primers in the PCR reaction mixture, were used for sequencing analysis as template DNA. Sequencing was performed with a DNA sequencer ABI PRISM 377 using a Taq Dye Terminator Cycle Sequencing Kit (Perkin Elmer). We have already sequenced 45 000 shotgun clones and will continue random sequencing up to 55 000 clones. A statistical coverage of sequenced region reached 7.1-fold at the stage of 45 000 clones, and the clones sequenced seemed to be almost saturated, as shown in Fig. 11.

11.3.3 Assembly of Whole Genome Shotgun Library

The DNA sequences determined by means of the ABI sequencer were assembled into contigs using Phrap (http://bozeman.mbt.washington.edu/phrap.docs/phrap.html)

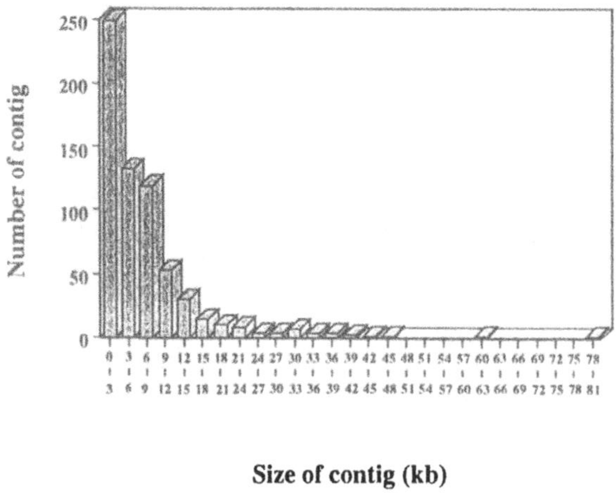

Fig. 13. Distribution of contig constructed at a statistical coverage of 7.1-fold. Total number of contigs is 592. Minimum and maximum lengths of contig were 5.98 and 0.53 kb, respectively. Total length of contig in this stage was 3.98 Mb

with default parameters and without quality scores. The conditions of contigs constructed at each statistical coverage are shown in Fig. 12-1,2. At a statistical coverage of 7.1-fold, the assembly using Phrap yielded 656 contigs (Fig. 12-1) and the average contig length was 6.73 kb (Fig. 13). The total length of each contig was 3.98 Mb, corresponding to 93.7% of the whole genome of *Bacillus halodurans* C-125 (4.25 Mb), as shown in Fig. 12-2. For finishing the sequencing of the whole genome, we are trying to fill gaps that occurred using the sequences of both end of large insert library such as λ and cosmid clone and by means of PCR with the primers designed based on the internal sequence in each contig.

References

Boylan SA, Suh JW, Thomas SM, Price CW (1989) Gene encoding the alpha core subunit of *Bacillus subtilis* RNA polymerase is contrascribed with the gene for initiation factor 1 and ribosomal protein B, S13, S11, and L17. J Bacteriol 171:2553-2562

Bult CJ, White O, Olsen GJ, Zhou L, Fleischmann RD, Sutton GG, Blake JA, FitzGerald LM, Clayton RA, Gocayne JD, Kerlavage AR, Dougherty BA, Tomb J-F, Adams MD, Reich CI, Overbeek R, Kirkness EF, Weinstock KG, Merrick JM, Glodek A, Scott JL, Geoghagen NSM, Weidman JF, Fuhrmann JL, Nguyen D, Utterback TR, Kelley JM, Peterson JD, Sadow PW, Hanna MC, Cotton MD, Roberts KM, Hurst MA, Kaine BP, Borodovsky M, Klenk H-P, Fraser CM, Smith HO, Woese CR, Venter JC (1996) Complete genome sequence of the methanogenic archaeon, *Methanococcus jannaschii*. Science 273:1058-1073

Carlson CR, Kolsto AB (1993) A complete physical map of a *Bacillus thuringiensis* chromosome. J Bacteriol 175:1053-1060

Carlson CR, Kolsto AB (1994) A small (2.4 Mb) *Bacillus cereus* chromosome corresponds to one conserved region of a larger (5.3 Mb) *Bacillus cereus* chromosome. Mol Microbiol 13:161-169

Chow KC, Wong JT (1988) Cloning and nucleotide sequence of the structural gene coding for *Bacillus subtilis* tryptophanyl-tRNA synthetase. Gene 73:537-543

Errington J, Appleby L, Daniel RA, Goodfellow H, Partridge SR, Yudkin MD (1992) Structure and function of the *spoIIIJ* gene of *Bacillus subtilis*: a vegetatively expressed gene that is essential for E-G activity at an intermediate stage of sporulation. J Gen Microbiol 138:2609-2618

Felsenstein J (1985) Confidence limits on phylogenies: an approach using the bootstrap. Evolution 39:783-791

Fujita M, Yoshikawa H, Ogasawara N (1990) Structure of the *dnaA* region of *Micrococcus luteus*: conservation and variations among eubacteria. Gene 93 73-78

Fujita M, Yoshikawa H, Ogasawara N (1989) Structure of the *dnaA* region of *Pseudomonas putida*: conservation among three bacteria, *Bacillus subtilis*, *Escherichia coli* and *P. putida*. Mol Gen Genet 215:381-387

Hansen, EB, Hansen FG, von Meyenburg K (1982) The nucleotide sequence of the *dnaA* gene and first part of the *dnaN* gene of *Escherichia coli* K-12. Nucleic Acids Res 10:7373-7385

Herfurth E, Briesemeister U, Wittmann-Liebold B (1994) Complete amino acid sequence of ribosomal protein S14 from *Bacillus stearothermophilus* and homology studies to other ribosomal proteins. FEBS Lett 351:114-118

Herwig S, Kruft V, Wittmann-Liebold B (1992) Primary structures of ribosomal proteins L3 and L4 from *Bacillus stearothermophilus*. Eur J Biochem 207:877-885

Honda H, Kudo T, Ikura Y, Horikoshi K (1985) Two types of xylanases of alkalophilic *Bacillus* sp. No. C-125. Can J Microbiol 31:538-542

Horikoshi K (1991) Microorganisms in alkaline environments. KODANSHA/VCH, Tokyo

Ikura Y, Horikoshi K (1979) Isolation and some properties of β-galactosidase-producing bacteria. Agric Biol Chem 43:85-88

Itaya M, Tanaka T (1991) Complete physical map of the *Bacillus subtilis* 168 chromosome constructed by a gene-directed mutagenesis method. J Mol Biol 220:631-648

Kang SK, Kudo T, Horikoshi K (1992) Molecular cloning and characterization of an alkalophilic *Bacillus* sp. C-125 gene homologous to *Bacillus subtilis secY*. J Gen Microbiol 138:1365-1370

Kimura J, Kimura M (1987) The complete amino acid sequences of the 5S rRNA binding proteins L5 and L18 from the moderate thermophile *Bacillus stearothermophilus* ribosome. FEBS Lett 210:85-90

Kimura M (1991) The nucleotide sequences of *Bacillus stearothermophilus* ribosomal protein S12 and S7 genes: comparison with the *str* operon of *Escherichia coli*. Agric Biol Chem 55:207-213

Kimura M, Kimura J, Ashman K (1985) The complete primary structure of ribosomal proteins L1, L14, L15, L23, L24, and L29 from *Bacillus stearothermophilus*. Eur J Biochem 150:491-497

Kromer W, Hatakeyama T, Kimura M (1990) Nucleotide sequence of *Bacillus stearothermophilus* ribosomal protein genes: part of the ribosomal S10 operon. Biol Chem Hoppe Seyler 371:631-636

Kudo T, Hino M, Kitada M, Horikoshi K (1990) DNA sequences required for alkalophily of *Bacillus*. J Bacteriol 172:7282-7283

Kunst F, Ogasawara N, Moszer I, Albertini AM, Alloni G, Azevedo V, Bertero MG, Bessières P, Bolotin A, Borchert S, Borriss R, Boursier L, Brans A, Braun M, Brignell SC, Bron S, Brouillet S, Bruschi CV, Caldwell B, Capuano V, Carter NM, Choi S-K, Codani J-J, Connerton IF, Cummings NJ, Daniel RA, Denizot F, Devine KM, Düsterhöft A, Ehrlich SD, Emmerson PT, Entian KD, Errington J, Fabret C, Ferrari E, Foulger D, Fritz C, Fujita M, Fujita Y, Fuma S, Galizzi A, Galleron N, Ghim S-Y, Glaser P, Goffeau A, Golightly EJ, Grandi G, Guiseppi G, Guy BJ, Haga K, Haiech J, Harwood CR, Hénaut A, Hilbert H, Holsappel S, Hosono S, Hullo M-F, Itaya M, Jones L, Joris B, Karamata D, Kasahara Y, Klaerr-Blanchard M, Klein C, Kobayashi Y, Koetter P, Koningstein G, Krogh S, Kumano M, Kurita K, Lapidus A, Lardinois S, Lauber J, Lazarevic V, Lee S-M, Levine A, Liu H, Masuda S, Mauël C, Médigue C, Medina N, Mellado RP, Mizuno M, Moestl D, Nakai S, Noback M, Noone D, O'Reilly M, Ogawa K, Ogiwara A, Oudega B, Park S-H, Parro V, Pohl TM, Portetelle D, Porwollik S, Prescott AM, Presecan E, Pujic P, Purnelle B, Rapoport G, Rey M, Reynolds S, Rieger M, Rivolta C, Rocha E, Roche B, Rose M, Sadaie Y, Sato T, Scanlan E, Schleich S, Schroeter R, Scoffone F, Sekiguchi J, Sekowska A, Seror SJ, Serror P, Shin BS, Soldo B, Sorokin A, Tacconi E, Takagi T, Takahashi H, Takemaru K, Takeuchi M, Tamakoshi A, Tanaka T, Terpstra P, Tognoni A, Tosato V, Uchiyama S, Vandenbol M, Vannier F, Vassarotti A, Viari A, Wambutt R, Wedler E, Wedler H, Weitzenegger T, Winters P, Wipat A, Yamamoto H, Yamane K, Yasumoto K, Yata K, Yoshida K, Yoshikawa H-F, Zumstein E, Yoshikawa H, Danchin A (1997) The complete genome sequence of the gram-positive bacterium *Bacillus subtilis*. Nature 390:249-256

Li X, Lindahl L, Sha Y, Zengel JM (1997) Analysis of the *Bacillus subtilis* S10 ribosomal protein gene cluster identifies two promoters that may be responsible for transcription of the entire 15-kilobases S10-*spc*-a cluster. J Bacteriol 179:7046-7054

Moriya S, Ogasawara N, Yoshikawa H (1985) Structure and function of the region of the replication origin of the *Bacillus subtilis* chromosome. III. Nucleotide sequence of some 10,000 base pairs in the origin region. Nucleic Acids Res 13:2251-2265

Nakasone K, Takaki Y, Fuji F, Takami H, Inoue A, Horikoshi K (1998) Cloning and expression of the gene encoding RNA polymerase a subunit from alkaliphilic *Bacillus* sp. strain C-125. FEMS Microbiol Lett 152:279-285

Ogasawara N, Yoshikawa H (1992) Genes and their organization in the replication origin of the bacterial chromosome. Mol Microbiol 6:629-634

Ogasawara N, Moriya S, Yoshikawa H (1985) Structure and function of the region of the replication origin of the *Bacillus subtilis* chromosome. IV. Transcription of the *oriC* region and expression of DNA gyrase genes and other open reading frames. Nucleic Acids Res 13:2267-2279

Ogasawara N, Nakai S, Yoshikawa H (1994) Systematic sequencing of the 180 kilobase region of the *Bacillus subtilis* chromosome containing the replication origin. DNA Res 1:1-14

Piggot PJ, Hoch JA (1993) The genetic map of *Bacillus subtilis*. In: Sonenshein AL, Hoch JA, Losick R (eds) *Bacillus subtilis* and other grain-positive bacteria: biochemistry, physiology and molecular genetics. American Society for Microbiology, Washington, DC, pp 425-461

Ramakrishnan V, Gerchman SE (1991) Cloning, sequencing and overexpression of genes for ribosomal proteins from *Bacillus stearothermophilus*. J Biol Chem 266:880-885

Sadaie Y, Yata K, Fujita M, Sagai H, Itaya M, Kasahara Y, Ogasawara N (1997) Nucleotide sequence and analysis of the *phoB-rrnE-groEL* region of the *Bacillus subtilis* chromosome. Microbiology 143:1861-1866

Saito H, Miura K (1963) Preparation of transforming deoxyribonucleic acid by phenol treatment. Biochim Biophys Acta 72:619-629

Sakamoto Y, Southerland KJ, Tamaoka J, Kobayashi T, Kudo T, Horikoshi K (1992) Analysis of the flagellin (*hag*) gene of alkalophilic *Bacillus* sp. C-125. J Gen Microbiol 138:2159-2166

Sambrook J, Frisch EF, Maniatis T (1989) Molecular cloning, a laboratory manual, 2nd edn. Cold Spring Harbor Laboratory, Cold Spring Harbor, NY

Smith CL, Condemine G (1990) New approaches for physical mapping of small genomes. J Bacteriol 172:1167-1172

Southerland KJ, Hashimoto M, Kudo T, Horikoshi K (1993) A partial physical map for the chromosome of alkalophilic *Bacillus* sp. strain C-125. J Gen Microbiol 139:661-667

Suh JW, Boylan SA, Oh SH, Price CW (1996) Genetic and transcriptional organization of the *Bacillus subtilis spc*-a region. Gene 169:17-23

Takami H, Horikoshi K (1999) Reidentification of facultatively alkaliphilic *Bacillus* sp. C-125 to *Bacillus halodurans*. Biosci Biotechnol Biochem 63:943-945

Takami H, Inoue A, Fuji F, Horikoshi K (1997) Microbial flora in the deepest sea mud of the Mariana Trench. FEMS Microbiol Lett 152:279-285

Takami H, Kobayashi T, Aono R, Horikoshi K (1992) Molecular cloning, nucleotide sequence and expression of the structural gene for a thermostable alkaline protease from *Bacillus* sp. no. AH-101. Appl Microbiol Biotechnol 38:101-108

Takami H, Kobata K, Nagahama T, Kobayashi H, Inoue A, Horikoshi K (1999a) Biodiversity in deep-sea sites niar the south part of Japan. Extremophiles 3:97-102

Takami H, Masui N, Nakasone K, Horikoshi K (1999c) Replication origin region of the chromosome of alkaliphilic *Bacillus* sp. strain C-125. Biosci Biotechnol Biochem 63:1134-1137

Takami H, Nakasone K, Hirama C, Takaki Y, Masui N, Fuji F, Nakamura Y, Inoue A (1999b) An improved physical and genetic map of the genome of alkaliphilic *Bacillus* sp. C-125. Extremophiles 3:21-28

Takami H, Nakasone K, Ogasawara N, Hirama C, Nakamura Y, Masui N, Fuji F, Takaki Y, Inoue A Horikoshi, K (1999e) Sequencing of three lambda clones from the genome of alkaliphilic *Bacillus* sp. strain C-125. Extremophiles 3:29-34

Takami H, Takaki Y, Nakasone K, Hirama C, Inoue A, Horikoshi K (1999d) Sequence analysis of 32 kb fragment including the major ribosomal protein gene clusters in alkaliphilic *Bacillus* sp. strain C-125. Biosci Biotechnol Biochem 63:452-455

Thuring RW, Sanders JMP, Borst P (1975) A freeze-squeeze method for recovering long DNA from agarose gel. Anal Biochem 66:213-220

Triglia T, Peterson MG, Kemp DJ (1988) A procedure for in vitro amplification of DNA segments that lie outside the boundaries of known sequences. Nucleic Acids Res 16:8186

Van der Zee A, Agterberg C, van Agterveld M, Peeters M, Mooi FR (1993) Characterization of IS1001, an insertion sequence element of *Bordetella parapertussis*. J Bacteriol 175:141-147

Xu Y, Kobayashi T, Kudo T (1996) Molecular cloning and nucleotide sequence of the *groEL* gene from the alkaliphilic *Bacillus* sp. strain C-125 and reactivation of thermally inactivated α-glucosidase by recombinant GroEL. Biosci Biotechnol Biochem 60:1633-1636

Yoshikawa H, Ogasawara N (1991) Structure and function of DnaA and the DnaA-box in eubacteria: evolutionary relationships of bacterial replication origins. Mol Microbiol 5:2589-2597

12 Future Scope

Kaoru Tsujii

As mentioned in the Preface, the first stage of our project research on deep-sea extremophiles finished September 30, 1998, and the second stage started October 1 of the same year. During the first stage we focused on microorganisms living in the extreme conditions of the deep-sea. During the second stage we are extending our research with a conceptually new point of view, that is, to the physicochemical environment surrounding the microorganisms. The underlying mechanisms of adaptation and response of the microbes to extreme conditions originate from interactions of the microbes with the surrounding environments, particularly deep-sea water. Research on the water structure of extreme conditions in the deep-sea is important to understand the adaptation mechanism of the extremophiles.

Our research strategy during the second stage is thus to study (1) the water itself under extreme conditions; (2) the interactions between seawater and extremophiles, including biopolymers, biomolecular assemblies, and supramolecular organs; and (3) the genome of such extremophiles to determine the adaptation mechanisms from the viewpoint of the gene. Finally, we continue to pay attention to useful microorganisms what produce important materials for human life and industrial applications.

Key words. Deep-sea, Subground, Environment, High pressure, Barophysiology, Piezophysiology, Supercritical water, Adaptation, Response, Extremophiles, Superbugs, Microbe, Microbiology, Exobiology, Origin of life

12.1 Physicochemical Environments in the Deep-Sea

Many physicochemical conditions in the deep-sea (i.e., pressure, temperature, salinity, organic solutes, particle dispersions) affect the nature of microorganisms. Among

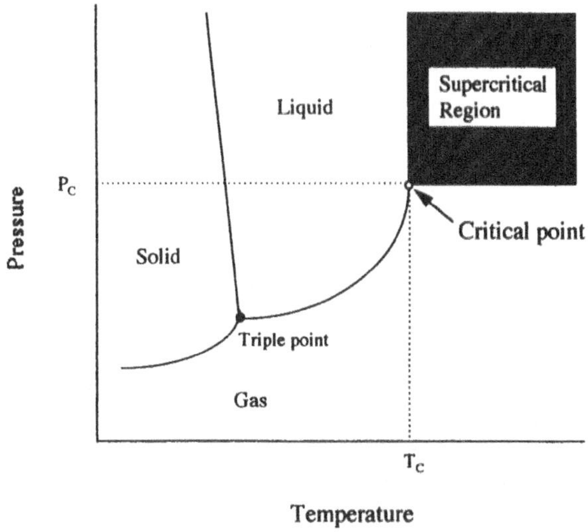

Fig. 1. Phase diagram of water. Supercritical water is present at a higher temperature and pressure than the critical point (374°C and 218 atm)

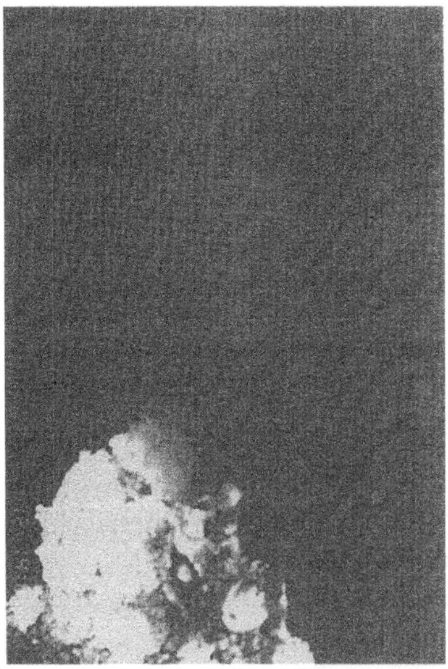

Fig. 2. Hydrothermal vent "black smoker" photographed at 1337 m depth in Okinawa Trough. (Courtesy of Japan Marine Science and Technology Center)

them we are particularly interested in water structure at high temperature and pressure, as this condition is a typical condition of the deep-sea. Water in the most extreme conditions is supercritical water, so our first target when studying water in the deep-sea is supercritical water.

Supercritical water is water that exists above the critical point, 647.14 K (374°C) and 22.064 MPa (217.8 atm); there is no distinction here between gas and liquid. Figure 1 shows the phase diagram of water with respect to temperature and pressure. Supercritical water is supposed to be present in the deep-sea, for example, over a volcano and a hydrothermal vent. Figure 2 is a photograph of a hydrothermal vent "black smoker" from which black hot water flushes out.

In the supercritical region, water shows unique properties.

1. Molecular interaction can be freely changed with changing pressure.
2. The dielectric constant is much smaller (10-20) than that of usual water (which is about 80).
3. Supercritical water behaves like a polar organic solvent and dissolves many organic compounds.
4. Supercritical water has no surface tension.

Taking these unique properties of supercritical water into account, two main research subjects have been established to study the deep-sea environment. One is "colloid and surface science in supercritical water," and the other is "chemical reactions in supercritical water." Both studies are to be done with special attention to the biological process in the deep-sea.

Observing matter in an extreme condition is useful for understanding the essential nature of it. Water structure in its normal condition of the deep-sea should be more profoundly understood when looked at from extreme conditions. The significance of studying supercritical water is based mostly on the above point. Another significant point is to observe the physical and chemical behavior of biomaterials in supercritical water, as it is proposed that the first living thing was born in a hydrothermal vent (Holm 1992; Ohshima 1995).

12.2 Adaptation and Response of Extremophiles and Multicellular Organisms to Extreme Conditions

Studies on the adaptation mechanism and response of extremophiles to extreme conditions in the deep-sea will be, of course, extended during the second stage of the project. The physicochemical aspects of deep-sea environments mentioned in the previous section are taken into account to elucidate the mechanisms. Some examples of research themes during the second stage of our project are enzyme activity in high pressure medium and water structure, protein and nucleic acid synthesis under high hydrostatic pressure, biomembrane formation under high pressure and high temperature environment, and cell structure and division under high hydrostatic pressure among others.

Barophysiology (piezophysiology) was proposed as a new scientific field during the first stage. This new concept was constructed from studies in yeast (Abe and Horikoshi 1995; 1998), *Escherichia coli* (Kato et al. 1994, 1995; Sato et al. 1996), and a barophilic (piezophilic) bacterium (Kato et al. 1994, 1995; Nakasone et al. 1998). During the second stage of our project, we extend the barophysiological (piezophysiological) study to multicellular organisms and combine it with the continued research on microorganisms. Many animals, such as fish, shellfish, sea cucumbers, and shrimp, are found in the deep-sea, even in the deepest place on Earth, the Mariana Trench. Figure 3 shows some of these unique animals. Their adaptation to the extreme conditions, especially the high hydrostatic pressure, may be accomplised in a more sophisticated way than it is by microorganisms. The adaptation mechanisms of microbes and multicellular organisms is being studied by both physiological and physicochemical approaches, taking the water structures into consideration.

The response of cells (microorganisms and multicellular organisms) to physical and chemical stimuli is also being studied as an extension of the previous studies (Koyama et al. 1996, 1997). If a cell shows some output responses when a physical or a chemical stimulus is applied, the responses must be reflections of physiological actions of the cell. So, at least in principle, analysis of these output responses can reveal the mechanism of adaptation to the stimulus. The physical and chemical stim-

A B C D

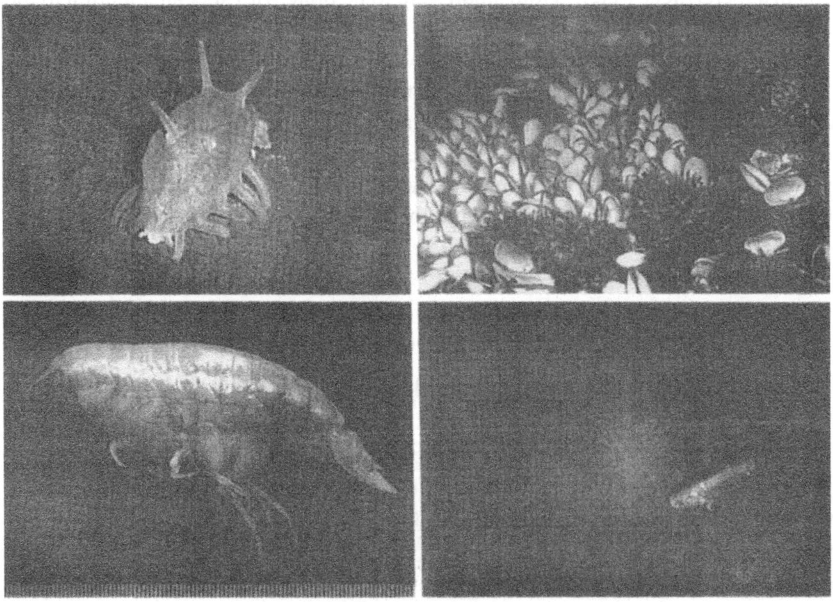

Fig. 3. Some animals living in deep-sea. **a** Sea-cucumber (at 6527 m depth in Japan Trench). **b** Clams, *Calyptogena soyoae* (at 1400 m depth in Okinawa Trough). **c** Amphipod, *Hirondellea gigas* (at 10 970 m depth in Mariana Trench). **d** Cod fish (at 3087 m depth in Enshu-nada Sea, Japan). (Courtesy of Japan Marine Science and Technology Center)

uli in which we are interested are pressure, temperature, chemicals, and even the electrical and magnetic fields. This research will contribute to our understanding not only of the adaptation mechanisms in deep-sea environments but also the reactions to stimuli of cells living in atmospheric conditions, including those of humans.

12.3 Genome Analysis of Extremophiles

Genome analysis is one of the best ways to understand the adaptation mechanism of extremophiles to an extreme environment. During the first stageof the project, we attempted to determine all the nucleotide sequences of the genome of facultatively alkaliphilic *Bacillus halodurans* C-125 to decipher the mechanism of adaptation to an alkaline environment (Takami et al. 1999a, 1999b, 1999c). Sequencing the whole genome of the alkaliphilic *Bacillus* strain provides the chance for further development of the enzyme industry, as several enzymes (e.g., protease, amylase, and cellulase) have already been commercialized. We selected *Bacillus halodurans* C-125, which is best characterized among alkaliphiles as a candidate for genome analysis, and routinized the sequencing process in May 1998 to complete it. After completion of the genome sequencing, we move this work to functional analysis of the gene to find some unique genes corresponding to alkaliphilic phenotype and to clarify the expression profile of the gene involved in enzyme production.

The most typical extreme condition of the deep-sea is the high hydrostatic pressure. Therefore, our next candidate of genome analysis is a barophilic (piezophilic) microbe. We expect that the genome analysis of a barophilic (piezophilic) extremophiles can elucidate the adaptation mechanism to high pressure and contribute to the development of some useful materials, such as enzymes acting at high pressure, lipids having polyunsaturated hydrocarbon chains, and so on.

12.4 Spreading Effects of Second-Stage Research and Further Scope

It is often said that the deep-sea is the final frontier on Earth. The final frontier will be made clear in large part at least in the field of microbiology after our research work during the second stage of this project is successfully completed. Is there already no "last frontier" on Earth?

We, JAMSTEC, have undertaken an extensive research project named "Ocean Drilling 21" (OD-21) (JAMSTEC 1998). We plan to build a ship equipped with a machine that can drill the ocean crust down to the mantle (about 7000 m below the deep-sea floor). A schema for the ocean drilling is shown in Fig. 4. The structure of the deep part of the earth surface, crust movement and change, the mechanisms of earthquakes, and other scientific puzzles are expected to be solved by this research. Another important feature of the OD-21 project is to search for novel underground extremophiles. The conditions affecting life inside the Earth's crust must be more

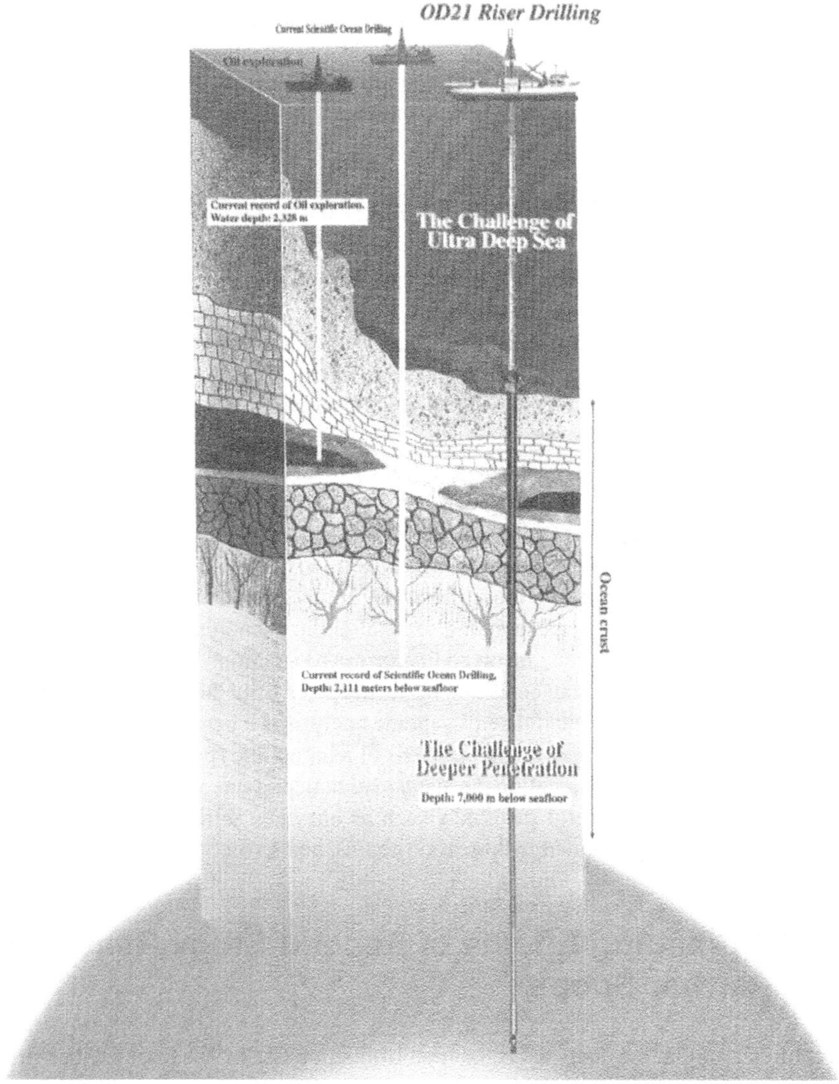

Fig. 4. Schema for the ocean drilling project. JAMSTEC has a large research project called "Ocean Drilling 21 (OD-21)" to search the ocean crust, earth mantle, (micro)organisms, and so on. (Courtesy of Japan Marine Science and Technology Center)

severe than those of the deep-sea. We are thus interested in the nature of extremophiles living in the crust.

The extremophiles living in the Earth's crust probably have a strong tolerance for such severe conditions as high pressure, little water and air, high or low temperature, poor nutrient, and so on. The severe conditions of life in the Earth's and the oceans'

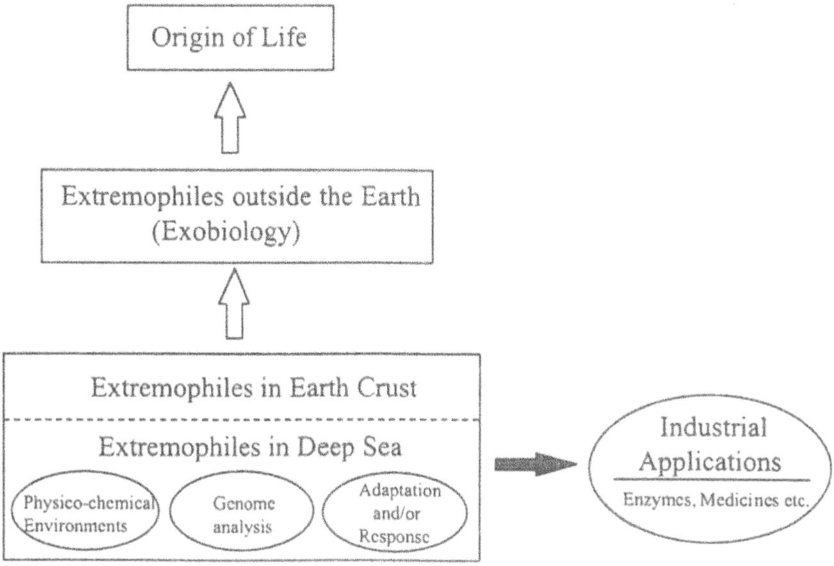

Fig. 5. Idealized research flow from "the extremophiles in a deep-sea environment" to "the origin of life"

crusts might be similar to those of a planet other than Earth, such as Mars (Brack and Pillinger 1998). If microbes do live in the Earth's crust, one could also imagine life on Mars. We propose here a new scientific field—exobiology—that is, biology away from Earth. Exobiology may elucidate the similarities and differences between life on Earth and life away from Earth. This comparative study might contribute to a clarification of the origin of life.

Origin of life is, of course, one of the most important and serious problems for humans to solve. Our research on the extremophiles in deep-sea environments could lead to knowledge about the origin of life; and it is an imaginative and romantic subject. Figure 5 illustrates the research flow that might lead to the answers.

References

Abe F, Horikoshi K (1995) Hydrostatic pressure promotes the acidification of vacuoles in *Saccharomyces cerevisiae*. FEMS Microbiol Lett 130:307-312

Abe F, Horikoshi K (1998) Analysis of intercellular pH in the yeast *Saccharomyces cerevisiae* under elevated hydrostatic pressure: a study in baro-(piezo-)physiology. Extremophiles 2:223-228

Brack A, Pillinger CT (1998) Life on Mars: chemical arguments and clues from Martian meteorites. Extremophiles 2:313-319

Holm NG (ed) (1992) Marine hydrothemal systems and the origin of life. Kluwer Academic,

Dordrecht

JAMSTEC (1998) Long term research program of JAMSTEC (in Japanese). Chap 4

Kato C, Sato T, Smorawinska M, Horikoshi K (1994) High pressure conditions stimulate expression of chloramphenicol acetyltransferase regulated by the *lac* promoter in *Escherichia coli*. FEMS Microbiol Lett 122:91-96

Kato C, Smorawinska M, Sato T, Horikoshi K (1995) Cloning and expression in *Escherichia coli* of a pressure regulated promoter region from a barophilic bacterium, strain DB6705. J Mar Biotechnol 2:125-129

Koyama S, Haruyama T, Kobatake E, Aizawa M (1997) Electrically induced NGF production by astroglial cells. Nat Biotechnol 15:164-166

Koyama S, Yanagida Y, Haruyama T, Kobatake E, Aizawa M (1996) Molecular mechanisms of electrically stimulated NGF expression and secretion by astrocytes cultured on the potential controlled electrode surface. Cell Eng 1:189-194

Nakasone K, Ikegami A, Kato C, Usami R, Horikoshi K (1998) Mechanisms of gene expression controlled by pressure in deep-sea microorganisms. Extremophiles 2:149-154

Ohshima T (1995) Life originates in hot sea (in Japanese). Tokyo Kagaku Dojin, Tokyo, Chap 4

Sato T, Nakamura Y, Nakashima K, Kato C, Horikoshi K (1996) High pressure represses expression of the *malB* operon in *Escherichia coli*. FEMS Microbiol Lett 135:111-116

Takami H, Nakasone K, Hirama C, Masui N, Fuji F, Takaki Y, Nakamura Y, Inoue A, Horikoshi K (1999a) An improved physical and genetic map of the genome of alkaliphilic *Bacillus* sp. C-125. Extremophiles 3:1-8

Takami H, Nakasone K, Ogasawara N, Nakamura Y, Hirama C, Fuji F, Takaki Y, Masui N, Inoue A, Horikoshi K (1999b) Sequencing of three lambda clones from the genome of alkaliphilic *Bacillus* sp. C-125. Extremophiles 3:29-34

Takami H, Takaki Y, Nakasone K, Hirama C, Inoue A, Horikoshi K (1999c) Sequence analysis of 32 kb fragment including the major ribosomal protein gene clusters in alkaliphilic *Bacillus* sp. strain C-125. Biosci Biotechnol Biochem 63:452-455

Appendix: International Congress on Extremophiles '98 at Yokohama

1. Outline of the Congress

The International Congress on Extremophiles'98 was held in Yokohama, Japan on January 18-22, 1998. This meeting was sponsored by Japan Marine Science and Technology Center (JAMSTEC), and supported by Science and Technology Agency of Japan and The Nippon Foundation, and was the first congress on Extemophiles in Asia.

The congress had two objectives:

First, the meeting provided a forum for the exchange of information between specialists in the rapidly growing basic field of Extremophile's research and promoted discussion and the exchange of ideas and applications between scientists from academia and industry.

Second, the scientific results of the deep-sea microorganisms research project (the DEEPSTAR program) during past 7 years had been published at the congress to emphasize the activity and quality of this Japanese project to international scientists.

About 380 researchers from 33 countries*, including Europe, America and Asia, attended the congress. The researchers who joined the congress had very exciting discussion about the microorganisms living under extremely environments, called "Extremophiles", from the surface of the earth to deep-sea floor at a depth of 11,000 m, including the possibility of life on Mars.

The DEEPSTAR group members presented the quantities of the papers in the

---

*; <Nationality of attendance>
Australia, Bangladesh, Belgium, Canada, China, Chile, Denmark, England, France, Finland, Germany, Ghana, Holland, Iceland, India, Indonesia, Israel, Italy, Japan, Korea, Malaysia, Norway, Pakistan, the Philippines, Poland, Portugal, Russia, Spain, Sweden, Switzerland, USA

A

B

Fig. 1. A The congress scene. **B** Congress poster with participants signs

field of deep-sea microbiology, like isolation of deep-sea extremophiles, study on metabolism and adaptation mechanisms in microorganisms, and genome analysis. These studies have been highly estimated within many researchers from many countries in the world.

This congress was in great success.

2. The Contents of the Congress, Extremophiles '98

The congress was composed of two special lectures, five sessions, and a panel discussion, including not only microbiology on the earth, but also a new fields of exobiology, subsurface microbiology, and the genome analyses. We focused on both fundamental and application studies of "Extremophiles" that were adapted to live in extreme environments.

Special lectures and the sessions were as follows:

Opening lecture by Prof. H.W. Jannasch
Session 1: Genetics and Molecular Biology
Session 2: Structure and Function of Proteins
Session 3: Physiology and Metabolisms
Session 4: Ecology and Diversity (including exobiology and subsurface microbiology)
Session 5: Genome
Panel Discussion: Future of Extremohiles
Closing lecture by Prof. Koki Horikoshi

The congress was also consisted of plenary seminars given by invited speakers and several oral presentations selected from the submitted abstracts.

List of the invited speakers was as follows;

Prof. A. Aguilar: European Commission, Belgium
Prof. G. Antranikian: Technical University Hamburg-Harburg, Germany
Prof. D. H. Bartlett: Scripps Institution of Oceanography, USA
Prof. A. Brack: Planetary Sciences Research Ins., CNRS, France
Prof. J. de Bont: University of Wageningen, The Netherlands
Prof. A. T. Bull: University of Kent, UK
Prof. D. A. Cowan: University College London, UK
Prof. R. H. Doi: University of California, Davis, USA
Prof. I. D. Dundas: University of Bergen , Norway
Dr. J. D. Farmer: NASA Ames Research Center, USA
Dr. S. Fitz-Gibbon: University of California at Los Angeles, USA
Prof. W. D. Grant: University of Leicester, UK
Dr. T. Gaasterland: University of Chicago, USA
Prof. M. H. Han: KRIBB , Korea

Dr. S. Ito: Biological Science Laboratories of Kao Co., Japan
Prof. H. W. Jannasch: Woods Hole Oceanographic Institution, USA
Dr. B. Jones: University of Leicester, UK
Dr. H. P. Klenk: The Institute for Genomic Research, USA
Prof. W. N. Konings: University of Groningen, The Netherlands
Prof. T. Krulwich: NY Mount Sinai School of Medicine, USA
Prof. J. P. Mckinley: Pacific Northwest National Laboratory, USA
Prof. N.Ogasawara: Nara Institute of Science and Technology, Japan
Prof. R. J. Parkes: University of Bristol, UK
Prof. D. Prieur: Marine Microbiology, Univ. of ParisVI, France
Prof. J. N. Reeve: The Ohio State University, USA
Prof. M. Rossi: University of Naples, Italy
Prof. N. Russell: Wye College, University of London, UK
Dr. C. W. Sensen: Institute for Marine Biosciences, Canada
Prof. K. O. Stetter: University of Regensburg , Germany
Dr. V. Swanson: Diversa Corporation, USA
Prof. A. Ventosa: University of Sevilla, Spain
Prof. W. M. de Vos: Wageningen Agricultural Univ., The Netherlands
Prof. J. Wiegel: University of Georgia, USA
Prof. B. Witholt: ETH, Switzerland
Prof. A. A. Yayanos: Scripps Institution of Oceanography, USA
Prof. G. Zeikus: Michigan State University, USA
Prof. W. Zillig: Max Planck Institut fur Biochemie, Germany

2-1. Science Program

Remarkable presentations that were published in this congress were as follows;

<Invited Presentations>

Opening Lecture by Prof. Holger W. JANNASCH (Woods Hole Oceangraphic Insitution, USA) **"The Deep-sea's Contribution to the Extremophiles"**

Session 1: Genetics and Molecular Biology

Wolfram ZILLIG (Max Planck Institut fur Biochemie, Germany) "Genetic Elements of Extremely Thermophilic Archaea in the Kingdom Crenarchaeota"
John N. REEVE (The Ohio State University, USA) "Histone Stability and Nucleosomes in Hyperthermophilic Archaea"
Garo ANTRANIKIAN (Technical University Hamburg-Harburg and Center for Biotechnology of Extremophiles, Germany) "Extremophiles as Cell Factories for Novel Thermostable Enzymes"
Douglas H. BARTLETT (University of California at San Diego, USA) "Genetic Analyses of Piezo-sensing and Piezo-adaptation in *Photobacterium profundum* SS9"
Chiaki KATO (Japan Marine Science and Technology Center, Japan) "Mechanisms

of Gene Expression Controlled by Pressure in Deep-sea Microorganism"
Roy H. DOI (University of California at Davis, USA) "*Clostridium cellulovorans* Cellulosome and Cellulases"
Toshiaki KUDO (The Institute of Physical and Chemical Research (RIKEN), Japan) "Phylogenetic Identification of the Intestinal Anaerobic Microbial Community in the Termite Hindgut without Cultivation"

Session 2: Structure and Function of Proteins

Wil N. KONINGS (University of Groningen, The Netherlands) "Bioenergetics of Extremophiles"
Nicholas J. RUSSELL (Wye College University of London, UK) "Towards a Molecular Understanding of Cold-activity of Enzymes from Psychrophiles"
Mose ROSSI (Istituto di Biochimica delle Proteine ed Enzimologia, Italy) "Domain Organization and Biochemical Features of *Sulfolobus solfataricus* DNA Polymerase"
Tairo OSHIMA (Tokyo University of Pharmacy and Life Science, Japan) "Adaptation of Isopropylmalate Dehydrogenase to Thermophilic and Mesophilic Temperatures"
J. Gregory ZEIKUS (Michigan State University, USA) "Thermozymes: Biotechnology and Structure-Function Relationships"
Susumu ITO (Biological Science Laboratories of Kao Corporation, Japan) "Alkaline Enzymes from Alkaliphiles: Enzymatic Properties, Genetics, and Application to Detergents"
Brian JONES (Genencor International BV, The Netherlands) "Microbial Diversity of Soda Lakes: Life in a Depression"

Session 3: Physiology and Metabolisms

Daniel PRIEUR (Station Biologique, France; Universite de Bretagne Occidentale, France) "Physiological Responses to Stress Conditions and Barophilic Behaviors of Deep sea Vent Thermococcales"
Willem M. DE VOS (Wageningen Agricultural University, The Netherlands) "Sugar Utilization and Its Control in Hyperthermophiles"
Dan A. COWAN (University College London, UK) "Biochemistry and Biotechnology of Nitrile Degrading Enzyme Systems from Mesophilic and Thermophilic Bacteria"
Terry A. KRULWICH (Mount Sinai School of Medicine, USA) "ATP Synthesis and pH Homeostasis: Studies of Two Processes that Necessitate Inward Proton Translocation in Extremely Alkaliphilic *Bacillus* Species"
A. Aristides YAYANOS (University of California at San Diego, USA) "The Influence of Temperature and Pressure Perturbations on the Dynamics of Bacterial Cell Processes"
Fumiyoshi ABE (Japan Marine Science and Technology Center, Japan) "Physiological Effects of Elevated Hydrostatic Pressure on Yeast -A Study in "Barophysiology"-"

Jan A. M. DE BONT (Wageningen Agricultural University, The Netherlands) "Mechanisms for Solvent Tolerance in *Pseudomonas putida* S 12"
Rikizo AONO (Tokyo Institute of Technology, Japan) "Improvement of Organic Solvent Tolerance Level of *Escherichia coli* by Overexpression of Stress-Responsible Genes"
Bernard WITHOLT (Swiss Federal Institute of Technology Zurich, Switzerland) "Organic Chemistry via Biocatalysis"

Session 4: Ecology and Diversity

Karl O. STETTER (Universitat Regensburg, Germany) "Novel Isolates and Cultivation Procedures of Hyperthermophiles"
Andre BRACK (CNRS, France) "Life on Mars: Chemical Arguments and Clues from Martian Meteorites"
Jack D. FARMER (NASA Ames Research Center, USA) "Extreme Thermophiles, Early Biosphere Evolution and the Exploration for Extraterrestrial Life"
Juergen WIEGEL (University of Georgia, USA) "Anaerobic Alkali-thermophiles, a Novel Group of Extremophiles"
James P. MCKINLEY (Pacific Northwest National Laboratory, USA) "Geochemical Investigations of Subsurface Microbial Ecosystems"
R. John PARKES (University of Bristol, UK) "A Deep Bacterial Biosphere in Marine Sediments"
Alan T. BULL (University of Kent, UK) "Taxonomy and Biotransformation Activities of Some Deep-sea Actinomycetes"
William D. GRANT (University of Leicester, UK) "Halobacteria from Ancient Salt: Evidence for Living Fossils"
Masahiro KAMEKURA (Noda Institute for Scientific Research, Japan) "Diversity of Extremely Halophilic Bacteria"
Antonio VENTOSA (University of Sevilla, Spain) "Moderately Halophilic Bacterial Diversity in Hypersaline Environments"

Session 5: Genome

Terry GAASTERLAND (University of Chicago, USA) "Integrated Analysis of Whole Genomes: Automated vs. Assisted Decisions"
S. FITZ-GIBBON (University of California at Los Angeles, USA) "Whole Genome Sequence of *Pyrobaculum aerophilum*"
Hans-Peter KLENK (The Institute for Genomic Research, USA) "The Genome of *Archaeoglobus fulgidus*"
Christoph W. SENSEN (Institute for Marine Biosciences, Canada) "Sequencing and Analysis of the *Sulfolobus solfataricus* P2 Genome"
Ronald V. SWANSON (Diversa Corporation, USA) "Complete Genome Sequence of *Aqmfex aeolicus*"
Naotake OGASAWARA (Nara Institute of Science and Technology, Japan) "An International Project for Systematic Analysis of Gene Function of *Bacillus subtilis*"

Hideto TAKAMI (Japan Marine Science and Technology Center, Japan) "Partial Genome Analysis of Alkaliphilic *Bacillus* sp. Strain C-125"

Panel Discussion: Future of Extremophiles

Chair person:
Prof. Koki HORIKOSHI (Japan Marine Science and Technology Center and Toyo University, Japan)
Panelists:
Prof. Alfredo AGUILAR (European Commission)
Prof. Ian DUNDAS (University Bergen, Norway)
Prof. Moon Hi HAN (Korea Research Institute of Bioscience and Biotechnology, Korea)
Dr. Hiroshi HOTA (Japan Marine Science and Technology Center, Japan)
Prof. James P. MCKINLEY (Pacific Northwest National Laboratory, USA)
Prof. A. Aristides YAYANOS (Univesity of California San Diego, USA)

Closing lecture by Prof. Koki HORIKOSHI

3. Organization of the Congress, Extremophiles '98 at Yokohama

Organizing Committee

Mr. T. Hirano: President, Japan Marine Science and Technology Center, Japan
Prof. K. Horikoshi: Japan Marine Science and Technology Center, and Toyo Univ., Japan
Prof. T. Oshima: Tokyo University of Pharmacy and Life Science , Japan

Local Organization

Dr. A. Inoue (Japan), Prof. R. Aono (Japan), Dr. M. Kamekura (Japan), Dr. C. Kato (Japan), Dr. T. Kudo (Japan), Dr. N. Nakamura (Japan), Prof. R. Usami (Japan)

Scientific Advisory Board

Prof. G. Antranikian: Technical University Hamburg-Harburg , Germany
Prof. D. H . Bartlett: Scripps Institution of Oceanography , USA
Prof. J. de Bont: University of Wageningen , The Netherlands
Prof. R. R. Colwell: Maryland Biotechnology Insitute , USA
Prof. R. H. Doi: University of California, Davis , USA

Prof. I. D. Dundas: University of Bergen , Norway
Prof. W. D. Grant: University of Leicester, UK
Prof. H. O. Halvorson: University of Massachusetts, USA
Prof. M. H. Han: KRIBB, Korea
Prof. H. W. Jannasch: Woods Hole Oceanographic Institution, USA
Prof. W. N. Konings: University of Groningen, The Netherlands
Prof. T. Krulwich: NY Mount Sinai School of Medicine, USA
Prof. H. Morgan: University of Waikato, NZ
Prof. D. Prieur: Marine Microbiology, Univ. of Paris VI, France
Prof. M. Rossi: University of Naples, Italy
Prof. N. Russell: Wye College, University of London, UK
Prof. K. O . Stetter: University of Regensburg, Germany
Prof. A. Ventosa: University of Sevilla, Spain
Prof. J. Wiegel: University of Georgia, USA
Prof. B. Witholt: ETH, Switzerland
Prof. A. A . Yayanos: Scripps Institution of Oceanography, USA
Prof. G. Zeikus: Michigan State University, USA

Sponsored by: Japan Marine Science and Technology Center
Supported by: Science and Technology Agency of Japan, and The Nippon Foundation

Appendix: List of Publications of the First DEEPSTAR Project: October 1, 1990 to September 30, 1998

1 Original Articles

1.1 Barophiles (Piezophiles) and Pressure Adaptation

1994

1) Kato C, Sato T, Smorawinska M, Horikoshi K (1994) High pressure conditions stimulate expression of chloramphenicol acetyltransferase regulated by the *lac* promoter in *Escherichia coli*. FEMS Microbiol Lett 122:91-96

1995

2) Kato C, Sato T, Horikoshi K (1995) Isolation and properties of barophilic and barotolerant bacteria from deep-sea mud samples. Biodiv Conserv 4:1-9
3) Kato C, Smorawinska M, Sato T, Horikoshi K (1995) Cloning and expression in *Escherichia coli* of a pressure regulated promoter region from a barophilic bacterium, strain DB6705. J Mar Biotechnol 2:125-129
4) Nakashima K, Horikoshi K, Mizuno T (1995) Effect of hydrostatic pressure on the synthesis of outer membrane proteins in *Escherichia coli*. Biosci Biotechnol Biochem 59:130-132
5) Abe F, Horikoshi K (1995) Hydrostatic pressure promotes the acidification of vacuoles in *Saccharomyces cerevisiae*. FEMS Microbiol Lett 130:307-312
6) Kato C, Suzuki S, Hata S, Ito T, Horikoshi K (1995) The properties of a protease activated by high pressure from *Sporosarcina* sp. strain DSK25 isolated from deep-sea sediment. JAMSTEC R 32:7-13
7) Sato T, Kato C, Horikoshi K (1995) Effect of high pressure on gene expression by *lac* and *tac* promoters in *Escherichia coli*. J Mar Biotechnol 3:89-92

1996

8) Kato C, Smorawinska M, Sato T, Horikoshi K (1996) Analysis of a pressure-regulated operon from the barophilic bacterium strain DB6705. Biosci Biotechnol Biochem 60:166-168

9) Sato T, Nakamura Y, Nakashima KK, Kato C, Horikoshi K (1996) High pressure represses expression of the *malB* operon in *Escherichia coli*. FEMS Microbiol Lett 135:111-116

10) Kato C, Smorawinska M, Horikoshi K (1996) Gene expression of aspartate β-D-semialdehyde dehydrogenese at elevated hydrostatic pressure in deep-sea barophilic and barotolerant bacteria. In: Trzeciakowski WA (ed) High pressure science and technology. World Scientific Publishing, Singapore, pp 869-871

11) Abe F, Horikoshi K (1996) Effect of hydrostatic pressure on the yeast vacuoles. In: Trzeciakowski WA (ed) High pressure science and technology. World Scientific Publishing, Singapore, pp 875-877

12) Kato C, Masui N, Horikoshi K (1996) Properties of obligately barophilic bacteria isolated from a sample of deep-sea sediment from the Izu-Bonin Trench. J Mar Biotechnol 4:96-99

13) Nakasone K, Kato C, Horikoshi K (1996) Molecular cloning of the gene encoding RNA polymerase a subunit from deep-sea barophilic bacterium. Biochem Biophys Acta 1308:107-110

14) Kato C, Tamegai H, Ikegami A, Usami R, Horikoshi K (1996) Open reading frame 3 of the barotolerant bacterium strain DSS12 is complementary with *cydD* in *Escherichia coli*: *cydD* functions are required for cell stability. J Biochem 120:301-305

15) Abe F, Horikoshi K (1996) Vacuolar acidification under high hydrostatic pressure in *Saccharomyces cerevisiae*. High Press Biosci Biotechnol 13:53-58

1997

16) Kato C, Ikegami A, Smorawinska M, Usami R, Horikoshi K (1997) Structure of genes in a pressure-regulated operon and adjacent regions from a barotolerant bacterium strain DSS12. J Mar Biotechnol 5:210-218

17) Kato C, Smorawinska M, Horikoshi K (1997) Comparison of the gene expression of aspartate β-D-semialdehyde dehydrogenase at elevated hydrostatic pressure in deep-sea bacteria. J Biochem 121:717-723

18) Abe F, Horikoshi K (1997) Vacuolar acidification in *Saccharomyces cerevisiae* induced by elevated hydrostatic pressure is transient and is mediated by vacuolar H^+-ATPase. Extremophiles 1:89-93

19) Abe F, Horikoshi K (1997) The yeast vacuoles may serve as proton sequestrants under high hydrostatic pressure. In: Heremans K (ed) High pressure research in the biosciences and biotechnology. Leuven University Press, Leuven, Belgium, pp 209-212

1998

20) Tamegai H, Kato C, Horikoshi K (1998) Pressure-regulated respiratory system in barotolerant bacterium, *Shewanella* sp. strain DSS12. J Biochem Mol Biol Biophys 1:213-220

21) Nogi Y, Masui N, Kato C (1998) *Photobacterium profundum* sp., nov., a new species isolated from a deep-sea sediment. Extremophiles 2:1-7

22) Li L, Kato C, Nogi Y, Horikoshi K (1998) Distribution of the pressure-regulated operons in deep-sea bacteria. FEMS Microbiol Lett 159:159-166
23) Abe F (1998) Hydrostatic pressure enhances vital staining with carboxyfluorescein or carboxydichlorofluorescein in *Saccharomyces cerevisiae*: efficient detection of labeled yeasts by flow cytometry. Appl Environ Microbiol 64:1139-1142
24) Kato C, Li L, Nogi Y, Nakamura Y, Tamaoka J, Horikoshi K (1998) Extremely barophilic bacteria isolated from the Mariana Trench, Challenger Deep at a depth of 11,000 m. Appl Environ Microbiol 64:1510-1513
25) Abe F (1998) A possible function of the yeast vacuole in cytoplasmic homeostasis under hydrostatic pressure. In: Bennett PB, Demchenko I, Marquis RE (eds) High pressure biology and medicine. University of Rochester Press, Rochester, NY, pp 38-43
26) Abe F, Horikoshi K (1998) Vacuolar acidification in *Saccharomyces* cells in response to elevated hydrostatic pressure. Rev High Press Sci Technol 7:1312-1314
27) Kato C, Li L, Nakamura Y, Nogi Y, Tamaoka J, Horikoshi K (1998) Properties of hyper-barophilic bacteria isolated from the Mariana Trench at a depth of 11,000 m. Rev High Press Sci Technol 7:1224-1276
28) Li L, Kato C, Nogi Y, Horikoshi K (1998) Two pressure regulated operons common in the "barophiles branch." Rev High Press Sci Technol 7:1315-1317
29) Nakasone K, Ikegami A, Kato C, Horikoshi K (1998) Analysis of the upstream element of the pressure-regulated operon from deep-sea barotolerant *Shewanella* sp. DSS12. Rev High Press Sci Technol 7:1318-1320
30) Qureshi MH, Kato C, Horikoshi K (1998) Purification of a novel *ccb*-type quinol oxidase specifically induced in a deep-sea barophilic bacterium, *Shewanella* sp. strain DB-172F. Extremophiles 2:93-99
31) Qureshi MH, Kato C, Horikoshi K (1998) Purification and two pressure-regulated *c*-type cytochromes from a deep-sea barophilic bacterium, *Shewanella* sp. strain DB-172F. FEMS Microbiol Lett 161:301-309
32) Nakasone K, Ikegami A, Kato C, Usami R, Horikoshi K (1998) Mechanisms of gene expression controlled by pressure in deep-sea microorganisms. Extremophiles 2:149-154
33) Nogi Y, Kato C, Horikoshi K (1998) Taxonomic studies of deep-sea barophilic *Shewanella* strains and description of *Shewanella violacea* sp. nov. Arch Microbiol 170:331-338
34) Nogi Y, Kato C, Horikoshi K (1998) *Moritella japonica* sp. nov., a novel barophilic bacterium isolated from a Japan Trench sediment. J Gen Appl Microbiol 44:289-295

1999

35) Abe F, Horikoshi K (1999) A new application of moderate hydrostatic pressure for efficient detection of living yeast by flow cytometry. In: Proceedings for international conference on high pressure bioscience and biotechnology (in press)
36) Masui N, Kato C (1999) New method of screening for pressure-sensitive mutants at high hydrostatic pressure. Biosci Biotechnol Biochem 63:235-237
37) Nogi Y, Kato C (1999) Taxonomic studies of extremely barophilic bacteria isolated from the Mariana Trench and description of *Moritella yayanosii* sp. nov., a new barophilic bacterial isolate. Extremophiles 3:71-77

1.2 Psychrophiles, Thermophiles, and Temperature Adaptation

1991

1) Hamamoto T, Horikoshi K (1991) Characterization of an amylase from a phychrophilic *vibrio* isolated from a deep-sea mud sample. FEMS Microbiol Lett 84:79-84

1994

2) Hamamoto T, Takata N, Kudo T, Horikoshi K (1994) Effect of temperature and growth phase on fatty acid composition of psychrophilic *Vibrio* sp. strain no. 5710. FEMS Microbiol Lett 119:77-82
3) Kobayashi T, Kwak YS, Akiba T, Kudo T, Horikoshi K (1994) *Thermococcus profundus* sp. nov., a new hyperthermophilic archaeum isolated from a deep-sea hydrothermal vent. Syst Appl Microbiol 17:232-236
4) Gonzalez JM, Sato T, Kato C, Horikoshi K (1994) Isolation and characterization of hyperthermophilic archaebacteria from southwestern Pacific hydrothermal vents. JAMSTEC J Deep Sea Res 10:471-480

1995

5) Hamamoto T, Kaneda M, Kudo T, Horikoshi K (1995) Characterization of a protease from psychrophilic *Vibrio* sp. strain 5709. J Mar Biotechnol 2:219-222
6) Takata N, Hamamoto T, Horikoshi K (1995) Low-temperature active lipase of deep-sea psychrophilic bacteria and effect of hydrostatic pressure on enzyme activity. JAMSTEC R 31:87-92
7) Kobayashi T, Higuchi S, Kimura K, Kudo T, Horikoshi K (1995) Properties of glutamate dehydrogenase and its involvement in alanine production in a hyperthermophilic archaeon, *Thermococcus profundus*. J Biochem 118:587-592
8) Kwak YS, Kobayashi T, Akiba T, Horikoshi K, Kim YB (1995) A hyperthermophilic sulfur-reducing archaebacterium, *Thermococcus* sp. DT1331, isolated from a deep-sea hydrothermal vent. Biosci Biotechnol Biochem 59:1666-1669
9) Gonzalez JM, Kato C, Horikoshi K (1995) *Thermococcus peptonophilus* sp. nov., a fast-growing, extreme thermophilic archaebacterium isolated from deep-sea hydrothermal vents. Arch Microbiol 164:159-164

1996

10) Nakashima K, Kanamaru K, Mizuno T, Horikoshi K (1996) A novel member of the *cspA* family of genes that is induced by cold shock in *Escherichia coli*. J Bacteriol 178:2994-2997
11) Gonzalez JM, Kato C, Horikoshi K (1996) Membrane-bound, sodium dodecyl sulfate-resistant proteolytic activity from the extreme thermophilic archaeon *Thermococcus peptonophilus*. J Mar Biotechnol 4:159-164
12) Gonzalez JM, Kato C, Horikoshi K (1996) Culturability and survival of an extreme thermophile isolated from deep-sea hydrothermal vents. Arch Microbiol 166:64-67

1997

13) Canganella F, Gonzalez JM, Yanagibayashi M, Kato C, Horikoshi K (1997) Pressure and temperature effects on growth and viability of the hyperthermophilic archaeon *Thermococcus peptonophilus*. Arch Microbiol 168:1-7

1998

14) Gonzalez JM, Robb FT, Ammerman JW, Maeder DL, Yanagibayashi M, Tamaoka J, Kato C (1998) *Pyrococcus horikoshii* sp. nov., a hyperthermophilic archaeon isolated from a hydrothermal vent at the Okinawa Trough. Extremophiles 2:123-130
15) Takai K, Nunoura T, Sako Y, Uchida A (1998) Acquired thermotolerance and temperature-induced protein accumulation in extremely thermophilic bacterium *Rhodothermus obamensis*. J Bacteriol 180:2770-2774

1999

16) Grote R, Li L, Tamaoka J, Kato C, Horikoshi K, Antranikian G (1999) *Thermococcus siculi* sp. nov., a novel hyperthermophilic archaeon isolated from a deep-sea hydrothermal vent at the Mid-Okinawa Trough. Extremophiles 3:55-62
17) Takai K, Inoue A, Horikoshi K (1999) *Themoaerobacter marianensis* gen. nov., sp. nov., an aerobic extremely thermophilic marine bacterium from the 11,000 m deep Mariana Trench. Int. J Syst Bacteriol 49:619-628

1.3 Organic Solvent-Tolerant Microbes

1993

1) Moriya K, Horikoshi K (1993) Isolation of a benzene-tolerant bacterium and its hydrocarbon degradation. J Ferment Bioeng 76:168-173
2) Moriya K, Horikoshi K (1993) A benzene-tolerant bacterium utilizing sulfur compounds isolated from deep sea. J Ferment Bioeng 76:397-399

1994

3) Fukumaki T, Moriya K, Horikoshi K (1994) Isolation of a marine yeast that degrades hydrocarbon in the prsence of organic solvent. Biosci Biotechnol Biochem 58:1784-1788
4) Komatsu T, Moriya K, Horikoshi K (1994) Preparation of organic solvent-tolerant mutants from *Pseudomonas aeruginosa* strain PAO1161. Biosci Biotechnol Biochem 58:1754-1755

1995

5) Moriya K, Yanagitani S, Usami R, Horikoshi K (1995) Isolation and some properties of an organic-solvent tolerant marine bacterium degrading cholesterol. J Mar Biotechnol 2:131-133
6) Moriya K, Horikoshi K (1995) Marine killer yeast isolated from deep sea and their properties. J Mar Biotechnol 2:135-138

7) Moriya K, Inada T, Kyo M, Horikoshi K (1995) Large-scale fermentation under high hydrostatic pressure using a newly developed deep-sea baro/thermophile collection and cultivation system. J Mar Biotechnol 2:175-177
8) Abe A, Inoue A, Usami R, Moriya K, Horikoshi K (1995) Properties of newly isolated marine bacterium that can degrade polyaromatic hydrocarbons in the presence of organic solvents. J Mar Biotechnol 2:182-186
9) Abe A, Inoue A, Usami R, Moriya K, Horikoshi K (1995) Degradation of polyaromatic hydrocarbons by organic-solvent tolerant bacteria from deep sea. Biosci Biotechnol Biochem 59:1154-1156
10) Li L, Komatsu T, Inoue A, Horikoshi K (1995) A toluene- tolerant mutant of *Pseudomonas aeruginosa* lacking the outer membrane protein F. Biosci Biotechnol Biochem 59:2358-2359

1998

11) Fukumori F, Hirayama H, Takami H, Inoue A, Horikoshi K (1998) Isolation and transposon mutagenesis of a *Pseudomonas putida* KT2442 toluene-resistant variant: involvement of an efflux system in solvent resistance. Extremophiles 2:395-400
12) Hirayama H, Takami H, Inoue A, Horikoshi K (1998) Isolation and characterization of toluene-sensitive mutants from *Pseudomonas putida* IH-2000. FEMS Microbiol Lett,169:219-225

1999

13) Kobayashi H, Takami H, Hirayama H, Kobata K, Usami R, Horikoshi K (1999) Outer membrane changes in a toluene-sensitive mutant of toluene-tolerant *Pseudomonas putida* IH-2000. J Bacterial 181:4493-4498

1.4 Microbial Diversity

1997

1) Naganuma T, Kato C, Hirayama H, Moriyama N, Hashimoto J, Horikoshi K (1997) Intracellular occurrence of ε-proteobacterial 16S rDNA sequences in the vestimentiferan trophosome. J Oceanographr 53:193-197
2) Naganuma T, Naka J, Okayama Y, Minami A, Horikoshi K (1997) Multiformity of the microbial population in a vestimentiferan tubeworm. J Mar Biotechnol 5:119-123
3) Kato C, Li L, Tamaoka J, Horikoshi K (1997) Molecular analyses of the sediment of the 11000 m deep Mariana Trench. Extremophiles 1:117-123
4) Takami H, Inoue A, Fuji F, Horikoshi K (1997) Microbial flora in the deepest sea mud of the Mariana Trench. FEMS Microbiol Lett 152:279-285
5) Tamegai H, Li L, Masui N, Kato C (1997) A denitrifying bacterium from the deep-sea at 11,000 m depth. Extremophiles 1:207-211

1998

6) Kobayashi H, Takami H, Inoue A, Horikoshi K (1998) Deep-sea, new resource for screening of biocatalysts. In: Kieslich K, van der Beek CP, de Bont JAM, van den Tueel WJJ (eds) New frontiers in screening for microbial biocatalysts. Elsevier Science, Amster-

dam, pp 259-264
7) Colquhoun JA, Heald SC, Li L, Tamaoka J, Kato C, Horikoshi K, Bull AT (1998) Taxonomy and biotransformation activities of some deep-sea Actimonycetes. Extremophiles 2:269-277
8) Kobayashi H, Takaki Y, Kobata K, Takami H, Inoue A, Horikoshi K (1998) Characterization of α-maltotetrahydrolase produced by *Pseudomonas* sp. MS300 isolated from the deepest site of the Mariana Trench. Extremophiles 2:401-407

1999

9) Yanagibayashi M, Nogi Y, Li L, Kato C (1999) Changes in the microbial community in Japan Trench sediment from a depth of 6,292 m during cultivation without decompression. FEMS Microbiol Lett 170:271-279
10) Li L, Kato C, Horikoshi K (1999) Bacterial diversity in deep-sea sediments from different depths. Biodiv Conserv 8:659-677
11) Li L, Kato C, Horikioshi K (1999) Microbial diversity in sediments collected from the deepest cold-seep area, the Japan Trench at a depth of 6,400 m. J Mar Biotechnol (in press)
12) Takami H, Kobata K, Nagahama T, Kobayashi H, Inoue A, Horikoshi K (1999) Biodiversity in deep-sea sites near the south part of Japan. Extremophiles 3:97-102
13) Kato C, Yanagibayashi M, Nogi Y, Li L, Horikoshi K (1999) Changes in the microbial community in the deep-sea sediment during cultivation without decompression. In: Proceedings for international conference on high pressure bioscience and biotechnology (in press)
14) Li L, Kato C, Horikoshi K (1999) Microbial communities in the world deepest ocean bottom, the Mariana Trench. In: Proceedings for international conference on high pressure bioscience and biotechnology (in press)
15) Grant S, Grant WD, Jones BE, Kato C, Li L (1999) Novel archaeal phylotypes from an east African alkaline saltern. Extremophiles 3:139-145

1.5 Genome Analysis Research

1998

1) Nakasone K, Takaki T, Takami H, Inoue A, Horikoshi K (1998) Cloning and expression of the gene encoding RNA polymerase a subunit from alkaliphilic *Bacillus* sp. strain C-125. FEMS Microbiol Lett 168:269-276

1999

2) Takami H, Nakasone K, Hirama C, Masui N, Fuji F, Takaki Y, Nakamura Y, Inoue A, Horikoshi K (1999) An improved physical and genetic map of the genome of alkaliphilic *Bacillus* sp. C-125. Extremophiles 3:21-28
3) Takami H, Nakasone K, Ogasawara N, Nakamura Y, Hirama C, Fuji F, Takaki Y, Masui N, Inoe A, Horikoshi K (1999) Sequencing of three lambda clones from the genome of alkaliphilic *Bacillus* sp. C-125. Extremophiles 3:29-34
4) Takami H, Takaki Y, Nakasone K, Hirama C, Inoue A, Horikoshi K (1999) Sequence analysis of 32 kb fragment including the major ribosomal protein gene clusters in alkaliphilic *Bacillus* sp. strain C-125. Biosci Biotechnol Biochem 63:452-455

5) Takami H, Masui N, Nakasone K, Horikoshi K (1999) Replication origin region of the chromosome of alkaliphilic *Bacillus halodurans* C-125. Biosci Biotechnol Biochem 63:1134-1137
6) Takami H, Horikoshi K (1999) Reidentification of facultatively alkaliphilic *Bacillus* sp. C-125 to *Bacillus halodurans*. Biosci Biotechnol Biochem 63:943-945
7) Takami H, Takaki Y, Nakasone K, Sakiyama T, Maeno G, Sasaki R, Hirama C, Fuji F, Masai N (1999) Genetic analysis of the chromosome of alkaliphilic *Bacillus halodurans* C-125

1.6 Others

1993

1) Naganuma T, Coury DA, Polne-Fuller M, Gibor A, Horikoshi K (1993) Characterization of agarolytic *Microscilla* isolates and their extracellular agarases. Syst Appl Microbiol 16:183-190
2) Coury DA, Naganuma T, Polne-Fuller M, Gibor A (1993) Protoplasts of *Gelidium robustum* (Rhodophyta). Hydrobiologia 260/261:421-427
3) Degnan BM, Degnan SM, Naganuma T, Morse DE (1993) The *ets* multigene family is conserved throughout the Metazoa. Nucleic Acids Res 21:3479-3484
4) Naganuma T, Horikoshi K (1993) Extremely halotolerant *Halomonas* sp. isolated from a tar ball and its cellular fatty acid composition. J Mar Biotechnol 1:151-157

1994

5) Kato C, Smorawinska M, Sato T, Horikoshi K (1994) Non-radioactive method for RNA analysis by primer extension. Toyobo Biochemicals 27:12-14
6) Naganuma T, Horikoshi K (1994) Cellular fatty acids of marine agarolytic gliding bacteria. Syst Appl Microbiol 17:125-127
7) Naganuma T, Degnan BM, Horikoshi K, Morse DE (1994) Myogenesis in primary cell cultures from larvae of the abalone *Haliotis rufescens*. Mol Mar Biol Biotechnol 3:131-140

1995

8) Naganuma T, Mizukoshi T, Tsukamoto K, Usami R, Horikoshi K (1995) Acclimatization of cultured chum salmon cells to high salinity. J Mar Biotechnol 2:227-229
9) Gonzalez JM (1995) Modelling enteric bacteria survival in aquatic systems. Hydrobiologia 316:109-116

1996

10) Naganuma T, Tsukamoto K, Mizukoshi T, Kato C, Nakasone K, Usami R, Horikoshi K (1996) In situ fluorescence PCR with cultured cells of chum salmon. J Mar Biotechnol 3:248-251
11) Naganuma T, Akutu T, Ishida T, Kato C, Horikoshi K (1996) In situ RT-PCR from the β-tubulin mRNA within the abalone ceels in primary cultures. J Mar Biotechnol 4:75-81
12) Naganuma T, Mizukoshi T, Tsukamoto K, Usami R, Horikoshi K (1996) Effect of hydro-

static pressure on the proliferation and morphology of the mouse BALB/c cells in culture. High Press Biosci Biotechnol 13:73-78

13) Hirayama H, Tamaoka J, Horikoshi K (1996) Improved immobilization of DNA to microwell plates for DNA-DNA hybridization. Nuleic Acids Res 24:4098-4099

1997

14) Ishikura M, Kato C, Maruyama T (1997) UV-absorbing substances in zooxanthellate and azooxanthellate clams. Mar Biol 128:649-655

1998

15) Shibuichi S, Yamamoto T, Onda T, Tsujii K (1998) Super water- and oil-reperellent surfaces resulting from fractal structure. J Colloid Interface Sci 208:287-294

2 Books and Review Articles

1993

1) Hamamoto T, Horikoshi K (1993) Deep-sea microbiology research within the Deepstar program. J Mar Biotechnol 1:119-124

1995

2) Bartlett DH, Kato C, Horikoshi K (1995) High pressure influences on gene and protein expression. Res Microbiol 146:697-706

3) Horikoshi K (1995) Discovering novel bacteria, with an eye to biotechnological applications. Curr Opin Biotechnol 6:292-297

1996

4) Kato C, Inoue A, Horikoshi K (1996) Isolating and characterizing deep-sea marine microorganisms. Trends Biotechnol 14:6-12

5) Horikoshi K (1996) Alkaliphiles—from an industrial point of view. FEMS Microbiol Rev 18:259-270

6) Kato C, Horikoshi K (1996) Deep-sea microbial research and its aspect to high pressure biotechnology. High Press Biosci Biotechnol 13:17-20

7) Kato C, Horikoshi K (1996) Gene expression under high pressure. High Press Biosci Biotechnol 13:59-66

1997

8) Kato C, Li L, Tamegai H, Smorawinska M, Horikoshi K (1997) A pressure-regulated gene cluster in deep-sea adapted bacteria with reference to its distribution. Recent Res Dev Agric Biol Chem 1:25-32

9) Kato C, Bartlett DH (1997) The molecular biology of barophilic bacteria. Extremophiles 1:111-116

1998

10) Kato C, Inoue A, Horikoshi K (1998) Characteristics of deep-sea microorganisms adapted to an extreme environment. In: Fingerman M, Nagabhushanam R, Thompson MF (eds) Marine biotechnology Vol 2: Environmental marine biotechnology. Oxford and IBH Publishing, New Delhi, pp 33-61

11) Abe F, Horikoshi K (1998) Analysis of intracellular pH in the yeast *Saccharomyces cerevisiae* under elevated hydrostatic pressure: a study in baro- (piezo-) physiology. Extremophiles 2:223-228

12) Kato C, Li L, Nogi Y, Nakasone K, Bartlett DH (1998) III. Marine Microbiology 1: Deep sea adaptations. In: Taniguchi Y, Stanley HE, Ludwig H (eds) In: Structure and function of biological systems under extreme conditions, pressure and temperature effects. Springer-Verlag, Heidelberg (in press)

1999

13) Kato C, Qureshi MH (1999) Pressure response in deep-sea piezophilic bacteria. J Mol Microbiol Biotechnol (in press)

14) Kato C, Nakasone K, Qureshi MH, Horikoshi K (1999) How do deep-sea microorganisms response to the environmental pressure? In: Storey KB, Storey JM (eds) Cell and molecular response to stress. JAI Press, Greenwich (in press)

Subject Index

A

α-maltotetraohydrase 20
α-Proteobacteria 61, 63, 73, 77, 78, 79
α-subdivision 68
acetonitrile 48
actinomycete
  phylogram of mycolate — 46
actinomycetes 39, 44, 75, 77
activity of amylase 20, 158, 160
ADD 193
alkaline phosphatase 127
  thermostable — 127
alkaliphiles 7
alkaliphilic *Bacillus* strains 249
alkaliphilic bacteria 249
Alvinella pompejana epibiont 82
amidase activity 48
amylase 18
  — activity 20, 159
  — producers 18
  — production 23
  — productivity 24
  stability of — 158
amylases 159
  characteristics of — 160
ancestral cells 114
androsta-1,4-diene-3,17-dione 193
archaea 30
archaeal 16S rDNA 30, 60, 64, 66
archaeal hyperthermophiles 117
Arcobacter 64

B

B. stearothermophilus 268
B. subtilis 268
β-Proteobacteria 78
β-subdivision 68
Bacillus halodurans C-125 249, 268, 277
Bacillus halodurans C-125 chromosome 253
barophiles 91, 227
  obligate — 36
barophilic bacteria 30, 83, 93, 95, 227
  moderately — 94
barophilic bacterial species 99
barophilic bacterium
  growth profiles of — 92
  moderately — 99
barophilic isolates 30
barophilic microorganisms 92
barophilic strains 97
barophysiology 227, 288
  — in eucaryotes 228
  — in prokaryotes 239

barotolerant bacteria 94
barotolerant bacterium 92
benzene-tolerant bacteria 177
benzene-tolerant bacterium
isolating of — 177
benzonitrile 48
BioLog 44
BioLog analysis 43
biotransformation 48
black smoker 286, 287

C

Calyptogena 55
Calyptogena communities
56, 60, 61, 63, 64, 65, 67
Calyptogena community 57, 70
carboxydichlorofluorescein (CDCF) 234
carboxyfluorescein (CF) 234
Challenger Deep 5, 7, 27, 99, 129
chemical reactions
— in supercritical water 287
chemoautotrophic animal 212
chemoautotrophic bacteria 212
cholesterol degradation 191, 192, 193
cholesterol-degrading bacterium 190
cold adaptation 172
cold-adapted enzymes 162
cold-seep 55, 212
colloid and surface science
— in supercritical water 287
Colwellia 99
Corynebacterium 47
Crenarchaeota 65, 66
Cytophaga 61, 63, 72, 73, 75, 76, 79, 83

D

δ- and ε-subdivisions
— of Proteobacteria 63
δ-Proteobacteria 61, 67, 70, 77, 79, 83
δ-subdivision 68
deep-sea actinomycetes 39
deep-sea isolates 15, 16
deep-sea mycolata 42
deep-sea rhodococci 48
growth of — 50
— Rhodococcus 49
degradation
— of *n*-alkanes 182
— by yeast 207
— of naphthalene 199, 200
— of sulfur compound 189
dendrogram 44, 45
DHA 35, 99, 167, 169, 171
diversity of actinomycetes
— in deep-sea sediments 51
DNA extraction 58
DNA libraries
— from alkaliphilic *B. halodurans* C-125 269
DNA sequencing 270
DNA–DNA hybridization analysis 34
docosahexaenoic acid 35, 99, 167, 169, 171

E

E. coli 100, 240
ε-Proteobacteria 61, 67, 70, 79, 82, 220, 221
ε-subdivision 68
eicosapentaenoic acid 35, 98, 169, 171
electron transport system
— for piezophilic *Shewanella benthica* 107
elemental sulfur 117, 119, 121
endosymbiont 212, 218
endosymbiotic bacteria 64
enzymatic activities of lipase 161
enzyme producers 18
EPA 35, 98, 169, 171
epibiont bacteria 70
Escherichia coli 101, 228
ether lipid membranes 127
Euryarchaeota 65, 66
exobiology 290, 291
extracellular proteolytic activity 142
extreme barophiles 32, 36
extremely barophilic bacteria 32, 35, 93
extremely barophilic bacterium 36, 99
extremely barophilic isolates 33
extremophiles 7

F

fatty acid composition 165, 166, 171
fatty acids 35
Flavobacterium sp. 180
flow cytometry 235
— analysis 237
fluorescent dye 236
fossil tubeworms 216

G

γ-Proteobacteria 61, 76, 77, 78, 79, 96
γ-subdivision 68
— of Proteobacteria 63
GDH 133, 135
GDH activity 135
GDH gene sequence 137
gene expression 240
effect of high pressure on — 245
effect of pressure on — 242
pressure regulated — 240
— by pressure 240
genetic map 103, 251
— of the chromosome of *Bacillus halodurans* C-125 264
genetic organization 268
genome 249
— of *B. halodurans* C-125 251, 269, 277
— of *B. subtilis* 250
genome analysis 250, 289
genus *Thermoccus*
physiological characteristics of — 123
glutamate dehydrogenase 133
— activity 134
— gene sequences 137
Gordonia 47
gram-positive bacteria 61, 63, 68, 72, 75, 78

H

Halophiles 7
hydrocarbon degradation 182
hydrolysis
— of nitrile and amide 49
hydrothermal vent 212, 286, 287
hyperthermophile 113, 145
phylogenetic position of — 114
hyperthermophilic
— archaea 128
— archaeon 142

I

Iheya Ridge 5
in situ hybridazation 220
Izu-Ogasawara Trench 5

J

Japan Trench 60

K

Kaiko 5, 27, 92, 129
Korarchaeota 66

L

λ phage library 269
lac promoter 240, 241
Lamellibrachia sp. 212
linkage map
— of *Bacillus halodurans* C-125 259
linking clone 253
lipases 161
log p 177
log p value 181
low-temperature adaptation 165, 169

M

M. yayanosii 99
maltotetraose 20
Mariana isolates 8, 11
phylogenetic tree of — 12
Mariana Trench 3, 6, 27, 99, 129
marine yeast 202
characteristics of — 206
membrane fluidity 165, 172
membrane permeability 172
methane seeps 211
Methanococcoides burtonii 65

microbial community 55
microbial diversity 56, 60, 67
— analysis 57
microbial flora 7
micromonosporae 40
moderately barophilic bacteria 94
moderately barophilic bacterium 99
moderately barophilic strain 98
Moritella 99
Moritella japonica 99
Moritella yayanosii 35
Morphological change of *E. coli*
— under elevated hydrostatic pressure 242
multicellular organisms 288
Mycobacterium 47
mycolate 40
mycolic acid 42

N

Nankai Islands 5
Nankai Trough 67, 68
nitrile hydratase activity 48
nitrile transforming activities 50
nitrile-metabolizing enzyme 50

O

obligate barophiles 36
obligately barophilic strain 30, 36
Ocean Drilling 21 289, 290
ocean drilling project 290
OD-21 289, 290
ORF analysis 270
— of *Bacillus* sp. C125 genome 272, 273, 274
ORF organization
— of *Bacillus halodurans* C-125 271
organic-solvent tolerant
— microorganisms 175
— microbes 176
organic sulfur compounds 189
utilization of — 190
oriC region 265
origin of life 114, 291

P

P. aeruginosa 81
P. fluorescens 81
P. horikoshii 127, 137
Photobacterium 99
Photobacterium profundum 97
phylogenetic position of *Thermoaerobacter marianensis* 130
phylogenetic techniques 57
phylogenetic tree
28, 30, 32, 60, 62, 66, 68, 95, 130
— based on gluta-mate dehydrogenase (GDH) gene sequences 136
— of bacteria isolated from the deep-sea 16
— of Mariana isolates 12
— of strain C-125 250
— of *Thermococcus* and *Pyrococcus* secies 128
phylogram
— of mycolate actinomycete 46
physical map 103, 251, 255
— of *Bacillus halodurans* C-125 251
— of the chromosome of *Bacillus halodurans* C-125 264
piezophiles 91
piezophilic microorganisms 92
piezophysiology 227, 288
— in eucaryotes 228
— in prokaryotes 239
Planctomyces 73
plasma membrane H^+-ATPase 233, 234
polyaromatic hydrocarbon degradation 199, 200, 201
polyaromatic hydrocarbon-degrading ability 194
polyaromatic hydrocarbon-degrading bacteria 194, 196
polyunsaturated fatty acid 35, 167
pressure regulated gene expression 240
pressure-induced accumulation 234
pressure-induced acidification
— of vacuoles 229
pressure-inducible proteins 100

pressure-regulated operon
30, 101, 102, 103, 104
pressure-regulated respiratory system 105
protease 162
purification of — 162
stability of — 162, 164
protease activity 138, 162, 164
effects of metal ions on — 165
Proteobacteria 68, 72
α-subdivision of — 73, 75
β-subdivision of — 73
δ-subdivision of — 68, 73
ε-subdivision of — 68, 75
γ-subdivision of — 63, 67, 68, 73, 75
proteolytic activity 138
highly thermostable — 139
— of *T. peptonophilus* 140
proton sequestrant 234
Pseudomonas 63, 67, 79, 80
Pseudomonas putida 175, 180
psychrophiles 7, 155, 156, 172
psychrophiles and psychrotrophs
properties of — 156
psychrophilic 92
psychrophilic and psychrotrophic bacteria 162
growth of — 156
psychrophilic bacteria 155
psychrophilic microorganisms 165
psychrotolerants 172
psychrotrophic bacteria 155
psychrotrophic enzyme 164
psychrotrophic isolates 159
psychrotrophic microorganisms 165
psychrotrophs 155, 156
PyMS 44
PyMS analysis 43
Pyrococcus 115
Pyrococcus horikoshii 124, 125, 126
electron micrograph of — 124
growth of — 124

R

16S rDNA 46, 59, 62, 68
bacterial — 28, 60, 67, 68
distribution of — 61
16S rDNA analysis 11, 15, 16
18S rDNA 28
resistance to antimicrobials 180
respiratory
— chain 107
— chain enzyme complexes 107
— system 107
response of cells
— to stimuli 288
restriction fragments 252
RFLP analysis 59
Rhodococcus 42, 48
Rhodococcus luteus 47
Rhodothermus marinus 83

S

S. benthica 98
Saccharomyces cerevisiae 228
Shewanella 103
Shewanella barophile 98
Shewanella benthica 34, 83, 101
Shewanella violacea 98, 101, 102
Shinkai 2000 5, 57
Shinkai 6500 5, 57, 92
shotgun library 277
assembly of — 280
shotgun sequencing 277
Solemya velum symbiont 81
solvent-tolerant microorganisms 176
stability of amylase 158
stability of protease 162, 164
streptomycetes 40
strict aerobic 130
strict anaerobe 117, 119, 126
sulfate reducing bacteria 70
sulfide oxidation 211
sulfur circulation 64, 70
model of — 70

sulfur-degrading ability 189
sulfur-degrading bacteria 187
sulfur-reducing bacteria 64
supercritical water 287
  chemical reactions in — 287
  colloid and surface science in — 287
symbiont bacteria 55
symbionts 55, 218
symbiotic bacteria 211, 218
symbiotic ε-Proteobacteria 221

T

T. peptonophilus 138
  pressure and temperature effects on — 144
T. profundus 117, 135, 137
T. siculi 124
tac promoter 240
taxonomy
  — of balophiles 96
Thermoactinomyces 40
Thermoaerobacter 115
Thermoaerobacter marianensis 129, 132
  growth of — 130
  light and electron micrographs of — 128
  phylogenetic position of — 130
Thermococcus 115
Thermococcus peptonophilus 117, 119, 138, 142, 145, 148
  effect of starvation on — 148
  electron micrographs of — 118
  growing characteristics of — 118
  growth, ATP content, and proteolytic activity in — 146
Thermococcus profundus 115, 133
  growth of — 116
Thermococcus siculi 122
  growth of — 122
  transmission electron micrographs of — 120
Thermococcus sp. 120
  growth of — 120
thermophiles 7, 113
  survival and viability of — 142
thermophilic proteases 139
thermostable alkaline phosphatase 127
tolerance of organic solvents
  — by marine yeast 205
  — by yeast 203, 204
toluene-tolerant strain 180
tubeworms 211, 212, 221
  vestimentiferan — 212

V

V-H^+-ATPase 229, 234
vacuolar acidification 229, 230, 231, 232
vacuolar pH 229
Vibrio sp. 160, 162, 165
Vibrio vulnificus 82, 83

W

whale carcass 213
whale skeleton
  — at the Torishima Seamount 215

Y

yeast 228
yeast vacuole 229, 234

The manufacturer's authorised representative in the EU is Springer Nature Customer Service Centre GmbH, Europaplatz 3, 69115 Heidelberg, Germany. If you have any concerns regarding our products, please contact ProductSafety@springernature.com

Printed and bound by CPI Group (UK) Ltd, Croydon, CR0 4YY
29/07/2026
02180634-0006